Astronomers' Universe

For further volumes:
http://www.springer.com/series/6960

W. M. Goss

Making Waves

The Story of Ruby Payne-Scott:
Australian Pioneer Radio Astronomer

W.M. Goss
National Radio Astronomy Observatory
New Mexico
USA

ISSN 1614-659X
ISBN 978-3-642-35751-0   ISBN 978-3-642-35752-7 (eBook)
DOI 10.1007/978-3-642-35752-7
Springer Heidelberg New York Dordrecht London

Library of Congress Control Number: 2013938945

© Springer-Verlag Berlin Heidelberg 2013
This work is subject to copyright. All rights are reserved by the Publisher, whether the whole or part of the material is concerned, specifically the rights of translation, reprinting, reuse of illustrations, recitation, broadcasting, reproduction on microfilms or in any other physical way, and transmission or information storage and retrieval, electronic adaptation, computer software, or by similar or dissimilar methodology now known or hereafter developed. Exempted from this legal reservation are brief excerpts in connection with reviews or scholarly analysis or material supplied specifically for the purpose of being entered and executed on a computer system, for exclusive use by the purchaser of the work. Duplication of this publication or parts thereof is permitted only under the provisions of the Copyright Law of the Publisher's location, in its current version, and permission for use must always be obtained from Springer. Permissions for use may be obtained through RightsLink at the Copyright Clearance Center. Violations are liable to prosecution under the respective Copyright Law.
The use of general descriptive names, registered names, trademarks, service marks, etc. in this publication does not imply, even in the absence of a specific statement, that such names are exempt from the relevant protective laws and regulations and therefore free for general use.
While the advice and information in this book are believed to be true and accurate at the date of publication, neither the authors nor the editors nor the publisher can accept any legal responsibility for any errors or omissions that may be made. The publisher makes no warranty, express or implied, with respect to the material contained herein.

Cover image: *The Great Wave, after Hokusai*, 1980, by Fiona Hall. Type C photograph, 33.0 x 25.1 cm. Courtesy of the artist and Roslyn Oxley9 Gallery, Sydney. In this work, Hall uses scraps of cloth, rotting banana peels, bits of plastic toys, a pair of children's shoes, matchsticks, perforated paper, old postcards, wire and more to reconstruct the old master print from c. 1831, *The Great Wave off Kanagawa* by Hokusai. The original print is considered by many to be a masterfully composed visual space with clear connections to the ideal mathematical proportions found in nature

Cover background image: Great Ball of Fire. This image from the Solar Dynamics Observatory of the news-making solar event on 2010 August 1 shows the C3-class solar flare (white area on upper left), a solar tsunami (wave-like structure, upper right), multiple filaments of magnetism lifting off the stellar surface, large-scale shaking of the solar corona, radio bursts, a coronal mass ejection and more. Credit: NASA/SDO/AIA

Printed on acid-free paper

Springer is part of Springer Science+Business Media (www.springer.com)

To Libby

Foreword

It is rare for a complete biography of an Australian scientist, particularly of an Australian woman scientist, to be published. It is rarer for such a book to be authored by an American.

Though scientists have written discourses on the history of their discipline, it is most unusual for a scientist to write a full-length biography of a colleague in his field. It is also uncommon for a man to write about an Australian woman scientist; most of the work on Australian women scientists has been done by other women. However, this author, a distinguished researcher in the field of radio astronomy, became so interested in the history of his discipline and in the career of the pioneer radio astronomer, Ruby Payne-Scott, that he spent some years bringing this book to fruition.

Until relatively recently, Ruby Payne-Scott had been the only woman scientist mentioned briefly in histories of Australian science or of Australian radio astronomy. This book will be an invaluable resource for anyone interested in these disciplines. Being a scientist himself, the author explains Payne-Scott's scientific work in detail; therefore, the value and importance of her contributions can, for the first time, be recognised, not only by historians but also by scientists.

After a brilliant academic career, with an M.Sc. in physics (the highest qualification then available at any Australian university), Ruby Payne-Scott worked as a science teacher, one of the few professional positions available to Australian women in the 1930s and especially in the Depression. However, the Second World War opened up opportunities for women science graduates. She was one of the first of the scientific staff members appointed to the new Radiophysics Laboratory of the Council for Scientific and Industrial Research (CSIR) from which radio astronomy developed and notably the first woman scientist in the Laboratory.

Ruby Payne-Scott was part of a pioneering group of radar scientists during the Second World War, led by J. L. Pawsey, whose scientific distinction and leadership qualities have been referred to by all writers in the field. Although it was mainly due to him that radio astronomy developed in Australia from 1944, she was one of the key people contributing to Australia's pre-eminence in the world in radio astronomy for many decades. Pawsey valued her judgment and experience so highly that when she was absent from a meeting, he would often not make a final decision until she had been consulted. She became the overall advisor to the group on scientific issues, engineering planning and mathematics; she also made major contributions to the development of radio astronomical techniques.

By 1951, when she left the discipline, she was set to be promoted to the highest research category short of the leader and to be paid the second highest salary on the scientific staff. Her standing was confirmed later by a member of this distinguished group who himself became an important radio astronomer but who disliked Payne-Scott; nevertheless, he considered her, as the author records, to have been "one of the best physicists at Radiophysics – no, one of the best physicists in Australia".

In telling Payne-Scott's story, the author highlights the inferior position of women in the workforce at that time. Married women could not become permanent employees in the public service. The practice of requiring women to choose between marriage and their careers inevitably deprived Australia of unknown talent.

Payne-Scott had to suffer the indignity of keeping her marriage secret from CSIR for some 5 years. When the marriage was discovered, she fought vehemently against the injustice of this regulation but was forced to become a temporary employee, losing all her superannuation entitlements in the process. She finally resigned in 1951 when pregnant with her first child, as there was no maternity leave at that time.

The War years provided some measure of equal pay for women. After the War, the old discriminatory practices returned. Payne-Scott, together with other colleagues, campaigned unsuccessfully for the recognition of the principle of equal pay.

"Women's rights" was not the only issue about which she felt strongly and for which she argued publicly and vigorously. During the War, it was natural that the type of work which the group was engaged in was classified; but after the War, she was bitterly opposed to secret research in the CSIR. She believed that it was impossible to do good research in the atmosphere of limitations imposed by a sponsoring body, particularly when that body was the military. She wrote to CSIR: "Frightened men do not produce great research."

The Australian Security Intelligence Organisation kept files on her that have only recently been made available to researchers. A subsequent media release by the National Archives of Australia, headed "The Secret Life of Ruby Payne-Scott", states that she was "passionate about both the independence of scientific research and human rights. These sentiments were deemed a security risk".

The work of pioneering Australian women scientists is gradually being recognised. CSIRO offers OCE Science Team Career Awards. One of these is the OCE Payne-Scott Career Award for researchers returning from family-related career breaks. The life and work of a feisty, brilliant woman is finally being recognised.

Nessy Allen

Preface

In the mid-twentieth century, radio astronomers revolutionised our view of the universe by observing new objects, ones that had not been previously known by optical astronomers. Even the sun is vastly different when studied by the radio astronomer, yielding clues about the outer regions of the sun that are difficult for the optical astronomer to observe. Ruby Payne-Scott was one of the first radio astronomers in Australia and the first female radio astronomer in the world. Due to her brief period of, only 6 years, employment as an astronomer, her contributions largely had been forgotten by the end of the twentieth century. In addition, she was an independent-minded, sometimes, confrontational woman, even suspected of being a communist in an era when communism was a precarious ideology for an Australian to embrace. Even so, the reader may wonder why the life of Ruby Payne-Scott is of significance to us almost 60 years after her retirement in 1951.

By studying the life and career of Ruby Payne-Scott, we gain a rare insight into the intersection between a woman, who was extremely passionate about research in the new field of radio astronomy, and societal pressures that caused her to leave the field and not realise her full potential as a research scientist. What great talent did this one Australian woman have and why was she only able to devote herself to it for about 10 years of her life? And why, for decades afterward, were her discoveries and work within radio astronomy partially erased from public knowledge?

I first heard about Ruby Payne-Scott when I was a young staff member at the Radiophysics Laboratory (RPL) in Sydney, in the decade starting in 1967. I had arrived with my wife from Berkeley, California, where I had completed a Ph.D. in radio astronomy. While using the Parkes radio telescope, John Bolton, the director, told me that I should have been there at RPL 20 years earlier, when the brightest staff scientist was a woman, Ruby Payne-Scott. He claimed she was the first person in Australia to recognise the importance of radio "confusion". I had never before heard of this great female scientist, and I admit I was not suddenly struck by the need to find out everything I could about Ruby Payne-Scott. Bolton's comment, however, adhered to my memory and resurfaced years later as I began hearing stories from the aging founders of the field about the early days of radio astronomy. Truly, the irony of his admiration in light of the tales of antipathy and confrontation between Payne-Scott and Bolton, which I heard many years later, was the catalyst for piecing together her story.

Even as a child, I always enjoyed reading biographies. It is a format for understanding the history of the world that I find accessible and meaningful. There are numerous histories of optical astronomy, yet understandably few about its younger cousin, radio astronomy. Radio astronomy is a relatively young branch of science after all, as will be explained in this book; it was born from radar research performed during the Second World War. My colleague, Woody Sullivan, has in fact published a few histories of early radio astronomy, and many conversations with him over the years piqued my curiosity about the instruments and the people who engineered them in the creation of the innovative new field. I also began to feel the pull of the historian and I felt that only someone who was a radio astronomer could effectively portray the unacknowledged achievements of someone about whom I had heard so much. The more I spoke with the contemporaries of Ruby Payne-Scott, such as Harry Minnett, Don Yablsley, Chris Christiansen, Bernie Mills, John Murray and Dick McGee, the more I felt that hers was a great story and an important one that needed to be told.

Starting in 1997, while visiting my daughter who was spending a half year at the University of Sydney, I started to collect information about her life, from both senior colleagues who had known her as well as the vast archival materials from the 1940s and 1950s in the National Archives of Australia. Of course many of these conversations occurred 45 plus years after the relevant events; frequent contradictions in recollections of the same events were often observed. The collections from the archives can be used in many cases to resolve these contradictions. In many cases these contemporary records from 1945–1955 can even be utilised to correct vague or even flawed memories.

A crucial element in efforts to recreate her life was meeting three key individuals: her children, Peter and Fiona Hall and her close friend Elizabeth Hall. I could never have predicted that, thanks to these friends, my quest for the past would become a bit of an obsession that constantly brought surprising discoveries. The more I learned, the more intriguing the mystery of her life became. In the end, this has been a joyful journey as I have unravelled the life of this remarkable scientist, woman and mother.

Ruby Payne-Scott's story is an inspiring saga of achievement and adversity. Fortunately, her successful career and life have now been recognised. A main motivation for this book was to make even more people aware of her important contributions to the field of radio astronomy and to further the quest for equality in the workplace.

This book is based on the 2009 Springer (Astronomy and Space Science Libraries series) *Under the Radar, The First Woman in Radio Astronomy: Ruby Payne-Scott* by W. M. Goss and Richard X. McGee. This volume is intended for a non-specialist reader. I do, however, hope that the astronomer reader will also enjoy this book. I have simplified the scientific text considerably; the detailed, technical appendices are not included. An additional chapter has been added at the beginning of the book to provide the reader with a basic background of both solar physics and radio telescopes.

Socorro, New Mexico W.M. Goss
September, 2012

Acknowledgements

Many colleagues and friends have contributed to this book. In the preface to *Under the Radar, The First Woman in Radio Astronomy: Ruby Payne-Scott,* I have thanked many people for their assistance in the preparation of the 2009 book. I am indebted to my late co-author of *Under the Radar*, Richard ("Dick") X. McGee (1921–2012) for extensive collaboration since 1967.

In the writing of this book, I have been ably assisted by my daughter-in-law Pax Bobrow as a conscientious editor of *Making Waves, The Story of Ruby Payne-Scott: Australian Pioneer Radio Astronomer*. The title of this book was suggested by Alison Muir (www.textualhealing.com.au) of Sydney, who also provided the title for the 2009 book. Alison also played a major role in the 2009 book launch at the University of Sydney. Loretta Appel has remained a valuable colleague, including the identification of Ruby Payne-Scott's image in several group photographs.

Dr. Elizabeth Hall has provided extensive assistance in updating the story of Payne-Scott's family. Betty Hall has been a steadfast friend and advisor for many years; without her, the two books about Ruby Payne-Scott would not have been possible. Betty provided the family history in Chap. 2.

The children of Ruby Payne-Scott, Peter Hall and Fiona Hall have continued to provide advice and support. A visit with Fiona in Adelaide in 2007 was an unforgettable experience for Libby Goss and me.

Jessica Chapman and Barnaby Norris have provided amazing support in the use of the CSIRO Radio Astronomy Image Archive. Any historical studies of radio astronomy in Australia would not be possible without the use of this 16,000 plus photo collection (https://imagearchive.atnf.csiro.au)

The staff of the National Archives of Australia (NAA) in Sydney have continued to provide excellent help since 2008. I thank Melanie Grogan, Kerrie Jarvis, Paul Wood, Simeon Barlow, Fiona Burn and Audrey Vintour-Cesar (and others) for their patience and diligence during my visits in 2008, 2009 and 2011 (twice).

Many comments for improvements have been provided by Tim Bastian, Stephen White, Aimee Norton, Harry Wendt, Claire Hooker and Don Melrose. Also I would like to thank Nick Lomb, John Deane, John Brooks, Gerrit Verschuur, Carolyn Little, Helen Sim Lakshmi Saripalli, Christine van der Leeuw, Pat Palmer, Minnie Mao, Anne Green, Robert Hayward, Caroline Baum, Gwen Anne Manefield, Barbara Manchester, Elizabeth Griffin, Robert Sault and Cornelia Lang. Again I would like to express my admiration and gratitude to Nessy Allen of the University of New South Wales, an expert on women scientists in Australia, for the detailed editing of the 2009 volume and especially for the forward to both books.

In addition, I thank Woody Sullivan, Ron Ekers, Bob Frater, David E. Hogg, Lory Wingate, Ken Kellermann, Ellen Bouton, Marsha Bishop, Lance Utley, Megan Nunemaker, Rodney ("Rod") D. Davies Roy MacLeod, Lois and Alan Whitney and Garry Tee for their help in recent years. Peter Murphy of the SEARCH Foundation in Sydney has provided advice for many years as well as a travel grant in 2009. The late John E. Baldwin was the host for my visit to the Cavendish Laboratory of the University of Cambridge in 2010. David Green of the Cavendish Laboratory of the University of Cambridge was my host in 2010 and 2012 for visits to the Churchill College Archives in Cambridge. Nan Dieter Conklin, the first female US radio astronomer, and author of *Two Paths to Heaven's Gate* (National Radio Astronomy Observatory, 2006) has continued to provide advice and support since 1965.

Thanks again to my children Prof. Andrew Goss of the University of New Orleans and Kate Goss, RN, MSN for their advice.

Numerous deceased colleagues of Ruby Payne-Scott have provided invaluable insights to her life and achievements: Joan Freeman Jelley, Bernie Mills, Paul Wild, Chris Christiansen, Don Yabsley, Ron Bracewell, Gordon Stanley, John Bolton and Harry Minnett. Additional insights have been provided by Rachel Makinson, Bruce Slee, Jan Christensen, Sue Brian, Inge Heleu of URSI, Lyn Brown, Glenys Edwards, Philip Edwards, Rita Nash, Joan and John Murray and Letty Bolton.

The Rosyln Oxley9 Gallery of Paddington, Sydney, has provided many of the images of Fiona Hall's work. Thanks to Roslyn Oxley and her staff, including Ivan Buljan and Jessica Maurer.

Finally I would like to thank W. Butler Burton, the chairman of the Editorial Board of the Springer *Astronomy and Space Sciences Library* series, for his initiative in the organisation of both volumes describing the life of Ruby Payne-Scott. In 2010, Butler suggested that I prepare a popular treatment of her life. Also I would like to express my gratitude to Harry Blom and Ramon Khanna of Springer for their support since 2009 in the preparation of this volume.

The National Radio Astronomy Observatory is a facility of the National Science Foundation operated under the cooperative agreement by Associated Universities, Inc.

Contents

1	**Introduction**	1
	Synopsis: The Life of Ruby Payne-Scott	1
	The Legacy of Ruby Payne-Scott	7
	Recognition	8
	Additional Note	10
2	**A Brief, Basic Guide to Terms and Concepts of Solar Radio Astronomy**	13
	The Astronomers' Sun	13
	Sunspots	17
	Flares	19
	Chromosphere	21
	Corona	23
	Tools of the Radio Astronomer	26
	Yagi Antenna	28
	Broadside Array	29
	Parabolic Reflector	32
	Michelson Interferometer	34
	Solar Grating-Array	35
	Mills Cross	36
	Culgoora Radioheliograph	36
	The Parkes Radio Telescope: The Dish	38
	Karl G. Jansky Very Large Array: The VLA	38
	Frequency Agile Solar Radiotelescope: FASR	41
3	**Ruby Payne-Scott's Ancestors and Her Early Childhood**	43
	Ruby Payne-Scott's Paternal Ancestors in the UK	43
	Emigration to Australia by Hubert and Agnes Scott	45
	Ruby's Early Childhood	49

4	**Ruby Payne-Scott's Education and Early Employment**	53
	Secondary School Education: Sydney	53
	Bachelor of Science in Physics at the University of Sydney	57
	Master of Science and the Cancer Research Committee at the University of Sydney	58
	Woodlands Glenelg Church of England Girls' Grammar School, Adelaide	61
	AWA- Amalgamated Wireless, Australasia: 1939–1941	64
	Additional Note	66
5	**Wartime Research by Ruby Payne-Scott at the Radiophysics Laboratory**	67
	New Career Opportunities for Australian Women, World War II	67
	Payne-Scott Joins the Radiophysics Laboratory in 1941	71
	Payne-Scott's War-Time Research	75
	Calibration of S Band Radar Receivers	76
	Equality of Pay for Women Ruling: September 1944	81
	Light Weight Aircraft Warning-Height 25 cm Radar Display	82
	Additional Notes	85
6	**1944–1945: Ruby Payne-Scott – The First Woman Radio Astronomer**	87
	Propagation Committee at Radiophysics Laboratory: 1944–1954	87
	Proto-Radio Astronomer, Ruby Payne-Scott	88
	Post-war Activities at RPL: 1945- How This Impacted the Role of Payne-Scott	91
	RPL's First Observations of Solar Noise- Role of Payne-Scott	95
	The First Summary Paper in Radio Astronomy: December 1945 Author Payne-Scott	96
	Symposium on Radar, 5–7 December 1945: Payne-Scott Was Present	99
	Additional Note	101
7	**1945–1946: Early Radio Astronomy at Dover Heights**	103
	The Collaroy Campaign: October 1945: First Solar Radio Astronomy in Australia	105
	Ground-Breaking Developments in Solar Noise Research and Techniques of Interferometry: February 1946, Dover Heights	113
	Publication of Dover Heights Research and Interferometry Techniques: Proceedings of the Royal Society, 12 August 1947	124
	Additional Notes	126
8	**1946–1947: Personal Tragedy and Professional Triumph**	129
	Payne-Scott as Solo Scientist at Dover Heights, Mid-1946	129
	Miscarriage and Missed Work: Late 1946–June 1947	133
	The Behemoth Type II Burst of March 1947: Payne-Scott, Yabsley and Bolton Observe an Amazing Event	137

	Plans for Pawsey's Departure Abroad: September 1947	142
	Payne-Scott Winds Down at Dover Heights: Late 1947	144
	Additional Notes	145
9	**1948: Hornsby Field Station: Daily Observations**	147
	Move from Dover Heights to Hornsby	147
	Payne-Scott Has Doubts at Hornsby: Advice from Pawsey	155
	Publication of the Hornsby Observations	158
	Additional Notes	165
10	**1949–1951: Radio Astronomy Blossoms as a Field, but Ruby Must Resign from the Radiophysics Laboratory**	167
	Payne-Scott's Career in 1948, Choice of the Swept-Lobe Interferometer Site	167
	Early Testing of the 97 MHz Interferometer	174
	Publication of the Results at 97 MHz: Paper I- The Instrument	180
	Paper II- Noise Storms (Type I)	182
	Paper III- Outbursts. Type IV	185
	Red Ruby: The Difficulty of Communist Ideology in Post-war Australia	190
	Transition from CSIR to CSIRO; Discovery by CSIRO of Marriage of 1944	192
	Payne-Scott's Resignation in July 1951	196
	Additional Notes	198
11	**1952: Ruby Payne-Scott's Last Experience as a Radio Astronomer at the International Union of Radio Science**	201
	Additional Notes	212
12	**The Married Life and Motherhood of Ruby Payne-Scott**	215
	A Remarkable Family: Bill and Ruby Hall	215
	Bill Hall	215
	Peter Hall: Mathematician	223
	Fiona Hall: Artist	225
	Additional Note	233
13	**1963–1974: Employment at Danebank School**	235
14	**The Last Years and Legacy of Ruby Payne-Scott**	239
	Last Years	239
	Ruby Payne-Scott's Legacy; Why Did She Not Return to RPL in the 1960s?	241
Bibliography for Making Waves		245
Biographical Sketch of the Author		253
Index		255

Abbreviations and Terminology

AASW	Australian Association of Scientific Workers
ABC	Australian Broadcasting Corporation
ACTU	Australian Council of Trade Unions
ANCORS	Australian National Committee of Radio Science
ANU	Australian National University, Canberra, Australia
ANZAAS	Australian New Zealand Association for the Advancement of Science
AORG	Army Operational Research Group, the group of J. Stanley Hey in the UK during the Second World War
ASIO	Australian Security Intelligence Organisation – earlier Commonwealth Investigation Service
ATNF	Australia Telescope National Facility of CSIRO, Australia
AUI	Associated Universities, Inc. Washington, DC. USA
AWA	Amalgamated Wireless, Australasia
AWAS	Australian Women's Army Service
BBSO	Big Bear Solar Observatory, California
Bursts	Early terminology for Type III bursts. These short timescale events were called "isolated bursts" by Pawsey and "unpolarized bursts" by Payne-Scott. See **Type III** bursts
CASS	CSIRO Astronomy and Space Science, Australia
CH	Chain Home; the chain of low-frequency aircraft radars in the UK, Second World War
CHL	Chain Home Low, warning radar for low-flying aircraft
CIS	Commonwealth Investigation Service, see ASIO
CME	Coronal Mass Ejections
COL	Chain Home Overseas Low – CHL used in New Zealand and Australia
CPA	Communist Party of Australia
CRC	Cancer Research Committee of the University of Sydney
CSIR	Council for Scientific and Industrial Research, 1926–1949

CSIR, OA	CSIR Officers Association
CSIRO	Commonwealth Scientific and Industrial Research Organisation, 1949-
CSIRO, OA	CSIRO Officers Association
CSO	Commonwealth Solar Observatory, Mt. Stromlo, Canberra, Australia, founded 1924. In 1957, it is transferred to the Australian National University, Mt. Stromlo Observatory – MSO
dB	Decibel, 10 times log (power ratio)
DSIR	Department of Scientific and Industrial Research, New Zealand
EAST	Eastern Australian Standard Time
Enhanced Radiation	Early terminology of Pawsey for Type I "noise storms" and associated bursts ("storm bursts")
FASR	Frequency Agile Solar Radio Telescope
IAU	International Astronomical Union
IRE	Institute of Radio Engineers (1912–1962)
LASCO	Large Angle and Spectrometric Coronograph - of SOHO
L band	Designation from the Second World War for the 15–30 cm range
LW/AW	Light Weight – Air Warning radar, the major achievement of RPL in the Second World War
LW/AWH	Light Weight – Air Warning-Height radar
MCA	Museum of Contemporary Art – Sydney
MHD	mageto-hydrodynamics
MSO	Mount Stromlo Observatory
nm	Nanometre (one-billionth of a metre) unit for wavelength X-ray, ultraviolet and optical radiation. For example, 500 nm is 5,000 Å (angstrom)
NRAO	National Radio Astronomy Observatory, USA
NSF	National Science Foundation, USA
NSL	National Standards Laboratory of the CSIR and the CSIRO, the Radiophysics Laboratory (RPL) shared this location from 1939 to 1968
NSW	New South Wales, Australian state
OBE	Order of the British Empire
Outbursts	Terminology invented by Allen (1947). Early term for Type II bursts. These were first discovered by Payne-Scott, Yabsley and Bolton (1947) during the giant outburst of 8 March 1947 at Dover Heights, Sydney
PC	Propagation Committee of the Radiophysics Laboratory during the last years of the Second World War until 1949, when the name was changed to the Radio Astronomy Committee
PMG	Postmaster-General's Department (Postal, Telephone and Telegraphic Services) – Australia

PPI	Plan Position Indicator; display of azimuth and range of radar targets
RAAF	Royal Australian Air Force
Rad Lab	The Massachusetts Institute of Technology Radiation Laboratory, the centre of US radar work in the Second World War
RADAR	The new term for RDF invented in 1940. Adopted by the US Navy in 1940 and by the Allied Powers in 1943 (**RA**dio **D**etection **A**nd **R**anging)
RDF	Radio Direction Finding, the first term for "RADAR"
RPL	Radiophysics Laboratory of the CSIR and CSIRO, sometimes RP
RRB	Radio Research Board, Australia
S Band	A designation from the Second World War for the 8–15 cm range. Microwave ovens operate at S band
SBW	Sydney Bush Walkers
SDO	Solar Dynamics Observatory, satellite
SEARCH	Social Education and Research Concerning Humanity Foundation, Sydney
SGHS	Sydney Girls High School
ShD	Shore Defence Radar
SOHO	Solar and Heliospheric Observatory, satellite
SWPA	South West Pacific Area – the Second World War
TRACE	Transition Region and Coronal Explorer, far UV satellite
Type I	Most common burst events on metre wavelengths from the sun. These are short, narrow frequency band events (storm bursts) that usually occur in great numbers together with steady or slowly varying background (noise storms) with broader band continuum. Noise storms may last for hours or days. Individual bursts last for a few seconds. Type I bursts have their origin from fundamental frequency plasma emission. They were discovered by Payne-Scott and colleagues in late 1945–1946 to be associated with sunspots with high degree of circular polarization (Wild 1951). See **enhanced radiation.**
Type II	Bursts with slow drift from high to low frequencies. They often show fundamental and secondary harmonic frequency structures. The drift rate is in the range 0.25–1 MHz/s, lasting some minutes (Wild 1950a). Type II bursts are excited by magnetohydrodynamic shock waves in the corona, and serve as a causal agency for magnetic storms and aurorae on the earth after 1.5–3 days. During sunspot maximum, the occurrence rate is about once per 2 days. See **outbursts.**
Type III	Bursts of short duration (duration a few seconds) with rapid drift from high to low frequencies. These bursts may exhibit harmonics. Often they accompany the flash phase of large

	flares. They were discovered by Payne-Scott at Dover Heights in 1946. Relativistic velocities are inferred due to radiation from plasma oscillations excited by discrete bunches of fast electrons. The drift rate is in the range of 20–100 MHz/s (Wild 1950b). During sunspot maximum, the occurrence rate is about once per 20 min. See **bursts**.
Type IV	Bursts discovered by Boischot and Denisse (1957), observed by Payne-Scott and Little circa 1959. They are flare-related broadband continua due to synchrotron emission. They have implied motions of about 1,000 km/s. They are also called Type IVM – "moving" in contrast to stationary component of Type IV continuum.
URSI	Union Radio Scientifique Internationale, International Union of Radio Science
VHF	Very High Frequency 30–300 MHz of 10–1 m
VLA	Jansky Very Large Array of the National Radio Astronomy Observatory
WAAAF	Women's Auxiliary Australian Air Force
WEB	Women's Employment Board, Australia, the Second World War and post war
WRANS	Women's Royal Australian Naval Service
WRNS	Women's Royal Naval Service (UK), known as the Wrens
Y Factor	Method widely used for measuring the gain and noise of a radio amplifier

Chapter 1
Introduction

Synopsis: The Life of Ruby Payne-Scott

Within this book I will explore the development of an intellectually gifted young woman during a time when women were discouraged from studying fields that were basically reserved for men. In *Irresistible Forces: Australian Women in Science* (2004), Claire Hooker has written

> ... [Ruby Payne-Scott] blazed her way to the peak of her profession, presaging the ambitions and successes that young women may have today.... Payne-Scott from the beginning was ready for anything she might encounter in science.

I will review Ruby Payne-Scott's educational career, her work both as an engineer and astronomer, as well as a school teacher, and her passion for the Australian wilderness and bushwalking. It was this passion for the outdoors that brought her to her husband, Bill Hall, and I will also describe their wonderful family life. There are some unknown aspects of Ruby's life that I will touch on: for example, how did a middle class Australian family produce such a remarkable woman? What were the factors that influenced her career choices?

With the support of her parents, Ruby attended schools that were meant to prepare her for further education at the University of Sydney. She was an outstanding student at university and spent some years in a cancer research laboratory following her graduate degree work. Figure 1.1 is one of the earliest photographs of Payne-Scott, as a student in the 1930s. When that work dried up, she taught mathematics to secondary school children. However, with the advent of World War II when able-bodied men were rounded up all over the world by the millions as solders, airmen when seamen, gaping holes were left in the workforce that could be filled by able-minded women.

Ruby took advantage of this and seized a research job at the Radiophysics Laboratory (RPL) in Sydney, Australia. One can only imagine the thrill that Ruby must have felt being able to work with a group of like-minded engineers and scientists, putting all of her considerable mental faculties to work on challenging problems to perfect an aircraft warning radar system for the defence of Australia.

Fig. 1.1 Photograph of Ruby Payne-Scott as a student in the 1930s, possibly while she was studying at the University of Sydney in 1929–1932, working on a B.Sc. degree in physics (Bill Hall family collection, used by permission of Peter Hall)

The RPL cohort was a cohesive and inquisitive group that respected all members based on the quality of their work. Ruby was a star performer in this group based on her experience and scientific skills. Unfortunately, during this era, women were not expected to play major roles in the workforce if they were married or having children. Were a woman worker to marry, she would lose her permanent employment status, be demoted to a temporary worker, and lose her pension. Further, when a woman became pregnant, there were no options for any paid leave and women were expected to abandon their jobs.

It was during this time that Ruby decided to get married and start a family with her bushwalking comrade, Bill Hall, but she was enjoying her work at the RPL. She was able to continue working for some time as a permanent employee by keeping her marriage secret from her highest ranked supervisors—those close to her at work certainly knew. By the end of the war, however, she was exposed to an increased scrutiny by bureaucrats, who discovered her marriage. She fought against the consequences of this discovery but was still demoted to the status of a "temporary" staff member and lost her superannuation. In 1951, when Ruby was only 39, she was forced to retire while pregnant with her first child, Peter Hall. After leaving her career in radio astronomy Ruby worked as a secondary school teacher for 9 years. Her two children grew up in a world that slowly was changed by Ruby and women like her, who championed the rights of women to combine a career and family.

The field of radio astronomy in Australia grew out of the radar research carried out during World War II at the Council for Scientific and Industrial Research (CSIR) Radiophysics Laboratory (RPL). Ruby Payne-Scott joined the new institute in 1941, as one of the founding scientific staff—she and Joan Freeman were the first female scientific staff. RPL played a key role in the War effort, producing numerous copies of the aircraft warning radars that were used so successfully in the Southwest Pacific Area by both US and Australian military personnel from 1942 to 1945 in the war against Japan. Payne-Scott made major contributions to this top secret radar research; she became the Australian expert on the theory of the detection of enemy aircraft using the display system that had been invented in the UK, named the PPI or Plan Position Indicator. She was also an experienced radio engineer; her work with B. Y. Mills to develop experimental, high-frequency (25 cm) aircraft warning radar contributed to her success as an experimental radio astronomer starting in mid-1945.

Late in World War II, women in the civil service in the Federal government were paid wages equal to that of their male counterparts, a great contrast to the previous convention of paying women only two-thirds of the male wage. In 1949, Payne-Scott was involved in a public controversy when the CSIR began to withdraw wage parity. It was only in 1969 and 1972 that Australian women were given wage equality based on rulings of the Australian Conciliation and Arbitration Commission. In addition, she and many of her male colleagues were strong proponents of non-military research in the newly established Commonwealth Scientific and Industrial Research Organisation (CSIRO), formed in 1949. She and many others participated in writing letters to both newspapers and to internal CSIRO publications in support of this cause, thus garnering the attention of the Australian Security Intelligence Organisation.

Payne-Scott started her career in radio astronomy, testing radar equipment with Joseph L. Pawsey in 1944. Her first observation was made from the RPL building on the campus of Sydney University in March 1944, during a test of military radar equipment at 10 cm. Thus Payne-Scott became one of the first radio astronomers, as well as the first woman radio astronomer. This observation also represented the first astronomical project with Pawsey, an association that became decisive in the years 1945–1951. Pawsey was the "father" of Australian radio astronomy. His research and recruiting of new, talented radio astronomers has had an effect on the astronomical world that continues into the twenty-first century.

An explosive growth of radio astronomy occurred in Australia starting in late 1945. RPL became one of the pre-eminent radio astronomy institutes in the world under the direction of Pawsey and Edward ("Taffy") G. Bowen. Within a few years, Australia established its international leadership in radio astronomy. Payne-Scott wrote one of the first summary papers in radio astronomy in December 1945 and participated in the first Australian publication of the budding field in early 1946, as the second author. From 1945 to 1952, the RPL radio astronomers published 62 papers in radio astronomy; Payne-Scott was a participating author in nine of these publications. During this period, radio astronomers in the rest of the world followed the Australian developments with great interest. A few Australian radio astronomers made visits to the US, Canada and Europe where the new results were shown during conferences and observatory visits.

In the short period from 1945 to her resignation in July 1951, Payne-Scott became a driving force in early radio astronomy in Australia; she was the first scientific leader in the solar radio group, directed by Pawsey. In 1946 she discovered Type III solar radio bursts, which originated at long radio wavelengths in the solar corona. She played an important part in the discovery of Type I bursts (1946) and Type II solar radio outbursts (1947). Payne-Scott, together with Alec Little, even detected Type IVM solar radio outbursts with the Potts Hill swept-lobe

Fig. 1.2 The most commonly published photograph of Ruby Payne-Scott. This was taken at the Potts Hill Reservoir, likely in late 1948. "Chris" Christiansen is to the right with Alec Little in the middle. Payne-Scott and Little were working on observations of the sun at 97 MHz using the newly constructed swept-lobe interferometer (Chap. 10) (CSIRO Radio Astronomy Image Archive B14315)

interferometer in 1949–1951, several years before the bursts were recognised as distinct physical entities by the French group of Boischot and Denisse. The photograph we see in Fig. 1.2 of Ruby Payne-Scott, W. N. "Chris" Christiansen and Alec G. Little, taken at Potts Hill sometime between 1949 and 1951, has become well known as it is the only photo of Ruby to appear in Australian publications in recent decades.

Payne-Scott also made major contributions to the development of radio astronomy techniques. Three prominent examples were: (1) the first ever interferometric measurements in radio astronomy on Australia Day, 26 January 1946, using the seacliff interferometer at Dover Heights- Sydney, (2) the mathematical development of "aperture synthesis", the technique utilised by many of the advanced radio astronomy instruments of the modern era (e.g. the Very Large Array, the Atacama Large Millimetre Array, the Multi Element Radio Linked Interferometer, the Australia Telescope Compact Array); and (3) the swept-lobe interferometer at Potts Hill, developed by her and Little, which could make a rudimentary ciné movie (25 frames a second), showing the motions of the solar radio bursts as the emitting gas moved outwards in the corona at high velocities. In addition, there is strong evidence that she was the first person in Australia to recognise the importance of confusion in radio astronomy—the necessity to achieve high angular resolution in detecting fine details, as well as good sensitivity to recognise distinct radio sources.

In these first years of growth in radio astronomy after World War II, Payne-Scott's auspicious career was marred by conflict with the bureaucracy of CSIR/CSIRO.

Ruby was subjected to discrimination against women, prevalent in Australian society in the 1940s and 1950s.[1] These controversies have been mentioned in a number of popular articles in the Australian press and also in books providing summaries of Australian astronomy. The latter have been correct in attributing her 1951 resignation to the birth of her first child in late 1951. The nature and the consequences of the discovery in 1950 of her "secret marriage" of 1944 have been described with considerable distortion in some popular articles and books.

The conflict with CSIR (Council for Scientific and Industrial Research) and CSIRO[2] occurred in the period February to May 1950, when she met in person with Ian Clunies Ross, the Chairman of the new CSIRO. Ruby Payne-Scott and William "Bill" H. Hall married in September 1944; this was known by most of her colleagues at RPL. The rule against married women at CSIRO maintaining permanent employment status was challenged head on by Payne-Scott. Clunies Ross wrote her a series of forceful letters, with equally strong replies from her side. Not surprisingly, Payne-Scott lost this battle and became a temporary employee of the CSIRO in 1950. As a consequence she had to forfeit her superannuation (pension) rights, the CSIR/CSIRO pension contributions (1946–1950), and the accrued interest on her own contributions. Ironically, some of the more productive research of her short astronomical career occurred after this demotion.

A year-long conflict with John G. Bolton was a part of Payne-Scott's life at the RPL, after he joined the CSIR in September 1946. Bolton was demobilised from the

---

[1] There is a temptation to evaluate these issues with the viewpoint of the more egalitarian society of the early twenty-first century; as a number of colleagues have pointed out to Goss, the draconian treatment of Payne-Scott in the mid-twentieth century was consistent with practices in many walks of life. The characteristic that distinguished Payne Scott was her resistance to these inequalities.

[2] The transition from the CSIR to the CSIRO (Commonwealth Scientific and Industrial Research Organisation) occurred in the period March to May 1949 with the passing of the Science and Industry Research Act 1949 by the Australian Parliament. The change from "Council" to "Commonwealth" was chosen to emphasise the national character of the new organisation and the word "Organisation" was used to highlight the changed character of the administration by the new CSIRO Executive of five members, including three scientists (Schedvin 1987).

British Navy in late 1945, having served as a radar officer on the Royal Navy aircraft carrier *Unicorn* for about a year in the East Indies and the Pacific. During the time of Pawsey's overseas trip, September 1946 to October 1947, the conflict reached a boiling point. Sharing the Dover Heights site by the two strong-willed scientists produced continual conflict and Payne-Scott was "exiled" to the Hornsby field station.[3]

An important gain for Payne-Scott's career at RPL was her interaction and the support of two prominent women colleagues during World War II: Joan Freeman Jelley and K. Rachel Makinson. Joan Freeman's autobiography (*A Passion for Physics*, 1991) preserves a number of famous anecdotes about Payne-Scott and others (see Additional Note, No. 1, end of this chapter).

Payne-Scott had been known as a "left winger" at the RPL in the 1940s. The Australian Security and Intelligence Organisation (ASIO) maintained a large dossier on her and suspected that she was a member of the Communist Party of Australia (CPA). ASIO had no proof of this affiliation at the time of a 1950 report. In 1999, however, Goss discovered that she had been a member of the CPA, possibly breaking with the Party later in the 1950s. Rachel Makinson has told Goss that Ruby was often referred to as "Red Ruby", even by her closest friends.

One of Payne-Scott's great loves was bushwalking. She met her husband through the Sydney Bush Walkers in 1941. They remained enthusiastic bush walker for many years. A typical picture of her during the period is shown in Fig. 1.3, a trip to the Blue Mountains west of Sydney.

In July 1951, Payne-Scott resigned from the RPL, with an advance notice of only 2 days. She was pregnant; her son Peter G. Hall—future Professor of Mathematics at the University of Melbourne and Fellow of the Royal Society of London—was born on 20 November 1951. There was no maternity leave at CSIRO. A daughter, the famous Australian artist Fiona Hall, was born 2 years later.

After the birth of her children, Ruby chose to remain at home in Oatley (a suburb of Sydney) to care for the two young children. After the children were about 10 and 12 years old, she became a mathematics and science teacher at Danebank Anglican School for Girls in nearby Hurstville; she was in this position from 1963 to 1974. It is likely that Ruby developed Alzheimer's disease at an early age, with signs of deterioration of her mental facilities in her last years at Danebank. She died in Sydney on 25 May 1981, a few days before her 69th birthday. Bill Hall died 21 July 1999.

---

[3] Melrose and Minnett (1998) have quoted one colleague at RPL who suggested that there was "a triangle of antagonism between John Bolton, Ruby Payne-Scott and Jack Piddington" [another prominent scientist at RPL] in the late 1940s. Minnett, himself, acknowledged these antipathies and described them as "creative tensions between very different personalities". RPL was blessed with some strong personalities!

Fig. 1.3 Bushwalking, a passion of Ruby Payne-Scott. Here she is probably in the Blue Mountains in the 1940s. Many photos of Payne-Scott show her eating or drinking at the time of the photographic session. Her daughter Fiona suggested that her mother likely thought that posing for photos was a waste of time; thus she could be more efficient when combining posing while eating or drinking! (Bill Hall family collection, used by permission of Peter Hall)

The Legacy of Ruby Payne-Scott

Ruby's legacy has two major components. First, she was a crusader for the rights of women in the scientific workplace in Australia. Other women had experienced discrimination; Payne-Scott complained loudly about the treatment. She helped pave the way for future generations. Secondly, Payne-Scott was one of the first three pioneers in the new field of radio astronomy, which burst into prominence at RPL in Sydney in 1944–1945. Within a few years, Australia and the United Kingdom became the leaders in this revolutionary new form of astronomy. Pawsey and Payne-Scott provided the key leadership for the rapid growth in solar physics that solar radio astronomy created in the first decade after World War II. After she retired in mid-1951, the Australian leadership role in solar physics was maintained by Paul Wild (Frater and Ekers 2012).

Recognition

Ruby Payne-Scott was hardly known outside the Australian astronomical community until the late 1990s. As an example, the influential popular book from 1956 by the *Scientific American* author John Pfeiffer, *The Changing Universe,* has a detailed description of many of the Australian achievements in radio astronomy. In particular, the Australian RPL achievements highlighted in a chapter titled, "The Sun in Action", included a description of the remarkable Type II outburst of 8 March 1947, with a whimsical cartoon of the effects of solar outbursts on terrestrial communication. This publication was authored by Payne-Scott, Yabsley and Bolton. No mention of Payne-Scott appears in the Pfeiffer volume, even though most of her Australian colleagues are explicitly named.

Goss and Dick McGee (co-author of the 2009 *Under the Radar, the First Woman in Radio Astronomy: Ruby Payne-Scott)* participated in a number of Australian radio programmes, an Australian Broadcasting Corporation television programme and numerous newspaper articles. The most successful and influential event was an Australian Broadcasting Corporation (ABC) Radio National Saturday broadcast on Valentine's Day, 2004. The broadcast was in the long running series, *The Science Show*, by Robyn Williams, directed in a thorough fashion by Pauline Newman Davies. Elizabeth (Betty) Hall, McGee, Claire Hooker, Fiona Hall, Carolyn Little and Goss were interviewed. The world wide web distribution of the transcript has led to numerous helpful comments to the authors. By contrast, the television programme in the *Rewind* series on 7 February 2005 by the Australian Broadcasting Corporation (ABC) was a disappointment. Originally it was taped as an episode for the *History Detectives*, a series which fell victim to internal infighting within the ABC. In spite of a heroic effort by the director, Laurie Critchley, the final version is a watered-down presentation that does not capture the essence of Critchley's original production. In particular the fascinating interview with Bruce Slee at Dover Heights, Sydney, was cut as well as a humorous interview with McGee at his home in Eastwood, Sydney.

Only two recent publications deal with Payne-Scott in detail. In her 2004 book, *Irresistible Forces: Australian Women in Science*, Claire Hooker has a thorough treatment in Chap. 11, "The Sun, Ruby Payne-Scott and the Birth of Radio Astronomy". In his 2009 book, *Cosmic Noise: A History of Early Radio Astronomy*, W.T. ("Woody") Sullivan has described her work in detail in a sub-section of his Chap. 14, "The Radio Sun: Payne-Scott's work".

Fig. 1.4 The Google Doodle of 28 May 2012, celebrating the 100th birthday of Ruby Payne-Scott. Note the PPI (Plan Position Indicator display), the headphones (which were actually used for the solar observations in late 1945–1946), and the solar Type I and Type III bursts (Used with permission of Google- Australia)

Within CSIRO there has also been recognition of Ruby Payne-Scott. "In Ruby's honour", states a recent description[4]:

> CSIRO, in 2008–09, initiated the Payne-Scott Awards which are designed to support researchers who have taken extended leave to care for a newborn child following birth. The grant provides support to researchers to re-establish themselves and re-connect with the research underway in their field and related fields of research.

Both women and men involved in newborn care are eligible, though the awards seem to have gone to women. In the first year of the new grant program, ten individuals applied and six awards were granted. Given the level of conflict that Payne-Scott had over her marriage in 1950 and the career-ending nature of her pregnancy in 1951, this grant in her name is appropriate.

The 28th of May, 2012 marked a century since the birth of Ruby Payne-Scott. It was a satisfying surprise for Goss and many others to find that Google Australia had used a cartoon ("Google Doodle") of Ruby at work as a radio astronomer as their logo on that day. The Doodle is shown in Fig. 1.4. This image highlights several achievements of her research, including Type I and Type III bursts and PPI detection. The Google Doodle attracted many web based articles in Australia; an impressive description of the life of Ruby Payne-Scott can be found at the CSIRO Staff Association web site (www.cpsu-csiro.org.au).[5]

To aid the reader in locating the various field stations of RPL in New South Wales, Australia, a schematic map of sites is shown in Fig. 1.5.

---

[4] http://www.csiropedia.csiro.au/display/CSIROpedia/Payne-Scott,+Ruby

[5] Professor Brian Schmidt of the Mt. Stromlo Observatory in Canberra has pointed out to Goss that Payne-Scott was competing for the 100th birthday "Google Doodle" with the Australian Nobel Laureate for Literature (1973) Patrick White, also born on 28 May 1912.

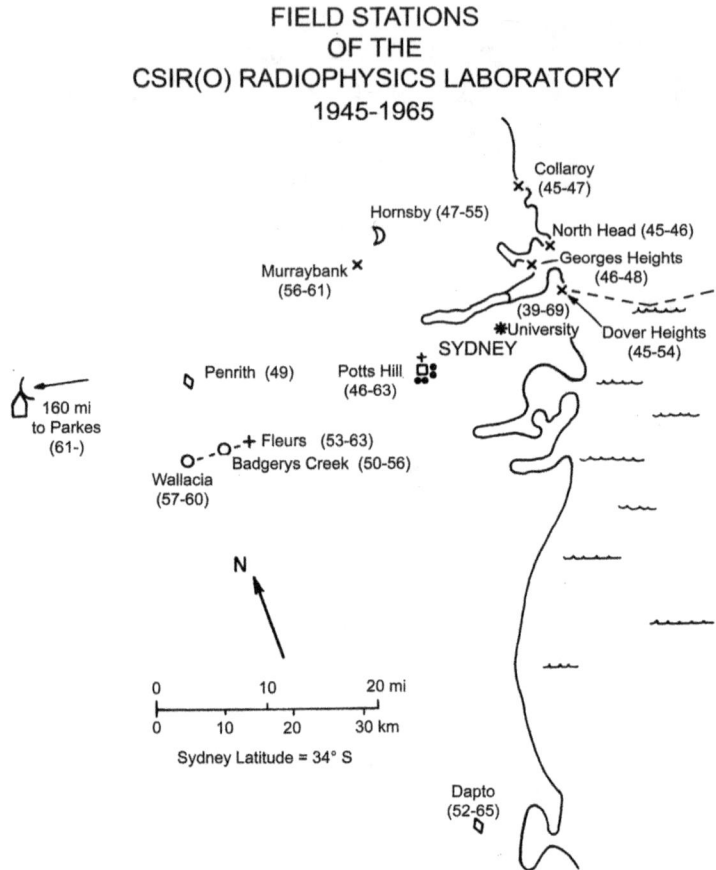

Fig. 1.5 A schematic map of the various RPL sites in New South Wales. The numbers represent the period in which each of the field stations was active. For example, Dapto (near Wollongong) was active from 1952 to 1965. The sites where Payne-Scott worked in 1945–1951 were Dover Heights, Hornsby and Potts Hill (CSIRO Radio Astronomy Image Archive)

Additional Note

1. The most famous colleague of the RPL during the War was certainly the late Dame Joan Sutherland, the famous opera singer (1926–2010). Although she may have heard of Payne-Scott while she was a clerical assistant at the RPL from April 1944 to January 1945, she never met Payne-Scott (letter from Sutherland to Goss, 23 April 2007). In 1944, Sutherland, at age 18, was a typist at RPL. She typed a report written by E. G. Bowen concerning meteorological effects on radar reception. The most famous story about Joan Sutherland at RPL concerns the Musical Revue, *Hush- Hush*, in late 1944, staged by Robert Coulson. This was a spoof of

the *Mikado* with a skit involving the "Lord High Clerical Officer" draped in red tape. Joan Freeman was in the chorus. The 18 year old Sutherland auditioned for a part and was rejected (confirmed by a letter from Dame Joan to Goss, April 2007). It appears that this rejection had no adverse effect on her career; already in 1947 she had made her stage debut in Sydney as Dido in Purcell's *Dido and Aeneas*.

Chapter 2
A Brief, Basic Guide to Terms and Concepts of Solar Radio Astronomy

The Astronomers' Sun

Humanity has observed the sun for many centuries. Among the millions of stars observable through optical telescopes, only the sun is close enough to be studied in all of its activity, in exquisite detail. But because it is so bright, it was difficult for anyone to see features on the sun's surface until more modern times. Chinese astronomers likely observed sunspots as early as 364 BC with naked eye observations; the observations would have been at sunrise or sunset when the solar radiation is attenuated by the earth's atmosphere or even through dense terrestrial clouds. In the early seventeenth century Galileo and others began detailed studies of the sun with some of the first optical telescopes; their detection of sunspots was a major discovery that impacted the understanding of the universe. Detailed telescopic solar studies began in the nineteenth century, including spectroscopic identification of many known and even some unknown elements in the solar spectrum. The sun emits a continuum of electromagnetic radiation from X-ray, to ultraviolet, optical, infrared and radio wavelengths. In the optical wavelengths (similar to the receptivity of the human eye at 400–800 nm), the **solar spectrum** shows absorption lines that enable astronomers to determine the chemical composition of the sun. The solar spectrum and associated discoveries were made by the German astronomer Fraunhofer in 1817 using the newly invented spectroscope.

Figure 2.1 shows a three-dimensional model of the sun with sections of the solar structure from the interior to the outer solar corona. In 1939, an understanding of the energy source of the sun was made by the German scientist Hans Bethe—later a prominent physicist at Cornell University in the US—who suggested that the energy source was the fusion of hydrogen nuclei (protons) into helium nuclei in a process known as the p-p (proton-proton) chain. This process releases vast amounts of energy and is, of course, a vital source for life on earth.

Astronomers have long realised that the sun is a common type of star in the Milky Way; it has a typical size, luminosity and temperature. Due to the proximity

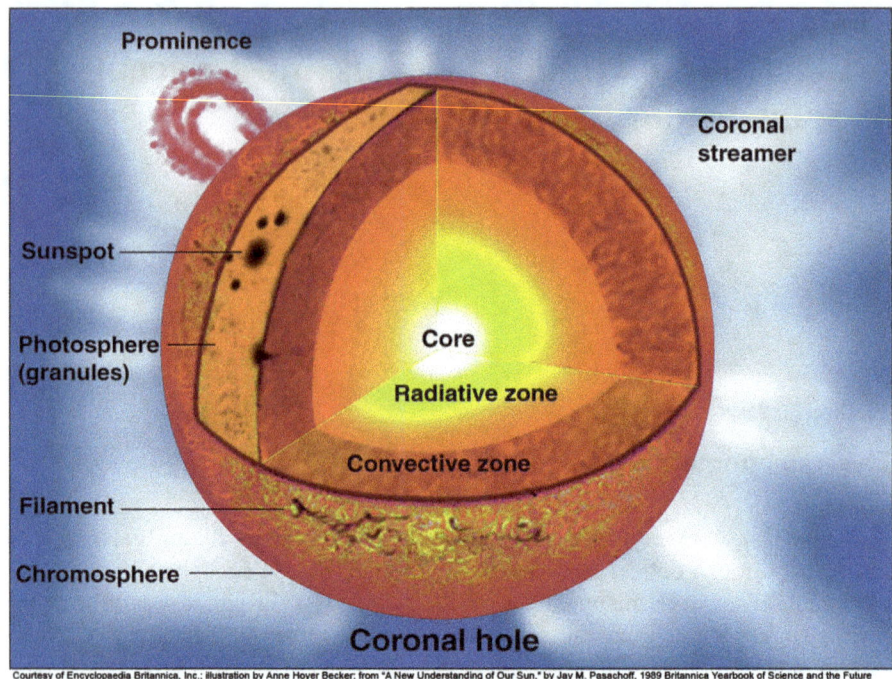

Fig. 2.1 A schematic 3-D model of the sun showing the interior and the solar atmosphere. The solar surface is shown as it would be observed through a hydrogen H-alpha filter (Courtesy of Encyclopaedia Britannica, Inc.; illustration by Anne Hoyer Becker; from "A New Understanding of Our Sun", by Jay M. Pasachoff, *1989 Britannica Yearbook of Science and the Future*)

of the sun to the earth compared to the nearest stars (about a factor of 200,000), detailed information gathered from the relatively nearby sun has allowed scientists to extrapolate information about the structure of far more distant stars. In the early twentieth century, telescopes were used to obtain images of the sun in the visible portion of the electromagnetic spectrum (e.g., from about 3,000 to 8,000 Å or 300–800 nm, from the nearby ultraviolet to the nearby infrared). These images revealed the existence of structures in the solar atmosphere such as flares and prominences. Some of these surprising features will be described below. Later in the twentieth century, totally new solar phenomena—many of them in the tenuous solar corona—were detected using ultraviolet, X-ray or gamma ray telescopes. These wavelengths are heavily attenuated by the earth's atmosphere and thus must be observed from space using rockets or satellite telescopes.

Based on these observations, astronomers and physicists have shown how prominent effects on earth are produced by solar activity. One example is the aurosa, produced when charged particles from the earth's radiation belts are driven into the atmosphere by geomagnetic storms that occur when coronal mass ejections—something like a large bubble of plasma erupting off the sun—strike the earth. The perturbed particles impinge on the earth's upper atmosphere at altitudes above 80 km, exciting molecules at these positions which radiate over a range of visible colours.

It was during and just after World War II that physicists and radio engineers discovered radio emission from the sun; in some cases, such as with both British and New Zealand military radars, the discoveries were serendipitous. This discovery of radio waves provided a method to investigate parts of the solar atmosphere that were difficult or impossible for the optical solar astronomer to detect. Thus new information about the sun could be obtained. For example, the properties of the solar corona were much more easily determined at radio wavelengths. For the ground-based optical astronomers of that era, the corona could only be observed during infrequent, total solar eclipses.

The optical and radio investigations of the sun's atmosphere have a double significance: (1) Physical processes can be studied on a very small scale of 100–1,000s of kilometres, and extreme conditions of high temperatures and low densities are observed that would never be possible to reproduce in the laboratory. These studies have advanced the knowledge of magneto-hydrodynamic (MHD) processes in a variety of situations. In addition, strong magnetic fields are observed at the solar surface and in the corona. (2) The study of conditions and changes in the earth's outer environment due to the propagation of energetic particles in the solar wind can now be studied on a routine basis. A prominent problem that has been addressed is how solar activity impacts radio communications as the ionosphere of the earth (at altitudes above 80 km) is disturbed by temporary increases in ultraviolet radiation from solar flares.

In the immediate post-war era, the physicists and engineers at the Radiophysics Laboratory (RPL) in Australia, including Ruby Payne-Scott, played a prominent role in solving these problems. They used the techniques of WWII radar to turn the military radar systems—the "swords"—into peacetime radio telescopes—"ploughshares". *Transmission* of radio pulses in the direction of enemy aircraft (with the subsequent reception of a reflected signal) was no longer necessary. Only the receiver and the antenna were used to *receive* the strong radio radiation from the sun. By a stroke of good fortune, a prominent period of high solar activity began in 1946, coinciding with the end of WWII.

Already in 1946, the Australians were joined in a competitive race to study the radio sun by two groups in the United Kingdom at Cambridge and Manchester. Both the Australians and the British physicists and engineers had little or no astronomical experience; yet within a few years all these groups became a part of the existing solar physics communities. In Australia, the Sydney group at RPL was fortunate that Clabon ("Cla") W. Allen, a well-known optical, solar physicist working at the Commonwealth Solar Observatory (later Mt. Stromlo Observatory of the Australian National University in Canberra) became a collaborator. Cla Allen was fascinated with this new method to investigate both "solar noise" and "cosmic noise"; the latter consisted of investigations of the newly discovered "radio stars" or radio nebula as well as the background radio radiation of the Milky Way. RPL even assisted in the construction of a simple radio telescope at the solar observatory in

Canberra. The term "radio astronomy" only began to be accepted in 1948, having been invented by J. L. Pawsey at RPL and Martin Ryle at Cambridge in that year.

In the early post-war era, the rapid growth of solar noise research contributed to the development of many techniques used by radio astronomers. The solar radio groups in the UK and Australia initiated many observing modes that had lasting importance for the growth of radio astronomy in the following decades. Due to the rapid variation in the radio signals from the sun, both in time and frequency, the pioneering radio astronomers created complex instruments in the late 1940s to follow the changes in the solar radio emission over time. In addition, principles needed to interpret the radio radiation of the sun were applied in studying the radio emission from other objects in the Milky Way as well as external galaxies. A vast breadth and depth of knowledge has been gathered by astronomers about the sun and its place in the galaxy.

The following summary is intended to provide a succinct description of the current knowledge of the sun's structure. Many details can be found in recent popular books about the sun; an excellent example is *Nearest Star, The Surprising Science of Our Sun*, by Golub and Pasachoff, Harvard University Press, 2001.

There are a few hundred billion stars in our galaxy, the Milky Way, and the sun is located in what is considered an outer suburb, not at all in the densely packed galactic centre. Its distance is in fact about 26,000 light years—or 8 kilo parsecs, to use the unit of distance adopted by astronomers—from the centre of the galaxy. The time for a total revolution of the sun around the centre of the Milky Way is 225 million years. Astronomers have determined that the age of the sun is 5 billion years, compared to the age of the universe which is 13.7 billion years. The mean distance between the sun and the earth is what astronomers call one astronomical unit or 1.5×10^{13} cm.

Our sun, with a surface temperature of 5,800 K, is a typical G2V star—the "G" signifies a class of moderate temperature, the "2" indicates being two-tenths closer toward the slightly cooler K class, and the Roman numeral "V" indicates a main-sequence luminosity class. (Stars on the main sequence have an approximate proportionality between temperature and luminosity, the hotter stars having a higher luminosity. Stars at birth and close to their death phase do not lie on the main sequence.) The mass of the sun is 2×10^{33} g, which is 300,000 times that of the earth. The radius of the sun is 700,000 km, more than 100 times the radius of earth. Most of the mass of the solar system resides in the sun; only about 0.13 % of the mass of the solar system is in its planets. As an example, Jupiter's mass is 0.10 % of the mass of the sun; in comparison, Jupiter's mass is 318 times the mass of the earth. Thus the motions of the planets, asteroids, and comets are governed by the gravitational pull of the sun. In rare cases, comets can come close to the massive planets, causing major changes in the cometary orbits.

1. The Interior and Photosphere of the Sun:

The interior of the sun can be divided into three zones: the core, from the centre to 0.25 of R_s (solar radius); the radiative zone, from 0.25 to 0.7 R_S; and then the outer convective zone, from 0.7 to 1 R_s. In the core the energy of the sun is generated by nuclear fusion while in the radiative zone the energy is carried outward by radiation. Above this region, the energy is carried by convection, a process in which the matter is heated from below, transporting energy as the matter moves outward against the pull of gravity.

The apparent, visible surface at the outer edge of the convective zone of the sun is actually a region about 400 km thick from which most of the sun's visible light is emitted. This region, called the photosphere, is where the density drops considerably and the scattering stops. The photosphere is a very small region since the radius of the sun is about 700,000 km; the photosphere extends to a point where a photon of light would experience on the average less than one scattering before leaving the star. Even though the gaseous sun does not have a solid edge, this edge is visibly well defined and considered by many as the solar surface. The density at the outer photosphere is only about 2×10^{-7} g/cm^3, or about 10^{17} protons/cm^3. In this region the effective temperature is 5,800 K.

On a scale of about one arcsec ($^1/_{1,800}$ of the solar diameter), the sun's photosphere is composed of short-lived convection cells with a typical size of 1,000 km, which produce a "salt and pepper" appearance on the solar surface. These granules carry energy from the hot interior to the base of the photosphere by convection, but only the tops of these granules are observed in the photosphere. They are dark at the edge where the cool material is flowing down and bright at the centre where the hot material is upwelling.

Major features in the photosphere are sunspots and flares.

Sunspots

Sunspots are cooler regions in the photosphere; the cooler temperature is a result of a strong magnetic field, which suppresses the upwards transport of energy by convective action and leads to decreased temperature. Different latitudes of the sun rotate at different rates (differential rotation) causing shearing. The subsequent eddies and other motions in the convective zone may give rise to the magnetic fields that cause sunspots. Sunspot temperatures are some 1,500–3,000 K cooler than the photosphere. Since they are cooler than the background they appear as dark spots (Fig. 2.2), though they would still be blindingly bright if viewed in isolation from the much hotter surrounding regions. Sunspots are regions of intense magnetic activity, usually appearing in pairs that have opposite polarity, similar to terrestrial magnets (Fig. 2.3 shows a sunspot from 4 August 2011). Sunspots occur as part of the 11-year solar cycle, with an increased number of sunspots at solar maximum and a decreased number at solar minimum. As an example a solar maximum occurred in 2001–2002, while the next predicted solar maximum will be in May 2013. The number of sunspots in the 2013 cycle is predicted to be about 30 % lower than the previous maximum. For example, on 9 May 2012 a prominent sunspot (AR1476) was detected in the new solar maximum (cycle 24). The diameter of this sunspot was about 160,000 km and its area 1,050 millionths of the solar area. This

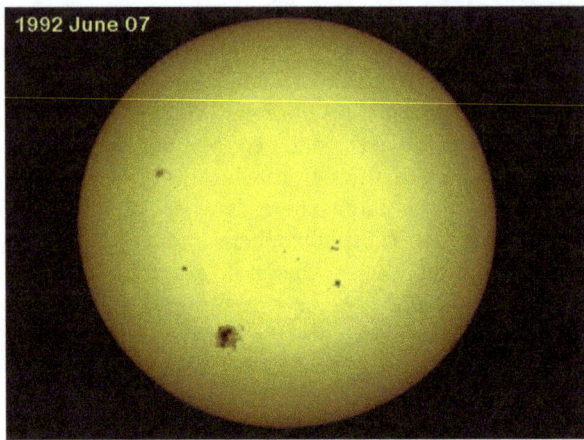

Fig. 2.2 The solar surface in visible light, near the maximum of the sunspot cycle of 1992. The small sunspots near the centre of the image are about the size of the earth (Marshall Space Flight Center Solar Physics web page "The Photosphere", http://solarscience.msfc.nasa.gov/surface.shtml)

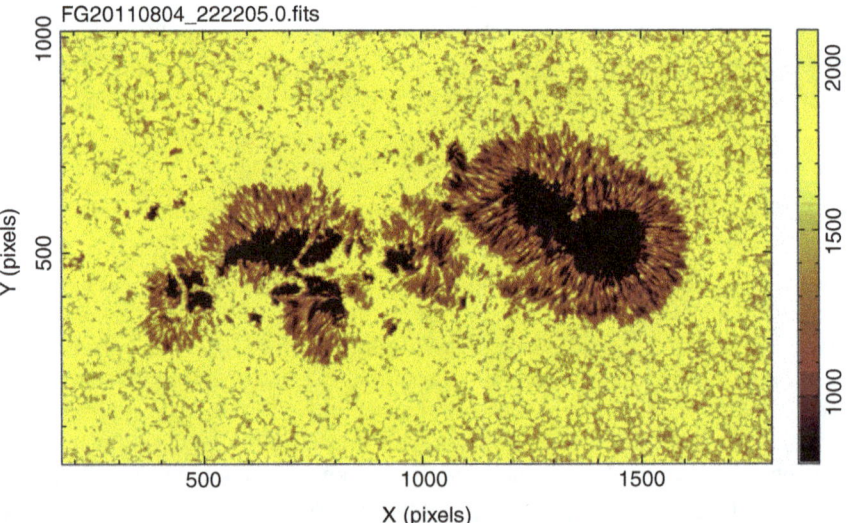

Fig. 2.3 High resolution image of a sunspot obtained with the Hinode (Japanese "sunrise") satellite, launched in September 2006. This is a cooperative mission between Japan, the US (NASA), Europe and the United Kingdom, consisting of a coordinated set of optical, extreme ultraviolet and X-ray instruments to investigate the interaction between the sun's magnetic field and its corona. This figure shows a high resolution optical (388–668 nm) image with a resolution of 0.2 arcsec. The pixel size is 0.08 arcsec; 500 pixels is thus 40 arcsec. The active region is AR 11263 from 4 August 2011. A few days later a prominent solar flare was produced (http://solarb.msfc.nasa.gov/news/12072012.html)

Fig. 2.4 Ultraviolet image of the entire solar surface facing the earth on 1 August 2010, obtained with the SDO (Solar Dynamics Observatory) satellite, launched on 11 February 2010 as part of NASA's "Living with a Star" program. The white area to the left centre shows a C-3 class solar flare. The colours in the image represent different gas temperatures. On 3 August 2010 prominent aurorae were observed in North America (NASA Image of the Day Gallery http://www.nasa.gov/multimedia/imagegallery/image_feature_1732.html)

size is five to six times smaller than the giant sunspots that Ruby Payne-Scott and colleagues observed during the prominent sunspot maximum of 1946–1947.

This 11-year cycle in sunspot activity was first observed by Samuel Heinrich Schwabe in the mid-nineteenth century. This German astronomer observed the solar surface for 17 years (1826–1843) hoping to discover a new planet, which was postulated to orbit the sun within the orbit of Mercury. The 11-year solar cycle of sunspots was found instead. The sunspot cycle is only a symptom of a more general activity cycle, driven by a magnetic dynamo operating in the interior of the sun.

Flares

Flares are an important constituent of solar behaviour. Figure 2.4 shows a prominent C solar flare as the white area in the upper left, while in Fig. 2.5 we see the famous "Seahorse" flare of 7 August 1972. A flare is a sudden, intense variation in brightness that occurs when magnetic energy is released. The temperature in a flare can reach 20 million K. Flare intensities are indicated—from weakest to strongest—in categories A, B, C, M and X. The scale is based on the peak rate of X-rays emitted by the flare and is logarithmic, like the Richter (earthquake) scale, thus B flares are ten times stronger than A flares, etc. Flares can originate in regions near sunspots, taking a few seconds to begin and lasting up to 4 hours. A typical flare lasts 20 min. Flares occur with rates from several per day when the sun is

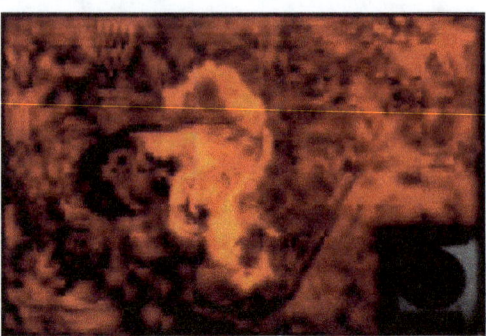

Fig. 2.5 The famous "Seahorse" flare as observed with the Big Bear Solar Observatory on 7 August 1972 in the H-alpha red line (656 nm) of hydrogen. This is an example of a "two-ribbon" flare in which the flare region appears as two bright lines threading through two sunspots (NASA "Solar Flares" http://solarscience.msfc.nasa.gov/flares.shtml)

active to less than one per week when solar activity is reduced. Many flares occur in conjunction with a coronal mass ejection (CME, see below), often likened to a large bubble of plasma erupting off the sun. (Solar fares can be observed without a CME and the latter can occur without the onset of a flare.) When the two occur simultaneously, often with large flares and fast CMEs, the event is called a "solar eruptive event" (Holman 2012).

Flares are observed at optical, radio and X-ray wavelengths. In the past, observations in the H-alpha line (the red line of the first Balmer line of hydrogen) were the most productive manner to detect solar flares. Energetic particles (electrons and protons) accelerated by the flare are detected at the earth after a delay of minutes to days following a strong flare on the sun. The radio burst connection with flares was established in the years 1946–1952 by the RPL group. The first flare in recorded history was discovered on 1 September 1859 by the English astronomer Richard Carrington. This observation was confirmed by another English observer, Richard Hodgson. About a day later a prominent geomagnetic storm was observed with auroras even at tropical locations such as Cuba and Hawaii. It is now widely believed that this flare may have been one of the most powerful flares ever observed.

2. The Outer Layers of the Sun

The solar radius extends from the centre of the sun to the top of the photosphere. The outer layers are found beyond the surface of the sun and consist of the chromosphere and the corona.

Most of the early Australian and UK observations of radio radiation arose from phenomena in the solar corona. This early, groundbreaking research made a major impact on the understanding of the outer layers of the sun. At the end of the exciting first decade of observations by solar radio astronomers, new physical processes were postulated to explain what they had found.

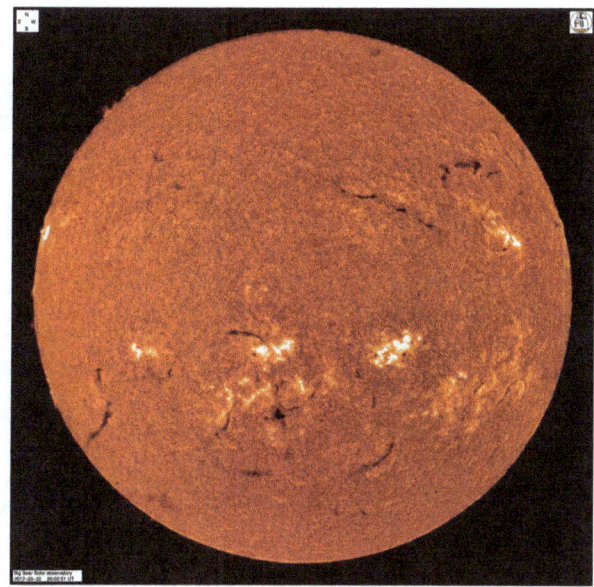

Fig. 2.6 H-alpha image of the sun's chromosphere on 30 June 2011 from the Big Bear Solar Observatory (BBSO) of the New Jersey Institute of Technology. Plages are the white area; solar filaments are also present (For latest H-alpha images from BBSO, http://bbso.njit.edu/Research/FDHA/)

Chromosphere

The **chromosphere** (the "colour" sphere) is a region 2,000 km thick that lies between the photosphere and the hotter, outer corona. In the chromosphere, the temperature rises from the photospheric temperature of 5,800 K to the one million degree temperature of the corona. The region is more visually transparent than the lower photosphere and is thus difficult to optically observe. At these higher temperatures hydrogen emits light that gives off a reddish colour (H-alpha emission); the name, chromosphere derives from this reddish colour.

Plages (French for "beaches") are bright regions of higher temperature and density within the chromosphere often close to sunspots (Fig. 2.6). In addition, plages can be present even in the absence of sunspots.

Prominences are a major component of the solar atmosphere; these are bright arch-like features that extend from the photosphere up to the corona. Prominences are stable structures with filamentary or braided shapes that appear to hang suspended above the surface (Fig. 2.7). Prominences are often associated with regions of sunspot activity, indicating that the sun's magnetic field plays a role in the formation of prominences. These features have typical time-scales of about a day and life-times of many days up to several solar rotations—one rotation being equal to about 26 days for a full rotation at the sun's equator as observed from the earth. The sizes are many thousands of km, with the largest prominences attaining sizes of up to 150,000 km. When a prominence is viewed from a different perspective, it has a different appearance. When the object is viewed face-on in the direction of the solar photosphere, the feature is darker than the surroundings (Fig. 2.6 and 2.8) and is called a **solar filament**. Prominences can become unstable

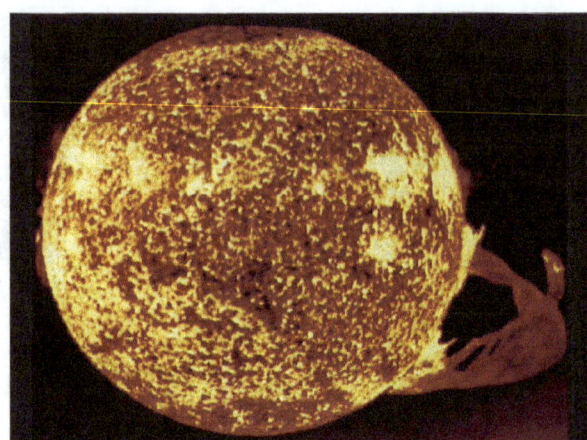

Fig. 2.7 The well-known image taken by the astronauts on Skylab 4 during the third and final mission. Image from 19 December 1973 showing a 600,000 km size prominence in the light of ionised helium (He II) in the extreme ultraviolet at 30.4 nm, provided by the U.S. Naval Research Laboratory. (Astronomy Picture of the Day, 30 August 1998, http://antwrp.gsfc.nasa.gov/apod/ap980830.html)

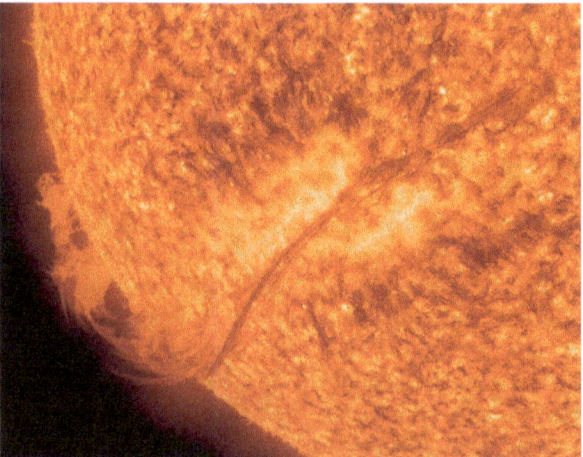

Fig. 2.8 Image of a prominence/filament from the Solar Dynamics Observatory on 6 December 2010, using the Atmospheric Imaging Assembly (AIA) at 30.4 nm in the light of ionised helium. It extends for over 700,000 km, comparable to the solar radius. This feature is called a prominence when seen bright against the dark sky above the sun's limb, and a filament when seen as a dark feature with the bright photosphere in the background, but both terms refer to the same physical feature. The SDO satellite (launched on 11 February 2010) is a part of NASA's "Living with a Star" program (http://www.nasa.gov/mission_pages/sunearth/news/News120610-filamentsnake.html)

and erupt, sending matter into space with velocities in excess of 600 km/s. These erupting filaments of gas are an impressive feature of solar activity due to their long lifetimes and energy release.

Corona

The **corona** is the extended outer atmosphere of the sun with a far larger volume than the photosphere. The typical height of the corona (depending on the 11-year solar cycle) is a few million km; in fact the corona is continuously expanding throughout the solar system forming the **solar wind**. The **heliosphere** is the region around the sun filled with solar plasma (ionised atoms); the size of the heliosphere is at least 100 AU in size based on the Voyager 1 spacecraft data. During periods of solar maximum the shape of the corona is roughly circular while during solar minimum the shape is elliptical with elongations along the solar equatorial regions.

Only in the nineteenth century was it realised that the corona belonged to the sun; earlier suggestions were made that this faint feature observed during eclipses might be a feature of the earth's atmosphere or that the features might arise from some phenomenon related to the Moon. The solar corona ("crown") is one of the more impressive components of the solar system. This extensive system occupied much of the attention of the new radio astronomers in the first decade after WWII. The temperature rises to about one million K at the base of the corona with particle densities in the range $10^{8-9}/cm^3$. This remarkable density decrease of a factor of 10 to 100 million occurs from the photosphere to the base of the corona. Within the corona, some regions with temperatures of several million K are also observed. The corona begins above the transition region, extending well beyond the solar surface. (Fig. 2.9).

Before the twentieth century the corona could only be studied during total solar eclipses. Spectra obtained in the nineteenth century showed unusual lines attributed to an unknown element, "coronium". This line at 530.3 nm is now known to arise from 13 times ionised iron (Fe XIV), (i.e., iron with 13 of the 26 electrons of the atom stripped away by intense collisions of protons in the corona). The hot corona of some millions of degrees was inferred by astronomers in the mid-twentieth century when these coronal spectral lines were identified with these highly ionised ions. Radio astronomers confirmed this hot corona a few years later by direct observations of the million degree corona.

Coronal Loops originate in the photosphere and extend through the chromosphere to the lower corona. Coronal Loops are observed in the ultra-violet (Fig. 2.10) and X-ray regimes (Fig. 2.11). These loops trace the magnetic field lines in the solar atmosphere with densities higher than their surroundings. These loops are associated with both active and quiet regions on the sun; the active regions produce the majority of the activity and are the source of flares and often a precursor of coronal mass ejections. The coronal loops are the closed magnetic flux field lines whereas open magnetic field lines result in the appearance of **Coronal Holes** (Fig. 2.12), regions of the corona that have a darker appearance due to the presence of lower density plasma.

Fig. 2.9 Solar eclipse photograph of 11 August 1999 taken by Luc Viatour (http://www.lucnix.be). The diffuse corona is clearly visible when the surface of the sun is blocked by the moon. Note the red (H-alpha) prominences around the limb of the moon (© Luc Viatour [CC BY-SA 3.0])

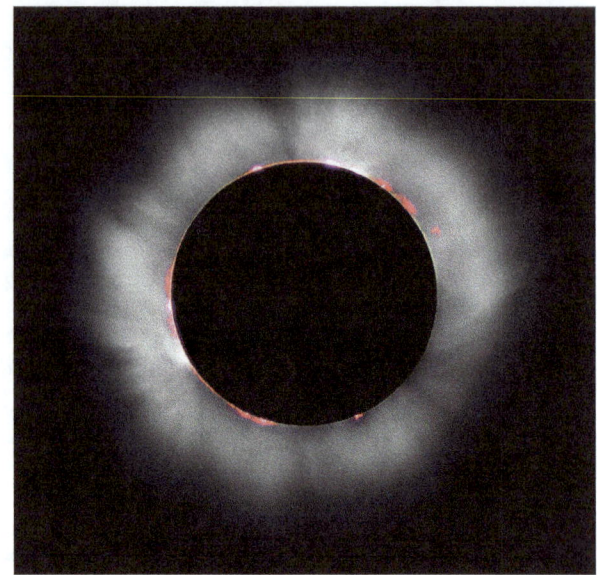

Fig. 2.10 Coronal loops observed with the Transition Region and Coronal Explorer (TRACE) at a wavelength of 17.1 nm, characteristic of hot plasma at 1 million K. Image from 6 November 1999. The satellite was launched on 2 April 1998. The Trace project is a mission of the Stanford-Lockheed Institute for Space Research and a part of the NASA Small Explorer program (http://soi.stanford.edu/results/SolPhys200/Schrijver/TRACEpodarchive.html)

Coronal mass ejections (**CME's**) represent a massive release of energy into the solar wind (Fig. 2.13). Recent research suggests that CME's are caused by magnetic reconnection, the rearrangement of magnetic field lines when magnetic fields of opposite polarity are brought together. This action leads to a large release of energy. The released energy and the associated matter may expand outward, causing a CME. Typical velocities are less than 100 to some thousands of km/s with a mean velocity of about 500 km/s; the energies are up to ten times that of flares. During solar maximum, CME's occur at a rate of about 4 per day, reducing to about one per 5 days in the solar minimum period.

Fig. 2.11 X-ray image from the Yohkoh ("Sunbeam") solar observatory from 24 January 1992. The joint project of Japan, US and UK was launched on 31 August 2000 in Japan. The image was taken with the Soft X-ray Telescope (SXT) with an angular resolution of 2.5 arcsec in the energy range 0.25–4 keV (wavelength range 0.3–5 nm). The coronal loops represent hot (above 2 million K) and large coronal magnetic structures (http://solar.physics.montana.edu/sxt/ under "Image Galleries" and then "High-resolution SXT Full-Sun images")

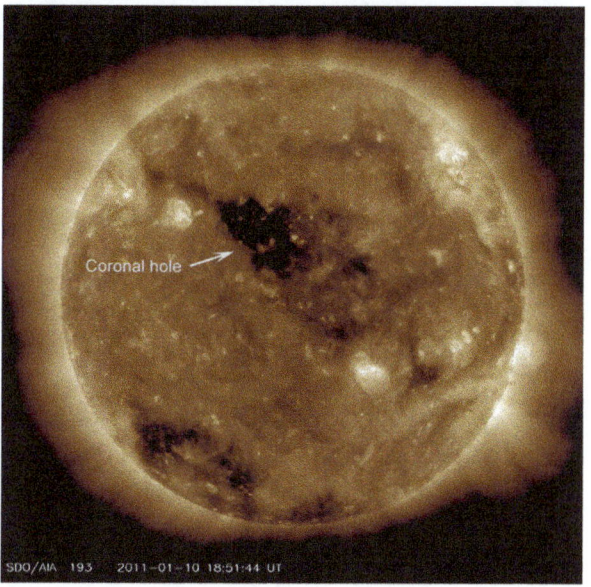

Fig. 2.12 An example of a Coronal Hole. Image from the SDO (Solar Dynamics Observatory), AIA, on 10 January 2011. The far ultraviolet image shows the dark coronal holes where the magnetic field lines are opening out to the interplanetary medium. These regions are also the sources of the fast solar wind of about 800 km/s. These particles will reach the earth in a few days, possibly causing aurorae (http://www.nasa.gov/mission_pages/sdo/news/news20110111-corona-hole.html)

As Ruby Payne-Scott and her colleagues began their exciting adventure in solar physics in 1945–1946, they could not have realised the important role they would play in the rapid advances in solar astronomy in the second half of the twentieth century. The knowledge that low frequency radio astronomy brought to understanding the physics of the solar corona had a far reaching effect. Payne-Scott's role in

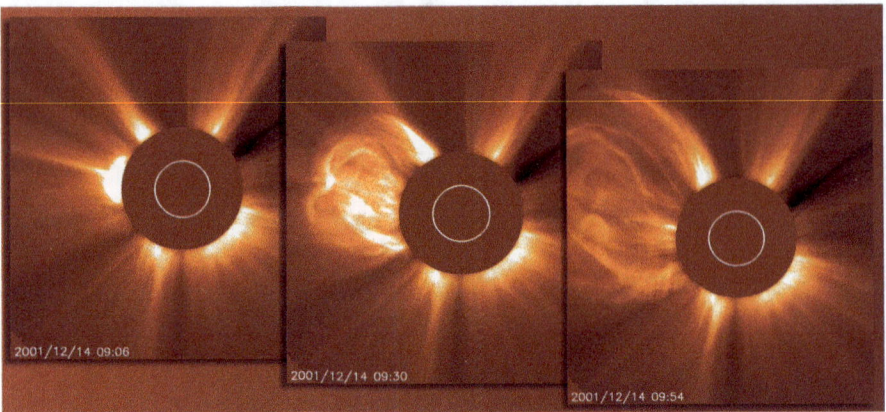

Fig. 2.13 Coronal Mass Ejection (CME). The SOHO satellite (Solar and Heliospheric Observatory, launched on 2 December 1995 and still operational in 2012) was used on 14 December 2001 to make a series of images of the sun. The LASCO (Large Angle and Spectrometric Coronograph) was used to create artificial eclipses of the surface of the sun (by blocking the bright solar image within the telescope on the satellite) enabling observations of the corona near the solar limb to a distance of 21 million km (about 1/7 of the distance from the sun to the earth). Using this instrument a number of sun grazing comets have been observed. These images were made over a 48 min interval and show a fast moving coronal mass ejection expanding at a speed of about 1,000 km/s. The sun is represented by the white circle. The field of view is 12 solar radii or 8.6 million km. This event was not directed towards the earth (http://sohowww.nascom.nasa.gov/pickoftheweek/old/17dec2001/index.html)

carrying out radio interferometry for the first time on Australia Day, 26 January 1946 has gone unrecognised. In her short period as an active radio astronomer (1944–1952), she made decisive contributions to the new field of radio astronomy. She discovered Type III bursts based on her work in 1947 at the Hornsby field station. Her understanding of these fast drifting radio bursts from the sun set the stage for the "most intensively studied form of radio emission in all of astrophysics" (Suzuki and Dulk 1985).

Tools of the Radio Astronomer

When Ruby Payne-Scott began her radio astronomy career in 1944, she was an experienced physicist who had worked on several radar research projects at the WW II Radiophysics Laboratory of the CSIR in Sydney, Australia. She was thus quite familiar with radio engineering techniques. The first radio telescopes were often radar antennas, altered only to receive radio radiation without the wartime practice of transmitting signals which were reflected from aircraft or ships. The new radio astronomers did not, in general, use the antennas as transmitters; instead they simply used the receiving equipment to detect the weak radio signals using the same antenna. There were, however, several groups working on lunar and meteor radar in the immediate post-WWII era, which did transmit signals to directly detect these objects with the radar technique.

What instruments did Payne-Scott use for her pioneering work of 1944–1951? How are these instruments related to modern twenty-first century instruments used by radio astronomers? The major research done by Ruby Payne-Scott in this era was in the VHF (Very High Frequency, 30–300 MHz or 10–1 m) range (wavelength and frequency are interchangeable; the wavelength is equal to the speed of light divided by the frequency). In this chapter, I will show a number of examples of the pioneering instrumentation used by Payne-Scott for the early solar radio noise research. I will show a few examples of twenty-first century radio telescopes that are the descendants of the post-WWII radio telescopes. The main achievement of Payne-Scott's career was the Swept Lobe Michelson Interferometer operating at a wavelength of 3 m (97 MHz) with which she recorded movies of the motions of solar bursts as they moved outwards in the solar corona. A major portion of the radio engineering planning was carried out by Payne-Scott.

Radio telescopes can be characterised by at least three attributes: (1) **sensitivity**, the ability to detect weak signals, (2) **angular resolution**, the ability to detect fine detail in the sky, and (3) **frequency response**, the determination of the intensity of the radio emission as a function of frequency or wavelength. For solar radio observations, high time-resolution (i.e. short time intervals) was essential. Payne-Scott's determination of the intrinsic frequency and time behaviour of the Type III solar bursts in 1946–1948 is an example of the latter property of a radio telescope.

The new solar radio astronomy of 1945–1951 had to address all of these attributes in order to decipher the mysteries of the radio radiation of the sun. By a stroke of good fortune, the beginning of 1946 coincided with a prominent solar maximum with large sunspots; these regions of the sun were associated with enhanced radio emission. Sensitivity was not a major limitation since solar bursts detected by Payne-Scott and colleagues had intensities in excess of a million Jansky. (**Jansky**, abbreviated as Jy, is the unit of intensity named after the American astronomer, Karl G. Jansky, who discovered radio emission from the Milky Way in 1933.) The sensitivity of her instruments was at the level of a few thousand Jansky, even enabling the quiet sun with intensities of 20–80,000 Jansky to be detected at wavelengths of 1–3 m (300–100 MHz). Modern radio telescopes such as the Jansky Very Large Array have sensitivities of a few millionths of a Jansky.

The relatively long wavelength of radio waves limits the angular resolution of the radio telescope. The radio wavelengths used in 1945 were a million times longer than optical wavelengths to which the human eye is sensitive. The typical angular resolution of a modest optical telescope is about an arcsec ($^1/_{1,800}$ of the sun's diameter, ½° as observed from the earth). The early radio telescopes had typical angular resolutions of tens of degrees, often referred to as the beam size. Thus in order to detect the details on the solar surface, the early radio astronomers needed to build special instruments called interferometers. By comparing different signal phases using an interferometer, much higher resolution (of the order of a fraction of a degree) could be achieved.

Fig. 2.14 Simple 65 MHz Yagi antenna used for routine solar radio astronomy monitoring. The antenna has an equatorial mount that enables continuous observations during daytime. The declination was set manually for each day. The antenna was located at the Potts Hill Reservoir (Chap. 10) of the Radiophysics Laboratory (RPL) of the CSIRO near Sydney Australia. The Yagi antenna was invented by Uda and Yagi in 1926 in Japan (CSIRO Radio Astronomy Image Archive B1465-1 from 26 July 1948)

In the following text, a number of radio telescopes (Figs. 2.14, 2.15, 2.16, 2.17, 2.18, 2.19 and 2.20) from the post-war era will be described. In addition, I will briefly discuss the Culgoora Radioheliograph (1967-1984); this instrument was the descendant of the ground breaking solar instruments constructed in Sydney in the post War era. In addition, two modern radio telescopes from the twentieth century will be discussed as well as a modern solar instrument planned for the second decade of the twenty-first century.

Yagi Antenna

In Fig. 2.14, we see the simplest element used by Payne-Scott in the late 1940s, the Yagi-Uda antenna, named after the discoverers of the device, Shintaro Uda and Hidetsugu Yagi of Tohoku University in 1926. It is commonly called a "Yagi" after the scientist who played the lesser role in the invention. This type of antenna is quite common for over the air VHF television and FM radio reception, whereas modern satellite television uses a high frequency, microwave dish antenna, much like modern radio telescopes.

Fig. 2.15 The 200 MHz shore defence radar at Dover Heights in Sydney during World War II. In the post-war era the antenna was converted to a radio telescope with no transmitter. In early 1946 this antenna was used for groundbreaking solar radio observations by Payne-Scott, Pawsey and McCready (Chap. 7). Payne-Scott carried out the first radio astronomical interferometry with this antenna on 26 January 1946 (Australia Day) at sunrise. See Fig. 7.2 (Copy obtained from the collection of W.T. Sullivan. Original from CSIRO Radio Astronomy Image Archive)

Broadside Array

In Fig. 2.15, the Shore Defence Radar (Sh.D.) at Dover Heights, in the eastern suburbs of Sydney is shown. This antenna was a WWII Australian Army radar used to detect enemy ships off the east coast of Sydney. J.L. Pawsey had been one of the major designers of this broadside array (200 MHz, 1.5 m wavelength) which consisted of 36 half-wave elements with a beam size of about 10°. The first use of interferometry in radio astronomy was carried out by Ruby Payne-Scott on Australia Day, 26 January 1946 at Dover Heights near Sydney using this radar antenna. This used the principle of **sea-cliff** interferometry. The antenna in Fig. 2.15 played a major role in the solar observations of 1946 and early 1947, but this instrument was scrapped by Bolton and Stanley in early 1947.

The principle of the sea-cliff interferometer is illustrated in Fig. 7.5 (see page 118). This interferometer is formed from the interference of the **direct** ray from the radiating source and the **reflected** ray from the sea; the waves from the two paths add in phase for some directions and cancel in other directions. The effective baseline of the interferometer, which determines the resolution of the interferometer, is twice the

Fig. 2.16 (**a** and **b**) The 100 MHz sea-cliff interferometer used by Bolton, Stanley and Slee at Dover Heights in the early 1950s at Dover Heights, used for the study of radio sources in the southern sky. This antenna was used for a survey that detected 104 discrete radio sources (Bolton et al. 1954) (CSIRO Radio Astronomy Image Archive)

height of the sea-cliff. This type of instrument is called a "Lloyd's Mirror". In late January 1946, Payne-Scott used this interferometer at sunrise to determine the size and position of the radio bursts associated with a major sunspot. Since the resolution was set by the height of the cliff above sea level, to change the resolution another location with a different cliff height would be required. John Bolton and Gordon Stanley, two colleagues of Payne-Scott, used this technique when they travelled to New Zealand in mid-1948 where the cliff heights were 300 m compared to the less than 85 m at the Dover Heights site in Sydney. At the Dover Heights site, the resolution of the sea-cliff interferometer was about 1/3 degree much less than the beam of the single broadside array of about 25 degree.

The second generation Dover Heights sea-cliff interferometer in 1952–1953 is shown in Fig. 2.16a, b. These show the last sea-cliff interferometer built for operation at Dover Heights at 100 MHz (3 m). The 6 by 2 array of Yagi's was

Fig. 2.17 (a) The 4.9 m reflector at Dover Heights used in 1950 by Stanley and Slee for an investigation of the properties of radio scintillation as a function of frequency. A striking star trail in visible light is seen during this long night time exposure from 16 July 1952. A similar photograph with John Bolton standing on the tower appears as a cover of Sky and Telescope, January 1953 (Bolton 1953). The star background includes the Southern Cross in the space between the reflector and the tower. The photograph is by the well known RPL photographer, Ken Nash who used a Rolleicord camera. A similar photograph appeared in Life magazine, November 1952 (CSIRO Radio Astronomy Image Archive, 2310-1 from 16 July 1952). (**b**) The 36 ft (11 m). transit parabola at Potts Hill. Sources were observed at transit (fixed east-west axis) with the telescope being moved in declination (north-south). The telescope was completed in 1952. The major use was for observations of the 21 cm hydrogen with an angular resolution of 2.8°. A map of the southern sky in neutral hydrogen was carried out as well as the detection of neutral hydrogen in the nearby galaxies the Large and Small Magellanic Clouds. The astronomers are left to right, Frank Kerr, Jim Hindman, Brian Robinson and Joe Pawsey. The 6 ft (1.8 m) reference antenna is in the right foreground (From W.T. Sullivan, originally CSIRO Radio Astronomy Image Archive, date circa 1953). (**c**) The 80 ft. (24 m) "hole in the ground" antenna at Dover Heights, Sydney, Australia (latitude -34° south) was completed in 1953. The survey of a limited part of the southern sky led to the radio detection of the galactic centre in early 1954. Dick McGee, one of the authors of the publication from 1954 (McGee and Bolton 1954), is shown in the photograph adjusting the mast of the telescope on 3 September 1953. With his adjustment, different regions of the sky near the zenith could be observed due to the earth's rotation. North Head at the entrance of Sydney Harbour is visible in the far distance, looking north (CSIRO Radio Astronomy Image Archive, B3150-1 from 10 Feb 1953)

mounted on an azimuth mounting and could only observe sources as they rose over the Tasman Sea. This publicity photo is somewhat ironic as this instrument was never used for solar research; likely the sun was only observed for testing purposes. Rather, a large survey of the sky from declination +50 to −50° (declination is the angular displacement with respect to the celestial equator, comparable to the latitude on earth) was carried out by John Bolton, Gordon Stanley and Bruce

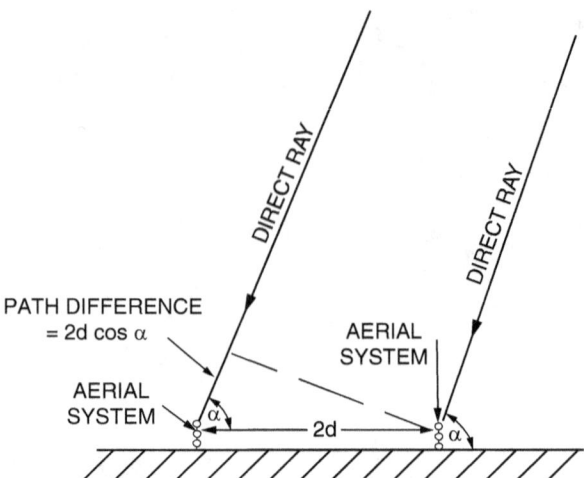

Fig. 2.18 A schematic diagram of the simple two-element Michelson interferometer, from Stanley and Slee (1950). Within a few years after 1950 most radio interferometry was done using Michelson interferometers (*Australian Journal of Scientific Research, Series A*, vol 3, page 234, 1950, "Galactic Radiation at Radio Frequencies, II. The Discrete Sources", Fig. 1b) (CSIRO Publishing, Copyright © CSIRO http://www.publish.csiro.au/nid.17.htm)

Slee. One hundred and four discrete sources (often called "radio stars" in this era) were detected. In reality radio nebulae were detected not the radio emission of stars.

Parabolic Reflector

For shorter wavelength observations (less than a few metres), the instrument of choice for the radio astronomer is the parabolic reflector, a paraboloid. This type of telescope was used in optical astronomy beginning in the seventeenth century; Isaac Newton is credited with building one of the first working optical reflectors in about 1668. In Fig. 2.17a, an early reflector telescope at Dover Heights is shown from 1952—a 16 ft. (4.9 m) reflector used to observe the intensity of strong radio sources at a number of frequencies; this type of observation enabled the determination of the **source spectrum**. At shorter wavelengths (for example 4 or 1.3 cm) it was necessary to use a solid surface; at 20 cm an open wire mesh was sufficient.

Also in 1952, a larger reflector was built at the Potts Hill reservoir site in the western suburbs of Sydney. This 36 ft. (11 m) transit telescope (Fig. 2.17b) was constructed mainly for 21 cm hydrogen line (the HI line arising from the neutral hydrogen atoms in the Milky Way) observations. The telescope could be positioned to detect most of the southern sky by moving the telescope only along the north-south meridian. The first HI line emission from an external galaxy was detected with this instrument. The southern Milky Way was mapped in detail, showing conclusive evidence for the spiral arm structure of the HI gaseous component of the Galaxy.

Fig. 2.19 (a) The 21 cm solar grating array designed by Christiansen, shown in the photograph, at Potts Hill, Sydney. The east-west array consisted of thirty-two, 1.7 m dishes providing a resolution of about 3 arcmin, a tenth of the solar diameter. The extent of the array was 213 m, and was completed in early 1952 (CSIRO Radio Astronomy Image Archive, 2976-1 from 14 January 1953). (b) Later in 1953, a north-south array was added to the 21 cm grading-array at Potts Hill. Sixteen elements were distributed perpendicular to the east-west array over a total extent of 160 m. Daily observations of the sun were made using both arrays from September 1953 to April 1954. The east-west array, comprised of solid surface antennas, is in the middle of the image, while the open mesh antennas extend northward, from the middle of the image to the bottom right (CSIRO Radio Astronomy Image Archive, B3475-1 from 25 October 1954)

A larger instrument was built in the early 1950s at Dover Heights with an ingenious design. This "hole in the ground" antenna could only observe close to the zenith (90° above the horizon) above Sydney, ideal for observations of the centre of the Milky Way which passed almost straight overhead. The 80 ft (24 m). diameter radio telescope (Fig. 2.17c) was used in January 1954 at 75 cm (400 MHz) to confirm that the radio source Sagittarius A was the centre of the Milky Way. This type of antenna had been invented at Jodrell Bank in the United Kingdom earlier in the 1950s. In the 1960s a large 1,000 ft. hemi-spherical "hole in the ground" antenna was built by Cornell University at Arecibo, Puerto Rico; this instrument has remained a major radio telescope into the twenty-first century.

Fig. 2.20 (**a**) The Mills Cross at Fleurs designed by B.Y. Mills, completed in 1954. At the time Mills was a staff member at the Radiophysics Laboratory of CSIRO; in 1960 he moved to the University of Sydney. This instrument operated at 80 MHz with the lengths of the arms measuring 450 m. The resolution was 0.8°; an all sky image was carried out with the detection of about 2,300 sources. North is to the top left and east to the top right (CSIRO Radio Astronomy Image Archive, B3476-4 from 25 October 1954). (**b**) Details of the construction of the Mills Cross at Fleurs. The view is to the north along the north-south arm. The building to the right (east) contained the receiver and control room of the array (CSIRO Radio Astronomy Image Archive 3454-1 from 7 October 1954)

Michelson Interferometer

The sea-cliff interferometer played a key role in high-resolution radio astronomy in the late 1940s in Sydney. The instrument, however, had a number of limitations. The major problem was the limited observing time of about an hour as the source—the sun—rose in the east over the Tasman Sea. A more flexible telescope of a two element interferometer was developed in 1946 at the University of Cambridge in the UK by Martin Ryle and colleagues. This Michelson Interferometer, named after the famous American physicist, Albert Michelson (1852–1931), could observe the radio sky over many hours a day. The radio astronomers in Sydney called this instrument a "vertical interferometer", a term that did not last long (Fig. 2.18).

The resolution could be varied at will simply by changing the physical location of the antennas on the ground. Thus the inflexibility of the sea-cliff interferometer was avoided. The main advantage of the Michelson interferometer has been summarised by Buderi (1996):

> By cabling two small aerials to a shared receiver, he could achieve the resolving power [the effective angular resolution] of a gigantic antenna with a diameter as great as the distance separating the two small arrays [sic]. With his interferometer, Ryle was able to narrow in on the solar region from which radio emissions arose...

The use of the Michelson interferometer was pioneered by Sir Martin Ryle (1918–1984) and his group at the Cavendish Laboratory at the University of Cambridge (UK). Ryle was awarded the Nobel Prize in physics (1974) for his developments in the field of aperture synthesis (simulating a large radio telescope by combing the signals from a number of small antennas spaced on the ground), based on the use of the Michelson interferometer to form an imaging radio telescope.

The first Michelson interferometer developed in Sydney was planned by Ruby Payne-Scott, Alec Little and Joseph Pawsey at the Potts Hill Reservoir (Fig. 10.5), page 176 in 1948–1949. The system was an ingenious swept-lobe interferometer that could follow and, using a movie camera, record the motions of the solar bursts. The instrument operated at 97 MHz (3 m wavelength) with three elements spaced up to 280 m apart. The group of Mills and Thomas used this instrument at night to observe the intense northern radio source Cygnus A. An individual element of the interferometer is shown in Fig. 10.4, page 176.

Solar Grating-Array

A completely new development in radio astronomy occurred in 1950–1953 at CSIRO in Sydney. W. N. "Chris" Christiansen and colleagues developed a "grating-array" consisting of thirty-two 1.8 m parabolic dishes spaced at equal intervals of 7 m over a total east-west length of 213 m. This instrument enabled the astronomer to make rapid images of the sun at a wavelength of 21 cm with a one-dimensional resolution of only 1/10 of a solar diameter of ½° (Fig. 2.19a); a year later a north-south array of 16 antennas was added (Fig. 2.19b). Using this instrument a two-dimensional image of the quiet sun was made during a period of low sunspot activity in 1953–1954 with a two-dimensional angular resolution of about 1/20°; the computing to form the image was done by hand over a 6-month period! (Wendt et al. 2008a).

Mills Cross

Bernard Y. Mills and Christiansen—both of the CSIRO—had a discussion in 1953 that led to the invention of the Mills Cross. A small prototype was built at Potts Hill in 1953, followed by the complete Mills Cross (Fig. 2.20a) at Fleurs (some 40 km west of Sydney) with both the east-west and north-south arms measuring 450 m in length and set to observe at 85.5 MHz (3.5 m wavelength). The details of this image synthesis instrument are shown in Fig. 2.20b, showing the intersection of the east-west arm and the north-south arm. With an angular resolution of 0.8°, the entire southern sky (declinations from +10 to −80°) was imaged. Over 2,200 radio sources were detected; with this number of radio sources it was possible to investigate the number of sources as a function of intensity, leading to some of the first conclusions of **radio cosmology**—the study of the structure and evolution of the universe using radio sources located at great distances from the Milky Way.

Later in the mid-1960s, Mills and his group (now at the University of Sydney) built the much larger Molonglo Cross with east-west and north-south arms, with each arm measuring 1.5 km in length, observing the sky at 408 MHz or 75 cm. The Molonglo instrument has been upgraded, remaining operational in the twenty-first century.

Culgoora Radioheliograph

The Culgoora Radioheliograph was opened in 1967, representing the most advanced solar radio telescope built by the CSIRO Division of Radiophysics. The instrument operated for 17 years—until 1984–in which a number of major modifications were made to extend the frequency coverage of the solar image synthesis (creating an instantaneous narrow beam by combining the signals of all the individual telescopes) instrument. This instrument provided a wealth of new information and insights into phenomena in the solar corona. The instrument consisted of a series of aerials arranged in a circular array with a 3 km diameter (Fig. 2.21a), located at a site about 600 km northwest of Sydney. The array consisted of ninety-six, 13 m diameter paraboloids with a simple, economical design (Fig. 2.21b). The initial operating wavelength was 3.75 m (80 MHz). Later wavelengths of 6.9, 1.9 and 0.9 m were added. At 80 MHz, the resolution was 3.8 arcmin over a field of view of 2°, creating a complete image in 1 only 1 second. The instrument could make separate images using different senses of circular polarization. The instrument was able to construct a two-dimensional image over a field of view corresponding to the entire solar corona (about 2° at 80 MHz) in an almost simultaneous manner and was ideally suited to observe time-variable phenomena over a wide frequency range. Many advances in solar physics were made in the lifetime of this instrument. During its lifetime, this instrument was also used to observe non-solar radio sources such as pulsars, radio galaxies and supernova remnants.

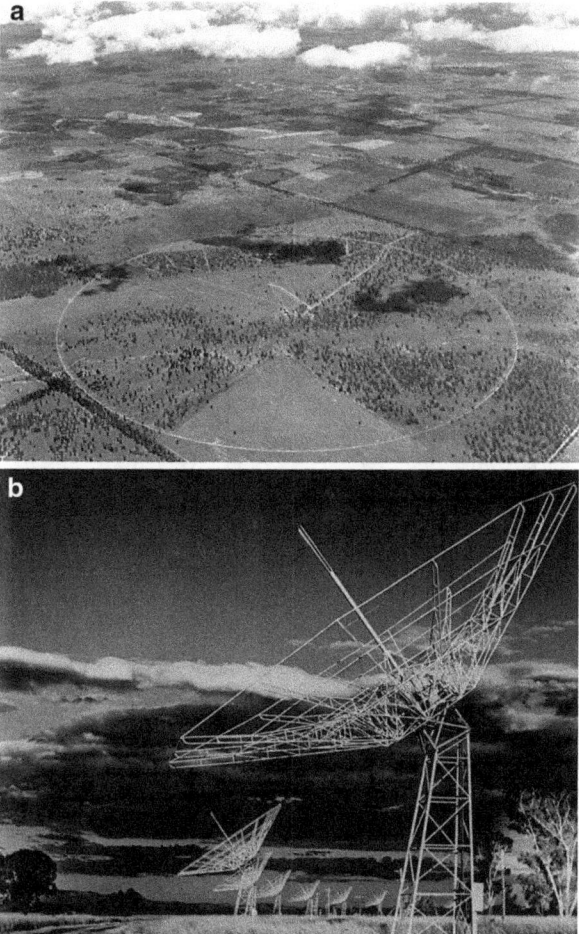

Fig. 2.21 (a) Aerial view from the southwest, of the Radiophysics Laboratory Culgoora Radioheliograph of CSIRO in north-central New South Wales, Australia. This instrument operated from 1968 to 1984. The 3 km diameter ring of 96 steerable 13 m paraboloids produced a beam of 3.8 arcmin over a field of 2° at 80 MHz. The instrument made images in circular polarization at intervals of 2 s, by scanning 48 beams across the sun. During the heliograph's remarkable lifetime, additional frequencies at 40,160 and 327 MHz were added. The driving force behind this project was J. P. Wild. The prominent road in this image to the centre of the array runs roughly east-west (CSIRO Radio Astronomy Image Archive, B7660-25 from 1964). (**b**) Close up of the Culgoora Radioheliograph antennas. The low-cost, 13 m antennas were characterised by simplicity. The Australia Telescope Compact Array of CSIRO (CASS) today occupies the site (J.P. Wild Observatory) near Narrabri, New South Wales, Australia. A few of the 13 m antennas remain (CSIRO Radio Astronomy Image Archive B8553-6 from 1967)

The Parkes Radio Telescope: The Dish

The Parkes 64 metre radio telescope is an icon in Australia; the dish celebrated its 50th anniversary on 31 October 2011, commemorated by Google with a Doodle of the Parkes telescope. The telescope is located near Parkes, New South Wales, Australia and is a facility of the CSIRO (Commonwealth Scientific and Industrial Research Organisation), CASS (Commonwealth Astronomy and Space Science). The importance of this instrument in the Apollo 11 lunar mission of NASA (National Aeronautics and Space Administration) during July 1969 is a well known story summarised by John Sarkissian at www.parkes.atnf.csiro.au/news-events/apollo11/. The actual moon walk by Neil Armstrong began at about 12.56 pm Australian Eastern Standard Time (AEST) on Monday 21 July 1969. The television images from the moon were shown around the world as received with the Parkes 64 m radio telescope. A fictional account of this event at Parkes has been captured in the 2000 Australian film, *The Dish*, directed by Rob Sitch and starring Sam Neill as the director of Parkes (a characterisation that bears little resemblance to the director at the time, John Bolton).

The telescope continues to maintain an outstanding research output. To date astronomers have detected about 2,000 pulsars in the Milky Way Galaxy. 1,250 of these pulsars were detected at Parkes since 1968, under the leadership of R.N. Manchester and colleagues. Numerous investigations of the interstellar medium in the Milky Way and in external galaxies have continued. The antenna was initially constructed for operation with an upper frequency of 1.4–2.3 GHz (21–11 cm wavelength). Presently the antenna is regularly used at 22 GHz (1.3 cm) for interstellar water maser and ammonia line observations; this remarkable change occurred as more accurate quality surface panels were added allowing the aerial to operate at shorter wavelengths. Fig..2.22a and c show the improved quality of the instrument from 1968 to 2011; both photographs were taken by Goss who started his career as a postdoctoral fellow at CSIRO in August 1967. Figure 2.22b is a striking publicity photo taken by Ken Nash in 1968.

Karl G. Jansky Very Large Array: The VLA

The Very Large Array of the National Radio Astronomy Observatory near Socorro, New Mexico, USA, is located on the Plains of San Augustin at an elevation of 2124 m. This aperture synthesis instrument is likely the most successful radio telescope built to date. The instrument was opened in 1980 and the completely renovated radio telescope was renamed and opened as the Karl G. Jansky Very Large Array on 31 March 2012. The updated array makes use of completely new electronics but reuses the original twenty-seven, 25 m antennas. These antennas can be moved to a number of configurations along the three railway tracks, each of length about 20 km. In 2012, the sensitivity has been increased by a factor of about ten. The renovation is a collaborative project between the USA, Canada and Mexico. The VLA covers a wavelength range from 7 mm (45 GHz) to 4 m (74 MHz)—a factor of about 600.

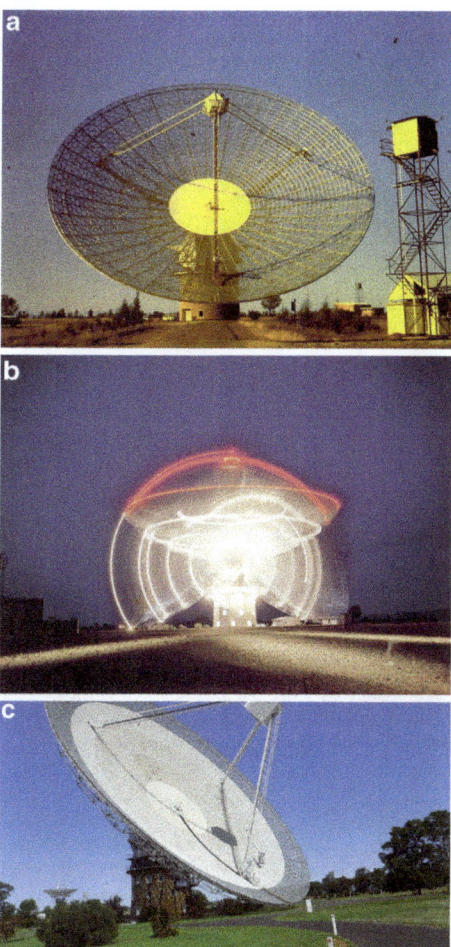

Fig. 2.22 (a) The Parkes 64 m radio telescope—The Dish—as photographed by Goss in 1968, after arriving in Australia as a NATO Postdoctoral Fellow in 1967. The solid inner surface has a diameter of 16.7 m. In 1968 the upper frequency limit of the telescope was in the range 6 to 3.6 cm. (b) The Parkes radio telescope in 1968. This Ken Nash publicity photo is a prominent icon in Australia (CSIRO Radio Astronomy Image Archive, 8886-1). (c) The Parkes radio telescope at the time of the 50th anniversary 31 October 2011. The new panels continue the progress over the last 50 years of improving the surface of the antenna to increase sensitivity at the higher frequencies (up to 1 cm). In order to support the NASA missions to Mars in 2003–2004 the mesh panels in the range 45 to 54 m were replaced with solid panels. After the refitting and extensive panel adjustments the roughness of the current surface of 54 m is only 0.8 mm, implying use to a wavelength of about 1 cm (Image taken by Goss, 2011)

The image in Fig. 2.23a shows the 'D' array, the smallest of the four configurations with a total extent of 1 km. In the C, B and A configurations the total length of the baseline is 3.6, 10 and 36 km respectively. The telescopes are moved from one configuration to the other about every 4 months; a double railroad

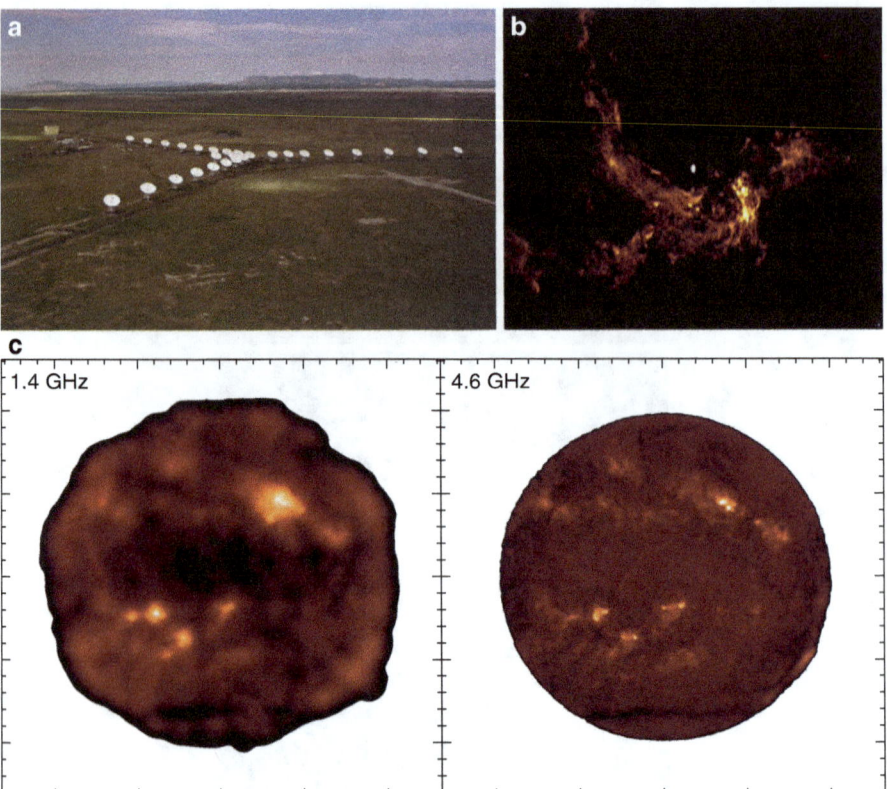

Fig. 2.23 (**a**) The Jansky Very Large Array of the National Radio Astronomy Observatory (NRAO) located on the Plains of San Augustin near Socorro, New Mexico, USA, a facility of the National Science Foundation operated under a cooperative agreement by Associated Universities, Inc. The location is at an elevation of 2,130 m. Each of the 25 m antennas can be moved along a railroad track to a number of different positions along three arms of a "Y". There are four possible configurations; A, B, C and D. The image here is the D array with a maximum baseline of about 1 km. Each array is progressively larger by about a factor of three. In the A array the total size is about 30 km. The frequency range of the radio telescope is from about 50 MHz to 50 GHz (Associated Universities, Inc). (**b**) During the last 30 years, the VLA has produced numerous images of new features in the Galactic Centre of the Milky Way. This image is a continuum image at 1.3 cm with a resolution of 0.2 by 0.1 arcsec of Sgr A West, a region of ionised gas that surrounds the centre of the Galaxy. The compact white region is the radio source Sgr A *—Sgr A "star"-associated with the four million solar mass black hole at the centre of the Milky Way Galaxy. A prominent feature is the spiral structure in the ionised gas; this is associated with a gaseous nebula Sgr A West. The total field size is 25 arcsec, about 1 parsec or 3 light years (Zhao et al. 2009; Associated Universities, Inc). (**c**) VLA images of the sun from 11 April 1999. Left is 1.4 GHz (21 cm) and the right is 4.6 GHz (6.5 cm), with angular resolutions of 30 and 12 arcsec respectively. The field of view is about 30 arcmin. A number of active regions are shown at a time close to sunspot maximum (Image provided by Stephen M. White)

track is used, with the location of each antenna site on a railway siding. In the A array at 7 mm, the resolution is 0.04 arcsec—the size of a golf ball as viewed from a distance of 150 km. The method of operation for aperture synthesis is the scheme proposed by McCready, Pawsey and Payne-Scott in 1947.

Since 1980, about 13,000 telescope proposals have been observed, written by scientists at universities and research institutes from around the world. The scientific merits were judged by a panel of fellow scientists. About 2,500 users have utilised the VLA in its 30 year history, including many from Europe, South America, Canada, Australia, and Asia. During this period, around 200 Ph.D. candidates have completed their doctoral research at the VLA, thus earning their degrees from their respective universities. Scientists have carried out numerous groundbreaking observations of radio emission from planets, the sun, galactic nebulae, stars and molecular clouds in the Milky Way Galaxy, the Milky Way centre, nearby galaxies and distant radio galaxies. One of the more exciting fields of research in recent years has been the detection of high red-shifted (due to the expansion of the universe) molecular gas from young galaxies as observed in the distant universe. An iconic image of the region near the Milky Way centre is shown in Fig. 2.23b. The intense white dot is the radio source associated with the four million solar mass black hole at the centre of the Milky Way. The spiral structure associated with the surrounding region is a gaseous nebula. In Fig. 2.23c, two radio images of the active sun obtained by Stephen White and collaborators with the Very Large Array are shown; the images were obtained on 11 April 1991. Radio emission associated with active regions near sunspots is clearly detected at both 1.4 GHz (20 cm) and 4.6 GHz (6 cm).

Frequency Agile Solar Radiotelescope: FASR

The next generation of advanced solar radio telescopes will likely be FASR, the Frequency Agile Solar Radiotelescope project. The proposed instrument would be constructed in the US by a consortium of universities and the National Radio Astronomy Observatory; the project was one of seventeen recommended for construction later in this decade by the National Academies of Science Astronomy and Astrophysics Decadal Survey in 2010.

The instrument would consist of three sets of antennas. The log periodic dipole array of 15 elements would cover the 50–350 MHz range, while the mid-frequency range from 0.3 to 2.5 GHz would consist of fifteen, 6 m antennas and the high frequency range from 2 to 21 GHz would consist of forty-five, 2 m antennas. The instrument will produce high quality images with a resolution of 1 arcsec at 20 GHz with high time resolution of 20 ms. Major goals are the study of the nature and evolution of coronal magnetic fields, the physics of solar flares, the driving forces of space weather and the physics of the quiet sun. Figure 2.24 shows an artist's conception of the high and intermediate frequency antennas in a log spiral configuration at the proposed site at Owens Valley Radio Observatory (of the California Institute of Technology) near Bishop California, USA.

Fig. 2.24 An artist's conception of the proposed FASR (Frequency Agile Solar Radiotelescope). This is a proposed ultra-wideband imaging array operating over the frequency range 50 MHz to 21 GHz and would be the most powerful radioheliograph in the world. The instrument will image solar radio emission from the middle chromospheres to the outer corona once per second. The proposed site is the Owens Valley Radio Telescope near Bishop, California, USA (California Institute of Technology). The high and immediate frequency (open structure) antennas are shown. Looking from the SW. The FASR Project is proposed to be managed under Associated Universities, Inc. in partnership with the National Radio Astronomy Observatory, the New Jersey Institute of Technology, University of Michigan, University of Maryland, University of California-Berkeley, California Institute of Technology and the Observatorie de Paris (Image provided by Tim Bastian, National Radio Astronomy Observatory)

Chapter 3
Ruby Payne-Scott's Ancestors and Her Early Childhood

Ruby Payne-Scott's family history reveals that her forebears believed in the importance of education; these ancestors included a number of independent-minded women. Payne-Scott, herself an educator, had many role models to follow as numerous women in her family tree were teachers. Similar to Ruby Payne-Scott, her female ancestors tended to have children later in life, while men in her lineage tried their hands at numerous professions.

Although little is known about Payne-Scott's maternal ancestors, her paternal ancestors can be traced back three generations to the United Kingdom, based on census data, birth, death and marriage certificates and the occasional public notice in a local newspaper. Utilising these sources, this chapter describes the history of Payne-Scott's ancestors in order to place her educational and professional achievements within the context of a family story of success and emigration to a new continent.[1]

Ruby Payne-Scott's Paternal Ancestors in the UK

In the early nineteenth century Ruby's paternal great-grandparents were living in the town of Tiverton, in the County of Devon, England. Tiverton is an old town on the River Exe, in the southwest of England. Once a flourishing centre of the wool industry, Tiverton, along with its wool trade had declined by the early nineteenth century—primarily as a result of industrialisation and competition from abroad. The economic conditions in Tiverton again improved with the opening, in 1815, of the John Heathcoat and Co. lace-making factory, which utilised new machinery and

---

[1] The details of Payne-Scott's ancestors were researched by Dr. Elizabeth Hall. Some details of the life of Hubert Payne Scott and his daughter Valerie Violet Payne-Scott have been modified since *Under the Radar, The First Woman in Radio Astronomy: Ruby Payne-Scott* was published in 2009. These changes have been provided by Dr. Hall. Many aspects of the Tiverton (Devon, UK) story were contributed by Cedric Ashton, Research Assistant, Tiverton Museum.

the recently popularised synthetic dyes. Within this factory Ruby's great-grandfather, John Scott, made a living as a dyer.

John Scott was born in 1801, well after the beginning of the Industrial Revolution. By 1830, when he married Margaret Payne, born in 1808, wages and conditions were such in Tiverton that the family could look forward to modest prosperity. John and Margaret had four children; Martha, Henry, Mary and Hubert. The elder son, Henry Thomas Scott[2] was born on 18 May 1835. Ruby's own grandfather, Hubert Payne Scott was born 7 years later in 1842, in the family home at 41 Peter Street in Tiverton.

At the time of the 1851 census, Hubert's oldest sibling Martha, aged 18, was single and still living at home, while Henry, aged 16, was a pupil-teacher,[3] and Mary and Hubert, aged 12 and 8 respectively, were still at school. By 1861, the census data indicate that John Scott had died, but that his wife was still living at the same address, along with her youngest child Hubert. At age 19 Hubert, Ruby's grandfather, was listed as the head of the household and a grain merchant by trade. From that same census we see that another "Payne" family was living next door, most likely Margaret's brother, Frederick.

The next appearance of Hubert in public records is in the 1871 census, which shows him living with a Sarah Payne, probably his maternal aunt, in Marylebone, a neighbourhood in central London. He may have been studying medicine at the time and courting his future wife, Agnes Duppuy. Agnes, born in 1847, was living with her widowed mother, Maria, at their home on George Street, in Croydon, a neighbourhood in south London. The exact circumstances of their meeting and courtship are unknown; in early 1872 they married in Hanover Square, which is in Mayfair, the neighbourhood adjacent and to the south of Marylebone. The same year of their marriage brought more delight to Ruby's grandparents, as Hubert became licensed to practice as a homoeopath by the London Society of Apothecaries in December, 1872 (L.S.A., London).

By 1881 Hubert was a general practitioner by trade. In their home at 8 Amherst Road, in New Cross, part of the borough of Lewisham in the southeast of London, Hubert and Agnes had three children: Marguerita Mary, born in 1875; Valerie

---

[2] Henry Scott had a most remarkable life. Like his younger brother, he became a physician. He was widowed in 1868 when his wife Sarah and daughter Margaret (born c. 1860) died in the Chincha Islands, off the coast of Peru. There is no information as to why the family was in South America. Henry then apparently remarried in the mid-1870s and became a clergyman in Swettenham in Cheshire. His name was still registered in the medical register. His second wife was Annie with a son, William, born c. 1878. In the 1881 and 1891 census records the family was living in Cheshire. He also appeared in a census record of 1901, visiting Mannington Bruce in Wiltshire. Henry was present at the death of his brother on 17 January 1917, signing the death certificate as H.T. Scott, M.D.

[3] A pupil-teacher would still be a student but assisting in the education of younger children. Henry was fortunate in attending the well-known Blundell's School, which had links to colleges at Cambridge and Oxford. In 1859, in his mid-20's he became a licentiate of the Society of Apothecaries, London and a Licentiate of the Faculty of Physicians and Surgeons, Glasgow. In 1861 he became a Licentiate of the Royal College of Physicians, London.

Violet, born in 1877; and Ruby's father, Cyril Herman, born in 1880. Agnes's mother, Maria, an individual with an interesting background, lived with the family. Not only was Maria born at sea but her maiden name, Duppuy—later Anglicised to Dupping—suggests a connection either with the Huguenots who fled from France to England in the eighteenth century or with refugees from the French Revolution. If she were still spry enough, she would have been ideally placed to help with the three children.

Emigration to Australia by Hubert and Agnes Scott

Homoeopathy had been popular in England for many years but by 1881 it was in decline. Although not rigidly enforced, the Medical Act of 1858, designed to control quackery, required health care practitioners to hold a medical degree.[4] Probably due to his early responsibilities this was something Hubert did not possess. Although it is unknown if these factors played a role in their emigration or if they were caught up in the search for a better life in Australia, Hubert and Agnes decided in the mid-1880s to emigrate to Australia from England. As was often the case with people emigrating, the move to Australia occurred in a two step process. In order to establish the family financially the father moved first, likely arriving in Sydney from London on 17 April 1885 on the steamship *Austral*.[5] Upon arrival he immediately began his medical career in Sydney as a "Medical Practitioner". He is listed as "Homoeopath", in the 1886 Australasian Medical Directory. Almost 2 years later he was joined by the other four members of the family—Agnes, and the three children, Marguerita, Valerie and Ruby's father, Cyril, aged 12 to 7. The four of them arrived in Sydney on the ship *Windsor Castle* from London on 13 January 1887.

Between 1886 and 1903, Hubert practiced at various addresses in Sydney, beginning with 181 Macquarie Street, which was and continues to be a fashionable location for the medical profession. This address on Macquarie Street appeared under his name in medical registries up until 1897, in addition to other prominent Sydney locations on King Street (1896, 1899) and George Street (1900, 1903).

During this time the exact family home address is not known. A common practice of the day was for the medical practitioner to work out of the family home. If this were the case for Hubert, the family would have moved many times until 1902. In 1897, Hubert Scott was listed in residence at 76 Edgeware Road in Newtown, near the University of Sydney. In 1898, a listing for his medical practice at 3 Trafalgar Street, Newtown appeared, while the next year a new address at 63

---

[4] Morrell, "A History of Homeopathy in Britain", 1998; Torokfalvy and Armstrong, "Homoeopathy in Australia—a Brief History".

[5] The passenger list contains a Mr. H. Scott. The identification with Hubert is not conclusive, but the date is consistent with events of the following 2 years.

London Street was published. In 1900 and 1902, a listing on Marrickville Road in Marrickville was published.[6]

Hubert was not shy about letting others know his opinions about issues of the day as illustrated through his argumentative correspondence to numerous Australian newspapers. As an example, in 1893, he wrote several letters to the *Sydney Morning Herald* with advice on how to manage the Australian economy and on the need to provide cinder-traps on steam trains to prevent bush fires. One can imagine that Hubert was very proud of his son, Cyril, Ruby's father, for being an accomplished chess player, often participating in the paper's chess tournaments.[7]

In 1891, with the full co-operation of his family, Hubert Payne Scott changed the family surname from Scott to Payne-Scott, possibly to avoid confusion with the very common name of "Dr. Scott". Hubert was now a respected member of the Sydney Homoeopathic community and his wife, Agnes, name occasionally appeared in connection with various charities. He became friendly with Dr. John Maffey, a well-known Melbourne Homoeopathic Physician and surgeon and worked with him towards developing a cure for cancer. In 1892, Hubert was also listed as providing services once a week to the newly-formed Homeopathic Dispensary in Sydney.[8]

Against this background of success, the events which followed are hard to understand. They may have been due to family friction or to a wish on Hubert's part to further his career. By 1903, the family was split apart by thousands of miles. Hubert left the family and returned to London from Sydney; he had been in Australia for 16 years.[9] Back in England, he resumed his activities as an active homoeopathic doctor. He is listed at various London addresses in 1907, 1911 and 1915. For example in 1915, his practice was at 21 Chilworth Street, London West (near Paddington Station). In 1917, Hubert died at the age of 76 in his brother's house (Dr. Henry Thomas Scott) on 17 January at 69 Mill Lane Hampstead. His brother Henry certified the death certificate; death was due to chronic bronchitis from which Hubert had suffered for about 10 years.

In the 1911 London census, Hubert still described himself as "married" although he had been apart from his wife and children for 8 years at that point. Back in Australia, Agnes listed herself in the 1903 postal directory as a music teacher, implying that she now needed to support herself after Hubert had left the family. By this time, the children were adults; Marguerita would have been about 28, Valerie about 26 and Cyril 23. Marguerita may have moved into her own home and Valerie

---

[6] The British and Australasian Medical Directories; the Australian Sands Street Directory, and the Victorian Government Gazette.

[7] "Master Cyril Payne Scott, one of the successful solvers in our first solution tourney, is hardly 16." *Sydney Morning Herald*, 3 August 1895, p.5.

[8] *Brisbane Courier* 12 December 1896, p.9 and Barbara Armstrong, *History of Homoeopathy in Australia* (www.historyofhomoepathy.com.au/people).

[9] The 1903 Medical, Dental and Pharmaceutical Directory of New South Wales listed Hubert with a practice at 50 George Street, while the UK Medical Register provided an address at 18a Ranelagh Road, London, W. (a location near Ealing).

Fig. 3.1 The house at 49 Liberty Street, Newtown in early 2007. This house was the home of Agnes Scott for some years into the twentieth century. After 1903 she was Agnes Payne-Scott, her new name. Her son (the father of Ruby Payne-Scott) lived nearby until moving to South Grafton, New South Wales, a few years later, perhaps circa 1906–1907 (Photo used by permission of Jan Christensen)

was a student at Hereford House, associated with Sydney Teacher's College.[10] There is a listing for only Agnes and Cyril living in a house named, "Pretoria", at 75 London Street in the neighbourhood of Newtown, but Valerie was probably still living at home.[11] Cyril and Agnes would remain physically close over the next 3 years, as Cyril remained a householder at 75 London Street, while Agnes took up residence with a housemate, Mrs. Sarah King, in the adjoining house around the corner, at 49 Liberty Street (Fig. 3.1). Agnes would remain at that address, even after Cyril moved on, until 1911 when she would move to 12 William Street in Ashfield, a neighbourhood to the west of "Pretoria" in Newtown. From 1919 to 1921 Agnes lived at 8 Holborow Street, Croydon, Sydney. On 6 February 1921, her death was registered in Croydon by her daughter Valerie. Hubert's marital status was not required on his death certificate and that of Agnes was left blank. In addition, her burial location remains a puzzle; the death certificate stated that burial was at Gore Hill, Sydney; however, the cemetery has no record of the grave.

Cyril's older sister, Marguerita Mary, became a nurse. Her name first appears in the electoral rolls for the Sydney suburb of Burwood in 1917 and again in 1921. In 1923 she married George Marks[12] in Petersham. George was 58 while his bride was 48 years old. No record of a shared address exists for Marguerita and George,

---

[10] Hereford House, an annexe of Sydney Teachers College. Later in 1919, while studying for a BA at the University of Sydney, Valerie's occupation was given as teacher and she was still living with her mother at 8 Holborrow Street, Croydon. By 1917, Marguerita was at 20 Edwin St., Croydon where she stayed until her marriage in 1923 (based on electoral rolls).

[11] Dr. Elizabeth Hall has written: "76 Edgeware Road, 3 Trafalgar Street, 63 London Street and probably Arundel Terrace, all rented by Hubert, are modest three-bedroom residences. 'Pretoria' at 75 London Street, is larger and more imposing. This was rented by Cyril. It is mentioned in a history of the area but there is no indication of whether or not it was used commercially. Agnes is shown as living there in 1903 but Hubert's name does not appear in connection with it or at any future address occupied by the family."

[12] George Marks (circa 1864–1951) emigrated from the UK to Australia sometime after 1891.

perhaps indicating that the marriage dissolved. This is further supported by the fact that in 1933 George is listed as living with his sister, Julia, in Lewisham, Sydney and that his death certificate in 1951 indicates "not married". Electoral rolls indicate that between 1930 and 1933, Marguerita was a nurse in the Croydon area; around 1936, at the age of 60 or so, she seems to have retired. During 1936 and 1937 she was a housekeeper in Lane Cove, another Sydney suburb and in 1943 her occupation was recorded as "home duties". Her last recorded address is 274 Addison Road, Marrickville—a "Home of Peace".[13] When she died there in 1963, the informant on her death certificate was her niece Ruby.

There was more contact with the other aunt, Valerie Violet; she was awarded a BA degree from the University of Sydney in 1925 at the age of 48 and a MA in 1932. Her thesis title was *The Medieval Woman in English Medieval Literature*. She was a teacher at North Sydney Girls High School until 1936. In 1937, she became a teacher at Girton Grammar School in Bendigo (Melbourne), an Anglican School (at that time) that had been founded in 1884. Possibly because of her own late educational development she was deeply interested in adult education. During a holiday overseas with a fellow teacher she spent time investigating continuing education and evening schools in England.[14] Valerie died in August 1948 and is buried in Rookwood cemetery in Burwood. Valerie's death certificate lists her niece Ruby Payne-Scott of 5 Fairleigh Street, Ashfield (Sydney) as the informant. Betty Hall has suggested that possibly Ruby had a greater level of contact with her Aunt Valerie than with Aunt Marguerita. In support of this, Ruby's son, Peter recalls his mother being very upset when Valerie died; also Valerie's name is written in a copy of Robert Louis Stevenson's *The Child's Garden of Verses* given to a young Ruby as a birthday present.

Around 1906–1907, Cyril moved some 600 km to the north, to an address in South Grafton, an area near the coast in northern New South Wales, 340 km south of Brisbane. The exact reason for this move is not known; perhaps it was to find employment. He was on the electoral roll in 1909 as "Cyric [sic] Payne-Scott" with profession "clerk". His mother, Agnes, and most likely his sisters, remained in Sydney. He met his wife, Amy Neale[15] in South Grafton, and they were married on 15 November 1910 at the Church of England there. The next day, an article in the local newspaper, *Grafton Argus*, under the heading "Wedding Bells", mentions that it "...was the scene of an interesting wedding. ... Mr. and Mrs. Payne-Scott are spending the honeymoon in the north". The article stated that Amy's older sister, Ruby Pearl Neale, was the bridesmaid while Cyril's occupation was "accountant".

---

[13] Braywood Private Hospital, 274 Addison Rd., Marrickville, a hospice for the dying, especially the underprivileged, based on a private communication from the archivist of Marrickville Council, 2012. Marrickville and Petersham are adjacent suburbs. Marguerita's death on 23 March 1963 is recorded as occurring at the "Home of Peace", Petersham.

[14] Passenger lists, *Sydney Morning Herald*, 18 January 1935.

[15] Amy was a school teacher born 1875 in Picton, NSW. Her parents were William Neale, an auctioneer born in 1840, who most likely died before 1910, and Ada Mary Moffitt, born 1846.

Fig. 3.2 The maternity hospital where Ruby Payne-Scott was born on 28 May 1912, Runnymede Private Hospital, South Grafton, New South Wales. Dr. Earle Page was the attending physician (Photo [circa 1910] used by permission of Frank Mack, President of the Clarence River Historical Society in Grafton)

Ruby's Early Childhood

Two years after the nuptials, our protagonist, Ruby Violet Payne-Scott, was born on 28 May 1912 in South Grafton. She was followed a year later by her younger brother Henry on 8 June 1913. Ruby was named after both her maternal and paternal aunts, Ruby Pearl Neale and Valerie Violet Payne-Scott. Ruby was born at Runnymede Private Hospital, a maternity clinic on Fitzroy Street in South Grafton owned by Dr. Henry. The birth certificate was signed by Dr. Earle Christmas Grafton Page, who later went on to be Australia's shortest serving Prime-Minister[16] and a Nurse Riordan. The hospital is shown, circa 1910, in Fig. 3.2. It is likely that the young Payne-Scott family lived with Amy's mother in "Uloom", the elegant Neale family house on Bent Street in South Grafton, shown in Fig. 3.3.[17] In this period, 1906–1914, there are numerous references in *The Clarence and Richmond Examiner* about Cyril's church work in the Church of England in South Grafton. He was an active Sunday school teacher and was elected to the South Grafton Parish Council; in October 1913, he was chairman of the local branch of the British and Foreign Bible Society.

By 1915, although several members of the Neale family continued to live in Grafton, the Payne-Scott family seems to have moved again. The details of this portion of their life remain quite uncertain. Ruby's children had memories of their

---

[16] Probably Page was a visiting physician at the Runnymede Private Hospital as he owned his own hospital, Clarence House on Through Street, South Grafton. Sir Earle Page (1880–1961) was born in Grafton; after finishing medical school in Sydney he opened the hospital in 1903. He was a member of the Australian Federal Parliament from 1919 to 1961. When Lyons died in 1939, Page was appointed Prime Minister and served for 20 days (7–26 April 1939). Page was the Minister of Health in the Menzies Government from 1949 to 1956.

[17] Details about South Grafton and the Neale family provided in 2007 by Frank Mack, President of the Clarence River Historical Society, founded in 1931 by Dr. Earle Page. Mack has published a feature article about Payne-Scott in the 18 November 2006 Clarence River Historical Society Newsletter (No. 94).

Fig. 3.3 A modern day view (circa 2007) of Ruby's maternal grandmother's house, "Uloom" in Grafton where the Payne-Scott family lived for a few years after 1911 (Photo used by permission of Frank Mack)

mother describing her life in Coonabarabran, NSW, for a few years before moving back to Sydney.[18] Recent investigations indicate that Cyril Payne-Scott and his family must have moved back to Sydney in about 1914. In fact, the local Grafton newspaper ran two notices in June 1914 which advertised the sale of furniture belonging to Cyril Payne-Scott owing to "early departure from the district". Possibly the memories that Ruby had discussed with her son, Peter, were related to childhood visits to relatives in Coonabarabran in the years 1919–1925.[19] At the time of the move to Sydney, Ruby was only 2 or 3 years old.

The Payne-Scott family was certainly established back in Sydney by 1914. The Sydney Diocesan Directories for 1915–1920 has a listing for C. Payne-Scott as a "Local Lay Reader" for the Church of England at Auburn, Sydney. The date of his appointment was 22 September 1914. He is also listed as a "Lay Reader" in 1916–1919. The Sands Directory notes that he was living at 37 Station Road, Auburn between 1915 and 1918. There is no address listed again until 1925, at which time the address is 118 Liverpool Street Sydney.[20]

Cyril tried his hand at running commercial businesses after moving back to Sydney. In 1925, Cyril was recorded as running a mixed business at 118 Liverpool

---

[18] The children of Ruby Payne-Scott have assumed that the Cyril Payne-Scott family lived for a few years in Coonabarabran, NSW (the future home of the Australian Astronomical Observatory, 450 km NW of Sydney); in fact the text in *Under the Radar, the First Woman in Radio Astronomy* is based on these memories of "living" in Coonabarabran.

[19] Interview with Peter Hall, February 2007. His mother had vivid memories of visiting Coonabarabran. from Sydney in the late 1920s on school holidays. She described the animals on their property (cats, dogs and rabbits) and visits from relatives from Grafton who complained about the summer heat in this inland community. These memories are consistent with Ruby visiting relatives in Coonabarabran instead of her parents.

[20] Due to possible street renumbering or even transcription errors in the records, the street number could be 118 or 156.

Street, Enfield.[21] The next year he was reported as having an additional address at 6a Burwood Road, Burwood, for a china, glass and earthenware shop, which existed for about a year. The Liverpool business continued until the mid-1930s.

---

[21] This building was later demolished; in 2007 there was no indication of the structure. When Payne-Scott entered Sydney Girls' High School in early 1926 the address listed on her enrolment form was her father's shop at 118 Liverpool Road. Ruby's Leaving Certificate of late 1928 listed the updated address, 156 Liverpool Street. Also her brother was pursuing the Intermediate School Certificate at a school in Summer Hill, only 5 km distant (National Archives of Austtralia-NAA: A9301, 20769).

Chapter 4
Ruby Payne-Scott's Education and Early Employment

Secondary School Education: Sydney

Ruby Payne-Scott must have been a noticeably bright child, and her parents must have planned to give her every opportunity they could for a good education, even if they could not afford to pay for her schooling at an elite private school. This goal is clearly based on the path her schooling took once she reached her pre-teens. Generally, a young person would attend school in Australia until the age of 14, at which point s/he would sit the examination for an Intermediate Certificate. One could not, however, enter university with an Intermediate Certificate. To do that, one would have to stay in school for another 2 years and pass the examination for a Leaving Certificate. Provided the requisite matriculation subjects had been taken, entry into university was then possible.

Based on her brother, Henry's discussions with a Royal Australian Air Force psychiatrist in late 1942,[1] it is likely that the Payne-Scott children were home-schooled by their mother in their early years. Amy Neale was a trained school teacher and Henry described being home-schooled by his Mother until he was about 10 years old, i.e. about 1923 or 1924.[2] Although Ruby was not specifically mentioned it seems unlikely that she would have attended school while her brother was home-schooled.

---

[1] National Archives of Australia- NAA: A9301, 20769 consisting of Henry Payne-Scott's war record. Henry had a troubled experience in the military in New Guinea which resulted in psychiatric care in 1942. From the mid to late 1930s to his death in 1970, he was estranged from his sister. The reason for the estrangement was likely caused by an event in the mid-1930s while she was still a student. She returned home from a vacation and found that Henry had sold her books.

[2] Amy Neale was first employed as a pupil-teacher at South Grafton Primary School in August 1889 at the age of 16; by April 1898 she had been promoted to a fully qualified grade IIIA teacher. Ruby was about 5 years old while the family was in Auburn. If she had been at school she would have been at either Auburn Public School or Auburn North Public School; unfortunately there are no records.

Starting in 1923, Ruby attended the Cleveland Street High School[3] in Sydney. She received her Intermediate Certificate[4] in 1925, making her 11 as she began Cleveland Street and reaching 13 by graduation, a younger age than average. This experience must have fostered independence. Ruby certainly did reach for her highest potential. She achieved excellent marks with A's in five subjects: English, Maths I, Maths II, French and Botany; and Bs in History, Geography and Latin. There was a report on 7 May 1934 in *The Sydney Morning Herald*, the major Sydney newspaper, that "Among girls who have gained noteworthy examination results [from Cleveland Street High School] were... Ruby Payne-Scott".[5]

With an Intermediate Certificate completed, Ruby was able to aim higher in the hierarchy of Sydney public schools. The Cleveland Street School would have acted as a feeder school for the prestigious Sydney Girls High School (SGHS).[6] Many of the public high schools in Sydney at the time were sex segregated; SGHS was one of the outstanding public high schools for girls in Australia. With a strong background from SGHS, Ruby was well-positioned to begin a university course in science.

The SGHS records indicate that Ruby entered the school on 9 February 1926 and finished her Leaving Certificate with honours in late 1928, when she was only 16. She was not quite 17 when she entered the University of Sydney in early 1929; again making her 1 year younger than many of her peers. A photo of her from this era is shown in Fig. 4.1. The final school report for Ruby at SGHS was indeed impressive with a pass at almost the highest level possible: first class honours in Maths I and Maths II and Botany. Her grades in English, Latin and Mechanics (Physics) were at the A level, while her only B was in French.[7]

Ruby left behind two articles in the SGHS publication, *The Chronicle*, in 1927. In the November 1927 issue she wrote an article about the "Pictures in the Library", in which she provided a piece by piece tour of the classic reproductions hanging in the school library, complete with the expectedly prosaic 15-year-old's opinions on each. Earlier, in June 1927, she had written a rather cute primer on how to succeed at SGHS, in rhyming verse:

---

[3] In 1912 or 1913, Cleveland Street High School became an Intermediate High School for both boys and girls. In 1929 (after Payne-Scott had left), the school became an Intermediate High School for boys only with the girls transferred to schools at Crown Street and Marrickville; the Cleveland Street name was maintained at the Crown Street site. Communication from Margaret Giannasca of Cleveland Street High School to Peter Hall, 8 March 1999.

[4] Susan Brian obtained a replacement Intermediate Certificate for Ruby Payne-Scott in August 2008 from the New South Wales Board of Studies.

[5] Newspaper article received from Margaret Giannesca, School Promotion, Cleveland Street High School, Sydney, February, 1999.

[6] Based on an interview in 1999 with Shirley Hoskin, the archivist of Sydney Girls High School.

[7] Much of the information about schools in Australia was provided by Rita Nash. Information about Sydney Girls' High School by Lilith Norman (*The Brown and the Yellow: Sydney Girls' High School 1883–1983*, 1983).

Fig. 4.1 Ruby Payne-Scott in the late 1920s (Bill Hall family collection, used by permission of Peter Hall)

TO A SYDNEY HIGH GIRL
(With apologies to Mr. Rudyard Kipling)
If you are always listening when you should be,
And keeping very quiet and very still;
If you can always see whate'er you should see
And tackle mathematics with a will;
If you can wait, and not be tired of waiting,
Although you're squashed in the Assembly Hall;
If you can work, and still not work be hating,
And yet on Wednesdays chase a hockey-ball;
If you are ready for examination
And need not cram your work right at the end;
If you can face that awful French dictation
As though it were your loving, kindly friend,
If you come first or last, although you've hard tried
And yet not let a change come o'er your face;
If you can lose just on the very last stride,
And yet can cheer the winner of the race;
If you can work, and yet find time for playing;
If you can play, and yet find time for work;
If you are always ready for essaying
Your task, and do not any duty shirk;
It does not matter whether you are clever,
It does not matter if you are a fool,

Fig. 4.2 The Payne-Scott family lived in Chatswood (10 Warrane Road) after about 1936; Cyril, Amy, Ruby and Henry were listed as living at this address in the late 1930s; Henry, an accountant, was estranged in later years from his sister. Ruby (University Demonstrator) was listed as living here from 1936 to 1938 (Photo used by permission of Jan Christensen, early 2007)

If you are all this, yet exams not weather,
Still will you be a credit to your school.
R. Payne-Scott (5B)[8]

By 1936, when Ruby was 24, her parents, Cyril and Amy moved to 10 Warrane Road in Chatswood (see Fig. 4.2), a Sydney suburb on the north shore. Cyril's occupation was listed[9] as homoeopathist, a return to the occupation of his father. There was no indication of a separate address for a medical practice. At this address, Henry (accountant) and Ruby (demonstrator at the University of Sydney) were also listed, Henry for the years 1936, 1937 and 1939, Ruby for the years 1936–1938. Ruby left for Adelaide in 1938 and returned to Sydney in mid-1939. Cyril died in 1942, followed by the death of his wife Amy in 1943. Thus at the time of the marriage of William Hall and Ruby Payne-Scott on 8 September 1944, both parents of the bride were listed as deceased. At age 31 years, Ruby had no surviving parents.

---

[8] The Leaving Certificate was taken in the fifth year of high school, thus the Fifth Form, now year 12 in NSW. '5B' was the level below '5A' of the Fifth Form classes.

[9] New South Wales Electoral Rolls of 1936–1939.

Bachelor of Science in Physics at the University of Sydney

Ruby began her studies at the University of Sydney in early 1929, before reaching her 17th birthday, with a merit-based bursary award and Science Exhibition scholarship from the university. She was an outstanding student, with an impressive record throughout her period as a student.[10] She completed the normal B.Sc. degree course in physics in 1929, 1930 and 1931 gaining outstanding marks; she achieved high distinctions in mathematics in all 3 years and in physics in 1930 and 1931. In 1931, she shared the Deas-Thomson Scholarship[11] with Reginald Healy for the "greatest proficiency in Physics III". She was also awarded the Walter Burfit scholarship for excellence in Physics III. Both awards required that the recipient be an honours student at the University of Sydney. In 1932 she spent the additional year completing Honours Physics, and was awarded an Honours degree, with first class honours, at the beginning of 1933.

Ruby was only the third woman to receive a degree in physics at the University of Sydney,[12] and in fact only the 90th individual to receive a honours degree in physics from the university.

Little of her experiences at the University of Sydney is known. From discussions with her son, Peter Hall,[13] we know that she found the mathematics lectures by H. S. (Horatio Scott) Carslaw (1870–1954) especially stimulating. Payne-Scott remembered that Carslaw read letters to the students during lectures from his former student John C. Jaeger (1907–1979), while Jaeger was a student at Cambridge (UK) in the early 1930s. During WWII, Payne-Scott and Jaeger became friends and colleagues, sharing a passion for cats.[14]

In the final year of Ruby's honours course, she published a short article in *Nature* entitled "Relative Intensity of Spectral Lines in Indium and Gallium" (Payne-Scott 1933). This short paper represented a portion of the research for her honours physics project. There is a single reference to a publication[15] with a discussion of similar results for thallium, by O. U. Vonwiller (1882–1972), who had suggested the project. He was a Professor of Physics at Sydney from 1923 to 1946. This paper by the 21-year-old Payne-Scott consisted of an analysis of photographic spectra of the two elements, indium and gallium, as well as alloys of indium with lead, in the

---

[10] Letter from Renata Mancini of the University of Sydney Archives to Peter Hall, March 1999.

[11] The Deas-Thomson Scholarship is named after the nineteenth century Vice-Chancellor and later Chancellor of the University of Sydney Sir Edward Deas Thomson.

[12] Edna Briggs (*née* Sayce) in 1917, Phyllis Nicol in 1926, Payne-Scott in 1933 and then Joan Freeman-Jelley in 1940 (see also Hooker 2004).

[13] Interview 12 February 2007.

[14] For a summary of Jaeger's life see Paterson (1982). Carslaw and Jaeger collaborated on a number of well-known applied mathematics books, including the well-known second edition of *The Conduction of Heat in Solids* (Carlslaw and Jaeger 1947). Ruby's and Jaeger's shared passion for cats is remembered by Peter Hall.

[15] *Physical Review*, 1930. "Intensity Measurements in the Arc Spectrum of Thallium", vol. 35, page 7.

range 4033–4511 Å. The short article presents the line ratios of the doublets of indium and gallium under various conditions; there is essentially no discussion of the results. The connection with Professor Vonwiller probably led to Payne-Scott's position at the Cancer Research Committee in the years 1932–1935.

Master of Science and the Cancer Research Committee at the University of Sydney

Ruby Payne-Scott began her association with the ill-fated Cancer Research Committee of the University of Sydney (CRC) in late 1932, where she continued to work for the following 3 years while finishing her M.Sc. degree.[16] Payne-Scott was awarded her M.Sc. degree in February 1936 and her thesis was published in *The British Journal of Radiology* in December 1937, having been submitted 13 months earlier; in the publication she lists her affiliation as Demonstrator in Physics, University of Sydney. While a M.Sc. student, she supported herself financially through positions at the Cancer Research Committee and as a Demonstrator/Tutor in the School of Physics.

The Cancer Research Committee had been formed in the early 1920s at the University of Sydney to foster research on the treatment of cancer. The history of the CRC has been described in detail by Hamersley (1988) and summarised by Hooker (2004). As Hamersley pointed out, from 1922 to the end of the enterprise in April 1938, its history was characterised by internal bickering and debate. Certainly the inexperience of this group of academics in funding, supervising and running such a research institute played a role in the ultimate failure of the CRC.

In the late 1920s, the opportunities for physics graduates in the cancer research field were enhanced by the need for an understanding of the physics of X and gamma rays. As Hamersley points out:

> For physicists this national ferment [new interest in the treatment of cancer by the use of X-rays] created some unique opportunities and challenges. Their participation in the national cancer effort encouraged the perception—new for Australians—that physics, traditionally one of the abstract sciences most remote from them, had something to contribute in matters of central human and societal concern. A limited number of new career openings for physicists also appeared, as a nation-wide scheme for providing physical services for radiotherapy, based on the physics departments of the state universities, gradually evolved. Finally, for physicists in the major centres of Sydney and Melbourne, there was the prospect of obtaining from cancer funds support for basic and applied research in radiation physics... [this] could bring significantly nearer realization of emerging ambitions to establish physics as a viable research discipline in Australia.

The CRC would have been an ideal place for a young physicist such as Ruby to gain professional experience and connections, earn a livelihood, and still work

---

[16] Based on the in-house journal, *The Journal of the Cancer Research Committee of the University of Sydney*, 1 October 1938, article by Vonwiller.

towards a higher university degree. Unfortunately, the CRC was almost completely devoted to researching the work of a single medical doctor, Wanford Moppett. In 1924, Moppett tried out a new theory of radiation therapy on chicken eggs that were in fact cracked open at the top to allow the extremely weak radiation into the embryo. After much criticism from the larger cancer research community, it was found that any effect upon the chicken eggs was probably caused by long exposure to a contaminated environment rather than the radiation. Hamersley (1988) pointed out that a number of investigators found that the changes might have been due simply to fungi or spore-forming bacteria in the exposure in the air, passing through the holes cut through the egg shells to allow the low energy X-rays to pass into the interior of the egg.

By the time Ruby Payne-Scott joined the CRC in 1932, the rationale for the research flagship of the enterprise was collapsing. What is more, in May 1934, while Ruby was working there the Director of Research, H. G. Chapman, committed suicide amid accusations of financial malfeasance and doubts increased about the reality of the Moppett Effect. In August 1934, O. U. Vonwiller and D. A. Walsh became the supervisors of research for the Cancer Research Committee. When the institute was closed in April 1938, Vonwiller wrote the apologia in the last issue of the in-house journal on 1 October 1938, placing much of the blame on the misdirection of Chapman.[17] As Hamersley has pointed out, the entire edifice established in the 1920s was flawed in contributing to this "collective folly". Even Vonwiller had played a role in committing resources to investigate the Moppett Effect.

Likely due to the fact that her previous advisor, Vonwiller, was the de facto director of the CRC during the final years (1934–1938), Payne-Scott was given a part-time appointment to support her research for her master's degree. On 10 December 1935, Payne-Scott and W. H. Love submitted a short research note, "Tissue Cultures Exposed to the Influence of a Magnetic Field", to *Nature* which was published on 15 February 1936.

The M.Sc. thesis presented by Payne-Scott was dated 28 February 1936 with a striking change to the title. The original title of "On the Amount and Distribution of the Scattered Radiation in a Medium Traversed by a Beam of X or Gamma Rays" shows that the words, "On the Amount and" are crossed out and the hand-written words "The Wave-Length" have been substituted. The corrected title is thus "The Wavelength Distribution of the Scattered Radiation in a Medium Traversed by a Beam of X or Gamma Rays". The paper (with the latter title) was submitted to the *British Journal of Radiology* on 17 November 1936 and published in Volume X, New Series, No. 120, December 1937 with essentially no changes from the thesis.

The publication summarised a theoretical treatment of the variation in the amount of scattered radiation as a function of primary wavelength and an investigation of the spectral distribution of the scattered radiation. Compton scattering, in which the incoming photons lost energy to the target electrons, produced a complex

---

[17] Ibid.

mixture of primary and scattered radiation. Payne-Scott wrote, "This softer radiation [Compton scattered] may, under certain circumstances amount to a larger proportion of the total radiation absorbed at a point, and thus may be of considerable importance in the study of the biological and other effects of the radiation." Since the scattering properties of water closely matched those of a biological tissue, Payne-Scott evaluated a number of the quantities derived for the scattering process for the particular medium of water. In addition, much of the experimental work available in 1936 was based on observations of radiation scattered from water phantoms (items substituted for real tissues in order to test the efficacy of the X-ray or gamma ray imaging). Strikingly none of the references in the paper referred to work done earlier at the CRC. Thus, there was no apparent connection with the earlier discredited research of this institute.

The mathematical skills that were to serve Payne-Scott well in her later scientific career at CSIR and CSIRO from 1941 to 1951 were clearly evident in this publication. "She was one of the very few people who had any association with the University of Sydney Cancer Research Committee who produced any good substantial work."[18] Although the official work of the CRC was shrouded in skepticism, Ruby Payne-Scott's research laid the foundation for much of her own future work on radar and later radio astronomy.

In the October 1938 article by Vonwiller, he specially praised Payne-Scott's research. After providing a detailed summary of the thesis, he added:

> Besides her routine duties Miss Payne-Scott did some original experimental work bearing on radiation measurement, and made an important and difficult theoretical investigation on the problem of quality and intensity of scattered radiation...Miss Payne-Scott must be credited with an important contribution, useful both for the results obtained and for the indication of methods of attack to be followed.

Ruby undoubtedly saw the end of the CRC on the horizon in the course of 1935–36. With her M.Sc. completed, and her current work situation crumbling, it would have made sense for the ever energetic Payne-Scott to go the extra mile and earn a Diploma of Education in 1937, from the Sydney Teachers College, a separate institution on the campus of the University of Sydney. Of course, in comparison to the majority of women seeking careers outside of family care in that era, getting a diploma of education would have been standard practice, while getting all the degrees with honours in physics would have meant going the extra mile. Ruby was not an average woman of her day. While it was conventional wisdom in those years for young men to obtain a safe, and if possible, permanent position of employment, women were more limited in their choices. Prior to World War II, women tended to work as nurses, teachers, typists, secretaries, clerks, nannies, and other jobs that could easily be left when family duties arose. While men could marry, have children and still maintain a career based on their inclination, luck and skill, a woman was generally forced to leave any upwardly mobile position as soon as she married, and certainly once she had children.

---

[18] Communication from H. Hamersley, 27 November 2006.

Ruby's own mother had been a school teacher, and it is doubtful that Ruby Payne-Scott was naïve about the career opportunities for women in the sciences at that time. Though her thesis advisor had recognised her brilliance and gotten her a position as a researcher at the CRC, Ruby would have been very aware that the demise of the CRC would mean one less in a limited pool of opportunities for a female physicist. The option of teaching was an extremely practical career choice on her part. Indeed she was to spend two periods as a secondary school teacher, 1938–1939 at Woodlands School and 1963–1974 at Danebank School.

Woodlands Glenelg Church of England Girls' Grammar School, Adelaide

As the Cancer Research Committee closed its doors in the late 1930s, there was a scarcity of permanent teaching positions at the University of Sydney; Payne-Scott must have been concerned about her future. During the course of 1937, she applied for a position as Science Mistress at Woodlands Church of England Girls' Grammar School (see Additional Note, No. 1, end of this chapter) in the sea-side suburb of Glenelg in Adelaide, South Australia. Payne-Scott completed the work for a Diploma of Education in 1937, and Sydney Teachers College awarded her the diploma in 1938. Obtaining a teaching post during those Depression years must have been an asset. Positions for physicists were indeed limited, especially for women. Employment outside universities was rare for any physicist. And so, at the age of 26, Payne-Scott joined the staff at Woodlands at the beginning of first term in 1938, remaining until the end of first term in 1939.

Payne-Scott only stayed at Woodlands for a year and a term, resigning in May 1939. Not only did she move outside New South Wales for the first and only time in her life, but she also moved from a research environment at the University of Sydney to that of a secondary school. In the 1960s she would return to teaching, at the Danebank School in Sydney. At Woodlands, Payne-Scott was a "boarding mistress", living in the school together with the student boarders. This meant that the school provided her accommodation but with the added responsibility of the supervision of pupils after school hours.

Jocelyn "Jock" Pedler (*née* Britten-Jones, deceased 11 February 2009)[19] from the class of 1939 was a boarder and remembered that Payne-Scott's room was on a mostly-enclosed balcony of the Law Smith building. This building is shown in 2007 in Fig. 4.3; the balconies were removed in the late 1930s. Although Jock Pedler was not especially interested in science, she was in the science club; she remembers that Payne-Scott was a reserved, serious teacher who suffered from poor eye-sight.

---

[19] Interview with Pedler, 23 March 2007 at the Archive Museum Woodlands Old Scholars' Association Inc. in the conference room at the Law-Smith Building at St. Peter's Woodlands Grammar School, Glenelg South Australia.

Fig. 4.3 Photo of the Law Smith building, St. Peter's Woodlands Grammar School, Glenelg, Adelaide, South Australia in 2007. (In 1938, Woodlands Church of England Girls' Grammar School.) In 1938–1939 Payne-Scott lived and worked here as a "boarding mistress" (Photo by Goss, March 2007)

Elizabeth Hallett, (*née* Brookman) (see Fig. 4.4), remembers that Ruby was "dedicated to the sciences and was a patient and inspirational teacher as well as having a quiet and pleasant personality...I cannot help thinking that her spell at Woodlands could only have been a...low spot in her career but the fact that three of the girls in the photograph (Fig. 4.4) gained their B.Sc.'s could have given her some satisfaction".[20]

This photograph shows Payne-Scott supervising a group of physics students, including Elizabeth Brookman (top), students performing physics laboratory experiments on the properties of inclined planes (middle) and on the density of matter with accurate weights (right). In Fig. 4.5, we see this same classroom used by a class for young children in 2007 at the renamed St. Peter's Woodlands Grammar School. In addition to the photo from 1939, a book celebrating the 50th anniversary of the Woodlands school[21] shows the entire school in 1938 just after Payne-Scott had joined the staff. Miss Monica Millington (OBE) had just become the second Headmistress of the school. Payne-Scott is clearly visible in this photograph together with approximately 23 other staff and about 220 students. As always, Payne-Scott can be identified by her blond hair and prominent glasses.

Payne-Scott was the president of the Senior and Junior Science Clubs in 1938 and the first term of 1939. In 1938, the club reported that "Miss Payne-Scott gave the Senior Club a fascinating talk on Evolution, and has helped members interested in photography to develop their own film". In 1939, the school magazine reported in the school notes section, "May 11–30. School closed for the vacation. We said good-bye to Miss Payne-Scott, who left us to take up a position in Sydney." The Science Club reported at the same time: "At the end of the first term Miss Payne-Scott left us to go to Sydney, and we would like to thank her for all she has done for us, and to welcome Dr. Gruenfeld as our new President... Later on in the first term

---

[20] Letter from Elizabeth Brookman Hallett to Goss, 22 June 2007.

[21] *Woodlands 1923–1973*, 1973, Chap. 4, "Problems and Changes".

Fig. 4.4 Photograph from 1939 of a physics class taught by science mistress, Ruby Payne-Scott, second from left. Elizabeth Brookman Hallett (*third from the left*) is the only student who was living in 2008. The unidentified student to the extreme left may be Jean Gooch (class of 1939) based on information from Jock Pedler in March 2007. From *Woodlands Reflections* (1999), on the occasion of the closing of the Woodlands School after 75 years (Photo provided with permission by Dawn Geyer of the Woodlands Old Scholars' Association Archives Museum, 2003)

Fig. 4.5 The science classroom used by Payne-Scott in 1939 in the Gillam Building, St. Peter's Woodlands Grammar School as photographed in March 2007. In 2007 this room was a Special Education Classroom with teacher Susan Bennett. In 2007, the school offered both remedial and gifted programmes for kindergarten to year 7 students (Photo by Goss)

[while Payne-Scott was still the Science Club President], we spent an instructive hour at the Brighton Cement Works. After becoming rather hot and dusty, we appreciated the ice creams which were so thoughtfully provided for us."

Kathleen Foy (*née* Bampton) was in Payne-Scott's botany and physiology class in 1938. This class was a requirement for the Intermediate Public Examination Certificate, which Bampton obtained. Bampton remembered a field trip to a local creek near the school during which both flora and fauna were collected.[22] Kathleen Foy crossed paths with Payne-Scott later, during the years 1963–1974, when both were teachers at Danebank School.

AWA- Amalgamated Wireless, Australasia: 1939–1941

The year 1939, with the approaching war in Europe and Asia, saw the expansion of the electronics industry in Australia; thus additional opportunities for physicists were available. At this point Payne-Scott left her job as a girls' school boarding mistress and joined Amalgamated Wireless, Australasia (AWA) for a period of slightly over 2 years, until August 1941. The approaching World War would have a major impact on her life and professional career.

Payne-Scott joined AWA as a librarian in mid-1939. Her colleagues were, among others, Chris Christiansen and A. L. Green.[23] Chris Christiansen wrote in October 1997[24]:

> I knew her before anyone else in R.P. because she came to work at AWA while I was there. She was put onto rather boring work and didn't last very long before applying for a job at R.P., where she was followed by Lindsay McCready, John Downes and a couple of others. I didn't go with her because I had a most interesting job in the overseas radio communication and also because I didn't want to get into 'secret' work particularly as I had a record from undergraduate days of being a left-wing militant and would probably be rejected for such government work (as indeed I later found that I would have been). I had no contact with Ruby until after the end of the war and could join the group at R.P. when they were starting work on radio astronomy.

With Marie Clark (Hooker 2004), Payne-Scott was only one of a small number of women on the professional staff at AWA. After a period of being in charge of the measurements laboratory, she was able to branch out into research, undoubtedly due to her expertise in physics and mathematics as well as her passion for electrical engineering. She was not pleased with the quality of the research atmosphere at AWA. Her son, Peter,[25] remembered his mother telling him that AWA had major difficulties finding qualified engineers in 1939. One striking story was about a

---

[22] Communication from Foy to Carolyn Little in 1999; communication to Dick McGee in June 2007.

[23] Green (1905–1951) was the Director of Research at AWA from 1941 to 1947. Green was a ionospheric scientist who had been recruited by the Australian Radio Research Board by J. P. V. Madsen in the early 1930s, having finished a Ph.D. with Appleton at King's College, London in 1934. In 1935, he joined AWA (Schedvin 1987). From 1947 to 1951 Green was in charge of the Ionospheric Prediction Service in Australia.

[24] Letter to Goss. Undated letter, received in Socorro, New Mexico, 7 October 1997.

[25] Interview with Goss, Socorro, New Mexico, 12 May 1998.

woman with a meagre background and no university qualifications who was recruited from the drawing office at AWA to join an engineering division. This individual expressed her misgivings about her lack of qualifications to Payne-Scott, who replied, "Never mind. Just put on a white coat and put a slide rule in your lapel pocket. No one will find out!"

An additional woman on the professional staff at AWA besides Marie Clark was Grace Noble. Grace Noble[26] and Ruby previously had a brief overlap as students at Sydney Girls' High School; she left in 1926, the same year that Ruby arrived. Noble had a geology degree from the University of Sydney. She worked initially at AWA doing mathematical calculations. When Payne-Scott left AWA in 1941, Noble was recommended by her as her replacement. Noble wrote, "I don't think I was replacing Ruby in any real sense, but I do know that radar... [was] being perfected." Noble worked on the problem of finding pure quartz crystals for radar and radio receivers, in Australia, as the supply source in Brazil disappeared during the War.

Payne-Scott was involved in two publications at AWA. In 1941 she wrote an article for the *AWA Technical Review*—it is likely that she was the editor of this journal for some period. The publication, "Superheterodyne Tracking Circuits-II"[27] was the second in a series of papers (the first was by Green 1941) on the problem of obtaining solutions for the tracking between the signal and oscillator circuits of superheterodyne receivers (as in FM and TV receivers); the wide band higher frequency was "mixed" (combined) with the local oscillator to produce a difference signal that was amplified and demodulated to receive the input video or audio signal. Lookup tables were presented to facilitate computations over a wide range of conditions.

The final publication by Payne-Scott was also published in the *AWA Technical Review*, but only after she had left AWA for CSIR in August, 1941. The paper, "Note on the Design of Iron-Cored Coils at Audio Frequencies" (Payne-Scott 1943), consisted of a detailed discussion of the calculations of the "Q" factor (magnification factor) of audio circuits for iron-cored coils at frequencies of 1–10 kHz (such as would be present in a modern stereo system).

These two publications indicated that Payne-Scott was certainly adept at finding solutions to practical electrical engineering problems. These skills were to serve her well as she began the next phase of her career, working on military radars necessary for the defence of Australia in the Second World War.

---

[26] Letter to Goss, 22 March 1999. Noble also knew Bill Hall and Ruby in the Sydney Bush Walkers organisation (Chap. 13).

[27] Ruby Payne-Scott and Alfred L. Green 1941; the article was later reprinted in *Wireless Engineer*, 1942, vol. 19, p. 290.

Additional Note

1. Glenys Edwards, a volunteer librarian in the Patchell Library Historical Centre at Annesley College in Adelaide, was the facilitator for much of the information received about Payne-Scott at Woodlands School. In March 2007, Miller and Libby Goss visited the school (now St. Peter's Woodlands Grammar School) with Edwards. Wendy Davis, Development Officer of St. Peter's Woodlands Grammar School provided tours of the school. In 1998, the original Woodlands Church of England Girls' Grammar School closed; the school had been founded in 1923. In that 75 year period over 7,000 students attended the school. The new co-educational school, St. Peter's Woodlands Grammar School, opened in early 1999 with a staff of 26 and 290 students. The Woodlands Old Scholars' Association Inc. has an office in the Law Smith Building, maintaining the archives of the previous school. W. M. Goss, Libby Goss and Edwards were hosted by the Woodlands Old Scholars' Association Archives Museum Group, including Dawn Geyer, Jill Colley and Mary Carver. Jocelyn Pedler (*née* Britten-Jones, formerly Cherry), class of 1939, was also present at Woodlands on this day, providing first-hand memories of Payne-Scott in 1938–1939. Pedler had been a boarder and thus provided additional information about Payne-Scott's role as a "boarding mistress".

Chapter 5
Wartime Research by Ruby Payne-Scott at the Radiophysics Laboratory

New Career Opportunities for Australian Women, World War II

Many of the male scientists of Australia had enlisted and were participating in the war in Europe beginning in September 1939. Thus as Australia became a major target for the Japanese armed forces in December 1941, women scientists had for the first time, an opportunity for employment outside the usual avenues in health services and education. In the 1930s, the University of Sydney Appointments Board conducted surveys of job advertisements in major Australian newspapers, and found that sex was a major factor for most employers in choosing an applicant. Hence female science graduates were urged to learn to type and perform shorthand (Carey 2002). Carey has quoted the 1941 Sydney Board:

> Never before in the history of the world has there been so great a demand for women with scientific knowledge.... Jobs which have...been the prerogative of men have opened their doors wide to women.... Employers who once refused to take women scientists are now begging for them.[1]

The number of women working in Australia in wartime scientific endeavours is hard to estimate. At RPL in 1945 there were two women working in a scientific staff of about 60.[2] In a publication in July 1943, *Australian Women at War*, edited by Mollie Bayne, the emphasis is on the 30,000 women in three branches of the 1943 armed forces—Army, Navy and Air Force[3]—and the estimated 200,000 in

---

[1] "Departing from their Sphere? Australian Women in Science, 1880–1960." In: Xavier Pons (ed.) *Departures: How Australia Reinvents Itself*, p. 181. (2002).

[2] Payne-Scott and Freeman. Rachel Makinson worked at University of Sydney, partially supported by RPL. The total staff at this time was about 300 (Evans 1970).

[3] The total number of women in the armed forces of Australia from 1941 to 1946 was about 65,000 (Thomson 1991) with 27,000 in the Women's Auxiliary Australian Air Force (WAAAF), 24,000 in the Australian Women's Army Service (AWAS) and only 2,000 in the Women's Royal Australian Naval Service (WRANS). About 4,000 women served as nurses in separate branches of the air force and army. The total number of males to serve in the Royal Australian Air Force (RAAF) in WWII was about 200,000.

industry. At that time it was estimated that a total of 580,000 women were involved in the war effort when the population of Australia was about seven million. Only 10 % of the women in the war effort worked for government entities, including local, state and commonwealth. Scientists did not appear as a separate category in this 1943 study by Bayne.

Bayne also provides a stirring call to arms for women in the post-war Australian society. She complains that there were no women in the Federal Parliament and only one woman on a Federal Government Board or Commission concerning female employment, at a time when most Australian working women were earning about half the wages of men. She compares Australia unfavourably with the achievements of the UK, the USA and the USSR in creating a more equitable work environment for women. She asks:

> Are they [the women] less capable? Or is the community [Australia] mediaeval-minded in the matter of a "woman's place"?

What about other prominent women in war-related research? Florence Violet McKenzie had become an electrical engineer based on course work at the University of Sydney and later Sydney Technical College in 1923.[4] In 1940, she formed the Women's Emergency Signalling Corps and succeeded in getting 14 girls, trained as wireless telegraphists, into the WRANS (Women's Royal Australian Naval Service) on Anzac Day (April 25) 1941. As a volunteer, she was in the end responsible for training 12,000 servicemen, as well as hundreds of women who had joined the women's armed services. She was made an honorary Flight Officer in the WAAAF (Women's Auxiliary Australian Air Force). Patsy Adam-Smith has written:

> After Japan entered the war, Mrs. Mac's [McKenzie] girls were already working as the nucleus of a service which, at the time of its greatest expansion, amounted to approximately ten percent of the Royal Australian Naval Service.[5]

There is little doubt that McKenzie, Payne-Scott, Rachel Makinson, Joan Freeman and many more Australian, American and British women all played significant roles in the Allied war effort (see Additional Notes, No. 1, end of this chapter).

The formation of RPL was based on the activities of the Radio Research Board (Minnett 1999), and its purpose was to coordinate wartime research into RDF (radio direction finding) or RADAR.[6]

---

[4] As was Payne-Scott, Florence McKenzie (*née* Wallace) was a graduate of Sydney Girls High School. She was the first woman radio amateur (ham) in Australia.

[5] In *Australian Women at War*, 1984, page 210 by Patsy Adam-Smith—an additional book with the same title as the Bayne volume.

[6] RDF (Radio Direction Finding) was the British term which evolved into the US acronym RADAR (**RA**dio **D**etection **A**nd **R**anging). This term was invented by S. M. Tucker of the US Navy in 1940 (Louis Brown 1999). As Hanbury Brown (1991) wrote: "Some years later we [the British] adopted the name Radar which, as Watson-Watt [one of the UK inventors of Radar] used to say in mildly disparaging tones was a 'synthetic palindrome invented by our friends the Americans'." The British adopted the new term RADAR after 1 July 1943; likely the Australians also changed terminology at about this time.

Efficient operations of the Australian WWII radars, designed by RPL, were essential during WWII. The air force, army and navy all used radars as an important weapon in the Pacific war. In early 1942, the rapid development of aircraft warning was essential in the north of Australia as the Japanese attacked the Australian mainland.

A complex push and pull went on in the military over the role that women would play in the armed defence of Australia during the war. Women served in many branches of the military, but the Women's Auxiliary Australian Air Force (WAAAF) was the largest of the women's auxiliaries with a total number of airwomen during the war of about 27,000 (Thomson 1991). The WAAAF had been formed in March 1941 after a long period of deliberation within the Australian government.[7] Based upon the experience of the UK,[8] which had shown that women made more reliable radar operators than their male counterparts, the command structure of the Royal Australian Air Force (RAAF), recommended to the Minister for Air, Arthur S. Drakeford, that women in the WAAAF be used extensively in the aircraft warning networks within Australia. By 1944, the goal from the Air Board of the RAAF was to substitute the WAAAF for male servicemen at the 10 % level for wireless (radio) assistants and store hands and fully 100 % of radar operators, signals clerks and wireless telegraphists.

These goals were never attained due to the reluctance of Drakeford to send airwomen to the north of Australia (Thomson 1991). Naturally, with such an attitude about service within the confines of their own nation, the Minister for Air would not allow women to be posted in the nearby South West Pacific Area (SWPA) such as Papua New Guinea where conflicts with the Japanese were raging.

Ironically the Australian Women's Army Service (AWAS) were posted to the Northern Territory in 1944, where WAAAF personnel were not allowed. Thomson wrote: "... the repeated refusals [to allow WAAAF to serve in the northern tropics] undoubtedly prevented the most effective use being made of the work experience they [WAAAF airwomen] had by then accumulated".

This policy was criticised by the military leadership in Australia as well as by the press and naturally by the WAAAF members. An example is an editorial in the Sydney Morning Herald in October 1941: "How could Cabinet reconcile this point of view with its pledge to wage war with the maximum of Australia's capacity?"

---

[7] Sources for this text are: *The WAAAF in Wartime Australia* by Joyce Thomson 1991(former WAAAF Squadron Officer 1941–1944), the Bowen archive in National Archives of Australia- NAA: C4661,1, "History No. 1 Radio School, Section 17, RAAF Station Richmond", written by an unnamed source from the RAAF Historical Section in 1987 and Simmons and Smith *Radar Yarns*, 1991.

[8] In the UK, women played major roles in the operation of the Chain Home radar stations, especially in the Battle of Britain (Rowe 1948). Rowe wrote: "All honour to the women who shared with the men the often primitive and isolated conditions at the radar stations and who carried on with their tasks when the stations were attacked by the enemy."

In spite of these limitations, by November 1943, 599 (440 in 1942 and 159 in 1943) WAAAF personnel had been trained as Radar Operators (for ground based aircraft warning) at the No. 1 Radar School at Richmond Aerodrome in New South Wales.[9] During this period, 1,242 male RAAF had received the same training. By the end of the war, 6,196 male and female personnel had been trained in all types of radar (ground and aircraft based operations and maintenance service) duties; the total also included 427 US airmen. The total number of women in the WAAAF represented about 10 % of the total RAAF and WAAAF personnel.

Even in mid-1942, the expectation was that about half of the radar station vacancies would be filled by WAAAF personnel. This goal was never attained due to the ministerial restrictions placed on the posting of women. Drakeford was even worried about setting up separate showers and latrines for the WAAAF as well as "likely immorality" (Thomson 1991). He also stated that the conditions for women in these northern areas were generally not "appropriate". By November 1942 only 17 stations on the main land (in Queensland south of Cairns and in Western Australia, South of Geralton) were deemed to be suitable for women. About a month later Drakeford did approve eight additional sites, including two near Cairns in the north of Queensland.

Thomson wrote:

> This limitation of the radar units on which WAAAF could work contributed largely to a surplus of female operators at a time when the expansion of the RAAF radar programme was being halted by the shortage of trained manpower.

Even though the 599 WAAAF radar operators had been trained, by late 1943, 200 of these had to be moved to other duties or even offered discharge. In spite of these short-sighted limitations, Australian women were actively involved in providing around the clock radar aircraft warning services in about 20 % of the roughly 120 stations on the Australian mainland.[10] An additional 90 solely RAAF-staffed radar stations were in the tropics north of Australia—Borneo, Dutch New

---

[9] At first there was quite some reluctance to train the women: "Naturally, instructors sometimes found it embarrassing, but, generally speaking the WAAAF's soon proved themselves capable operators." (Bowen archive NAA: C4661,1, author unnamed source from the RAAF Historical Section). The first school for women was in June 1942 with a 4 week course followed by a 12 week trainee course at a radar station in Eastern Australia. Most of the commanding officers of radar units (with rank of Pilot Officer or Ground Radar Officer) were trained in longer courses at the University of Sydney, the "Bailey Boys" after Professor Victor A. Bailey (Fielder-Gill et al. 1999). The courses lasted 3–5 months and produced 249 graduates from September 1941 to March 1944. There were 160 RAAF, 14 in the Royal Australian Navy and 75 in the Army. At the end of the war there were about 300 radar officers among the roughly 5,000 men and women working in ground, air and naval radar.

[10] Excellent reminisces of RAAF and especially WAAAF veterans of WWII can be found on the web with the two series of *Radar Yarns* collections by Ed Simmons and Norm Smith www.radarreturns.net.au/archive/Radar%20YarnsRRWS.pdf (originally from 1991). *Then More Radar Yarns* www.radarreturns.net.au/archive/More%20Radar%20YarnsRRWS.pdf *(from 1992)*. In addition their excellent history of the use of Allied air warning radar in the Pacific war, *Echoes over the Pacific* (1995), can be found at www.radarreturns.net.au/archive/EchoesRRWS.pdf.

Guinea, Papua New Guinea, the Solomon Islands, New Britain and the Admiralty Islands.

In summary, "WAAAF fortunate enough to fill the approved operators' positions were fully manning shifts on many of the radar units, under the charge of a RAAF radar mechanic [a male]. Hindsight shows that the morale and morals were high on these and other small units where airwomen felt their work was of immediate value to the war effort" (Thomson 1991).

There is no doubt that servicewomen made contributions to the efficiency of Australians in the war of 1941–1945 in the Pacific, but their work was curtailed by the social mores of the day. It required involvement in WWII to shift, ever so slightly the prevailing views on what a woman should be allowed to do. Subsequent years have shown that such an incremental shift was tenaciously widened by waves of women like Ruby Payne-Scott.

Payne-Scott Joins the Radiophysics Laboratory in 1941

The Council for Scientific and Industrial Research (CSIR) was created in 1926 as an institution for federally funded and guided scientific research. The CSIR Radiophysics Laboratory (RPL) was founded on 22 August 1939, just before the beginning of World War II- Australian followed the United Kingdom into war with Germany on 3 September 1939. In the first year of operation, RPL had a small staff of only five research officers and eight assistant research officers, with the Chief of the Division D. F. Martyn (who was to be Chief from 1940 to 1942). By the end of the war, there were over 300 staff members at RPL with 91 research staff and 97 engineering staff (Evans 1970). The success of the RPL in providing radar designs for the SWPA allied forces was substantial; more than 20 major radar development projects were undertaken during World War II, with RPL being responsible for the manufacture of more than "2,000 complete items of Service radar equipment" (Evans 1970).

In the course of 1941, additional scientific staff members were required at the RPL due to the impending threats to Australia from Japan; an advertisement was placed in early 1941 for eight new scientific officers. The deadline for the receipt of the application was 28 April 1941. The advertisement stated:

> Applicants should posses a University degree in science or engineering.... A knowledge of, or experience in, high frequency oscillations would be an additional qualification.

The salary range was to be in the range £A344 to £A588 per annum.

Well after the deadline, on 4 June 1941, Ruby Payne-Scott sent in her letter of application. She got off on the wrong foot by misspelling the name of the Chief of Radiophysics. The letter began:

> Dear Dr. Martin [sic] [Martyn was correct]: I should like an opportunity to discuss with you the possibility of my joining the staff of the Radiophysics Laboratory. I am a graduate of Sydney University with first class honours in mathematics and physics, and obtained my

MSc. degree in physics in 1936 ... Mr. Macready [sic][in fact McCready], who is a member of your staff, has known me at the University and at AWA and could be referred to for further information. I would be grateful if you would consider this letter as confidential ... [return address listed as 5 Farleigh Street, Ashfield].

On 17 June 1941, David F. Martyn wrote to the CSIR Secretary with a description of eight individuals previously proposed to the Minister of the CSIR for approval; however, two of them could not be released from their current positions. One was from the Patents Office in Australia and the other (Walker) from the New Zealand Department of Scientific and Industrial Research (DSIR).[11] Two late applications were therefore to be considered as replacements for these individuals; these were from A. Richardson and R. Payne-Scott, both from AWA.

Martyn had heard favourable reports about both of them and on 17 June 1941 recommended that an offer be made to Richardson; he planned to interview Payne-Scott personally before making a recommendation. Two days later he met Payne-Scott and was impressed. Then on 20 June 1941, the Secretary of the CSIR wrote to the Minister for Scientific and Industrial Research, Mr. H. Holt (later a short-serving Prime Minister of Australia in 1966–1967), pointing out that two late applicants for the RPL positions should be appointed, Richardson, an electrical engineer, and Payne-Scott, a physicist.

Both appointments, as Assistant Research Officer, at starting nominal salaries of £A434 and £A384 respectively, were made "without commitment"—i.e., there was no commitment for future superannuation (pension) rights. In other words, these were not permanent positions. In subsequent years, the issue of superannuation was to be the source of increased conflict between the CSIR/CSIRO and Payne-Scott.

It is worthwhile to note that Ruby's starting salary was only 88% of Richardson's. In addition, Payne-Scott suffered a slight salary reduction in the new position as her salary at AWA would have been £A400 per annum on 1 July 1941. Clearly, she was willing to take a reduction in pay for the opportunity to work at RPL.

On 17 July 1941, Payne-Scott wrote to G. Lightfoot, Secretary of the CSIR in Melbourne, accepting the position which had been offered to her on 26 June; she said that she could only begin the new assignment on 18 August 1941 due to the pressure of "urgent jobs" at AWA. Apparently, she was somewhat late in responding to the offer. She apologised for the delay as being due to her absence from Sydney on a holiday.

Payne-Scott began her new career on Monday, 18 August 1941. This was close to 4 months before Japan's Pacific War was launched against the Allies, Australia

---

[11] Based on the personnel file of Payne-Scott, this deliberation was likely the first occasion that F. W. G. White and Payne-Scott had an interaction. White had just arrived in Sydney a few months earlier (March 1941) on secondment from the Department of Scientific and Industrial Research in New Zealand. In a 17 June letter Martyn reported: "Professor White informs me that Dr. [E.] Marsden [the Director of Scientific Developments at the DSIR in Wellington, New Zealand] has told him that he does not propose to release Walker." Thus the position for Payne-Scott opened up since Walker was to remain in New Zealand.

and the USA, in early December 1941. The importance of secrecy was highlighted on her first day at work as Payne-Scott signed the "Secrecy Declaration", promising not to reveal any of the details of the work of RPL, except as authorised by the Chief of the RPL. The penalty for any offence would be administered under the Crimes Act of 1914–1937. The signature was witnessed by H. J. Brown, who had been an early appointment (1939) from EMI (Electric and Musical Industries) in England, and by the Assistant Chief, J. L. Pawsey, who was to play a major role in her subsequent career in radio astronomy at RPL.[12]

The work hours were listed as a nominal 36.75 hours per week with extra time (war effort) extended from 17:06 to 17:18 daily and from 9:00 to 12:00 on Saturdays.

Payne-Scott joined a cohort who would have illustrious careers in disparate disciplines in the coming decades.[13]

One noteworthy colleague, Harry Minnett, who served as chief of RPL from 1978 until 1981, was an invaluable source concerning Payne-Scott, as he knew her at the University of Sydney in the years 1936–1938, while he was studying Physics II and Maths II, as well as at RPL. Minnett joined RPL in April 1940 (Thomas and Robinson 2005). He wrote to Goss in 1998:

> To my surprise, Ruby turned up [at RPL] in mid-1941, just after Joan Freeman had joined... They were the only two women scientists on the staff and made a considerable contribution to the work, as well as adding a much needed feminine touch to the community. It was a great opportunity for them to work on state-of-the-art techniques, particularly as employment was not easy for women scientists before the war. Both went on afterwards to distinguished research careers.... I worked in association with Joan from time to time on microwave projects [they both developed the TR-transmit-receive switch for the LW/AHW Mk II 25 cm system, Minnett having developed the TR switch for the famous 200 MHz radars in 1941–1942] ... but never with Ruby.

Joan Freeman,[14] shed some light into the unique experience of being a female research scientist in the RPL during these early years. Freeman, working on radar,

---

[12] Pawsey (1908–1962) was originally from Victoria and had studied at the University of Melbourne. He completed a Ph.D. degree with J. A. Ratcliffe at the Cavendish Laboratory of Cambridge University in the years 1931–1934. He joined the RPL (also from EMI in the UK) as one of the first scientific appointments from late 1939, starting work on 2 February 1940. Pawsey was the major driving force (along with Bowen) for the development of radio astronomy in Australia at the CSIR/CSIRO starting in 1944.

[13] Other prominent appointments at RPL in mid-1941 were B. F. C. Cooper (graduate in 1941 from Electrical Engineering, Sydney University, who played a major role in the Darwin radar response to the Japanese attacks in early 1942, and had a long career at RPL working on radio astronomy instrumentation and aircraft landing systems), F. J. Kerr (from Melbourne University and the Radio Research Board with a later career at CSIR/CSIRO and the University of Maryland as one of the pioneers in 21 cm HI research), and Joan Freeman (from Physics, University of Sydney and later at the Atomic Energy Research Establishment at Harwell in the UK).

[14] Joan Jelley, née Freeman, left Australia on a CSIR Fellowship in August 1946; after gaining a Ph.D. in Nuclear Physics at the University of Cambridge, she had an illustrious career at the Atomic Energy Research Establishment, in Harwell, UK. In 1976, she shared the Rutherford Medal of the Institute of Physics with R. J. Blin-Stoyle. Freeman married the physicist John Jelley (of Cerenkov radiation fame) in 1958; the two had met in Cambridge in 1948. John Jelley died in

joined the RPL group in June 1941, barely 2 months before Payne-Scott began in August 1941.

In late 1941 or early 1942, both of the female scientists found themselves in a battle with Mrs. Enid Eastman the head librarian of the National Standards Laboratory, a facility shared with RPL. Rachel Makinson reported to me that Mrs. Eastman was known as "a bit of a dragon" at RPL and she apparently saw herself as the arbiter of female behaviour at the National Standards Laboratory and RPL. The head librarian was very upset by the fact that Payne-Scott wore shorts and that Freeman was a smoker. Payne-Scott found the shorts quite appropriate attire since she was climbing up ladders to work on radar antennas; naturally many of the men also wore shorts, especially during the hot summers before the days of air-conditioning. Freeman had taken up smoking while she was at university. Mrs. Eastman wrote a peremptory note to the two telling them to shape up! Payne-Scott reacted in an angry fashion and continued to wear shorts and urged Freeman to continue with her smoking since most of the men did so. Both were finally summoned to attend a meeting in the library with Mrs. Eastman at which time the proper behaviour of the females in the laboratory would be discussed. On that day Payne-Scott deliberately changed into shorts for the inquisition and Freeman refused to attend.[15] Payne-Scott must have had a shouting match with the librarian since she soon returned to her office; she had been dismissed by the disciplinarian. A stand-off continued and was only resolved when a new librarian arrived, the famous Australian novelist, Marjorie Barnard (see Additional Notes, No. 2, end of this chapter). As Freeman put it, Barnard, an enlightened colleague, was not at all "concerned about Ruby's shorts or my cigarettes".

A second story also involves Freeman and Payne-Scott; in this case it was Freeman who was more upset. Every few weeks during the early period of WWII, secret radar reports would arrive by air from Britain. The medium was microfilm; since there were no copying facilities at RPL, these were taken to the Sydney Public Library for photocopying. Since the reports were top secret, someone with security clearance had to take them to the Library and be present during the entire copying procedure as well as ensure that all work, including spoiled sheets, was brought back to the laboratory.

Payne-Scott and Freeman, ostensibly because they were women with the proper security clearance, and thus seen as appropriate staff for menial, "secretarial" chores, were told to alternate in accompanying the films. Each session lasted several hours while the attendant had to watch the copy process in an almost darkened room. Freeman despised the tedium and the lost time for her own projects; she felt that some of the junior men should also have been given this task. The worst part of the chore for Freeman was the endless chatter of the young woman doing the

---

1997, preceding his wife's passing by 8 months; Joan Freeman died on 18 March 1998 in Oxford. Nessy Allen (1990) has written a fascinating account of the lives of Freeman and Makinson, "Australian Women in Science—A Comparative Study of Two Physicists".

[15] In a few publications in Australia during the last 10 years, the claim has been made that Payne-Scott was also a smoker; this claim is quite unlikely based on comments from her family and friends.

copying. Payne-Scott would not join Freeman in making a fuss; she did not mind the respite as she could knit in the dark, a skill that Freeman had no intention of acquiring.

Through these stories we can see that women were hired as engineers and physicists, but had to battle many set notions of how they should approach the workplace differently from the men working beside them.

Goss heard from an RPL colleague that practical jokes were played on Payne-Scott during the War; her work-issued mat (foot rest) and radiator would be "stolen" on cold winter days, possibly in retaliation for her aggressive attitude to women's issues.

Of course, as much as Ruby challenged the gender norms of the day, at this stage in her life she was as yet completely oblivious to the plights of working mothers. K. Rachel Makinson (see Additional Notes, No. 3, end of this chapter), Payne-Scott's friend and colleague, was a physicist working at the University of Sydney in 1942, when her son David was only a year old. She was unable to come to work on a Saturday when David was ill (the work day ended at 12:30 on Saturdays). Payne-Scott had organised an informal lecture at RPL that was attended by both professional and workshop personnel. Later on, Payne-Scott criticised Makinson for her absence. Makinson bit her tongue and said nothing, though some years later after Payne-Scott had started a family, with disastrous effects on her career, Makinson was tempted to say, "See, I told you so", referring to the difficulties of working full time with a small child at home.[16] Kindly, she held her tongue.

Payne-Scott's War-Time Research

The years 1941–1945, the last 4 years of World War II, were a turning point in the life of Ruby Payne-Scott. Her abilities as a radio engineer were perfected as she learned a number of techniques that would serve her well in the post—war years as she became one of the early solar radio astronomers. There is no doubt that her association with J. L. Pawsey during these years was decisive in developing her research skills. The groundbreaking 10 cm radio astronomy project of March 1944 (Chap. 6) arose from Payne-Scott's wartime activities testing the new S band (10 cm) radar receivers. In particular, this project set the stage for the exciting astronomical endeavours of late 1945 and early 1946.

Payne-Scott's research activities during this period at RPL were summarised in a number of internal reports written by her between 1942 and 1945, as well as the two page report of a colloquium held on 30 January 1945. All of these reports were originally classified as "most secret", "secret" or "restricted".

---

[16] Allen (1990, 1993) has described how "devastatingly difficult" Makinson found her life during this period. She had to pay out more than her own salary for child care. To take the baby to child care and then go to the University of Sydney campus, she had to use five different public buses.

Payne-Scott's war time research can be divided into two categories: (1) establishment of accurate standards for the radio engineering test equipment used at RPL and (2) the detection of weak radar signals described both theoretically and in practical applications. The initial research projects at RPL continued the radio engineering investigations that Payne-Scott had carried out at AWA in the period 1939–1941, and had a major influence on her ability to make substantial contributions in the field of radio astronomy from 1944 to 1951.

Calibration of S Band Radar Receivers

As stated above, Payne-Scott began at a salary of £A384. By the end of December 1942, the new Chief of RPL, F. W. G. White, recommended that her salary be increased to £A402 per annum; at this time a male colleague on the same classification but with more experience was earning £A428 per annum. The three highest salaries at the time (not including that of the Chief) were those of H. J. Brown, J. H. Piddington and J. L. Pawsey at £A675. It is clear from the raises she was given that her work was valued.

In a 1998 letter to Goss, Minnett remarked on the relationship between Pawsey and Payne-Scott:

> ... Pawsey appreciated Ruby's abilities as a physicist, evident in her research and in colloquia. As well, her wartime work on small-signal visibility on radar displays and the accurate measurement of receiver noise factor was attractive to Pawsey, who had become keenly interested in the thermodynamics of external noise fields, as expressed by antenna temperature, and its relationship to the internal noise of receivers, as expressed by the noise factor. It was therefore natural for Ruby to assist him in early wartime searches for solar radiation.

Early on in World War II, from 1939 to 1942, extensive radar research was focused on 200 MHz radars, which required cumbersome antenna sizes of several metres. Such a large antenna proved impracticable for use with airplanes, and thus the RPL refocused research onto higher frequencies that would utilise smaller, more portable and practical antennas. Interactions with both British and US groups (especially the MIT Radiation Laboratory) were frequent and led to testing and prototyping of higher frequency radar systems (Brown 1999; Minnett et al. 1999a). Payne-Scott's research was mainly involved with evaluating various S band (10 cm or 3,000 MHz) components. When Mills joined the RPL in December 1942, his main task was to develop low noise preamplifiers in the receiver research group directed by Lindsay McCready.

A radio engineer seeking consistent data, no matter what the variances from one piece of equipment to the next, must find a way to calibrate each, individual piece of equipment before use in order to extract standardised data. In radar research, this means that every receiver needs to be tested by connecting to a microwave component that is cooled or heated to a known temperature to measure its exact response. This process is called "cold load" vs. "hot load" calibration.

Mills found that Payne-Scott was already establishing accurate standards for much of the test equipment. Encouraged by Pawsey, Mills was then tasked with determining the sensitivity of these systems using hot loads. "This naturally led to comparisons with signal generators calibrated by Ruby and much discussion about differences which showed up."[17]

In early 1944, Payne-Scott (1944c) wrote a report on the effects of high physical temperatures on the sensitivity of S band (10 cm) radar receivers; at times there were temperature extremes of more than 20 C. The heated crystals caused a severe deterioration in the noise factor (sensitivity) of the radar receivers and thus a major decrease in the sensitivity of the overall radar system. The report contained an analysis of the level of increase in the noise factor and the manner in which the receiver recovered after the crystals were replaced in the receiver. The concern was the rapid deterioration of the radar system in the tropics during the final years of WWII. Her procedure for removing the heated crystals and remounting them during periods of extreme heat was an effective solution.

Two classified reports were written in 1943 and 1944 describing an S band signal generator and an S band noise "thermal noise generator", the latter was described in *RP211*, dated 29 May 1944 (Payne-Scott 1943b, 1944a). In modern terminology the latter device would be called a "noise tube". A noise tube is a device that generates microwave energy. This report described an elegant device to produce a known amount of S band radio frequency power that was then used to determine in a precise manner the "noise factor"[18] (current terminology is "noise figure") of S band receivers. Knowledge of the sensitivity of the S band radar receiver for an active radar station was always essential. A rapid deterioration in the noise factor indicated that something was malfunctioning in the receiver; in this case the ability to detect incoming aircraft would be seriously impaired. The resultant errors in the location of incoming aircraft could be substantial.

During this period, receiver noise factors were usually determined with what is now referred to as the "signal generator twice-power" method.[19] The technique required two pieces of laboratory test equipment. A signal generator was used to feed a continuous wave (CW—a fixed frequency) signal of known frequency and strength into the receiver. A power meter was then used to measure the level of the amplified signal at the output of the receiver.

As the name implies, the twice-power method was a two–step process. First the output power of the receiver was measured with a matched "room temperature" load connected to the input. This presented an effective noise power signal of 290 K

---

[17] Letter from Mills to Goss, 14 September 1997.

[18] This term was invented by Harold Friis of Bell Labs in 1944. Friis was the supervisor of Karl Jansky (Sullivan 2009). In the current era the common usage is "noise temperature" to describe the sensitivity of the radio astronomy or radar receiver. As an example, the typical 200 MHz receiver of the 1945 era would have a noise figure of about 6 dB or 860 K. A modern 3 GHz receiver in 2009 would have a noise figure of only about 0.22 dB, a noise temperature of only 15 K.

[19] The following six paragraphs have been provided by Robert Hayward.

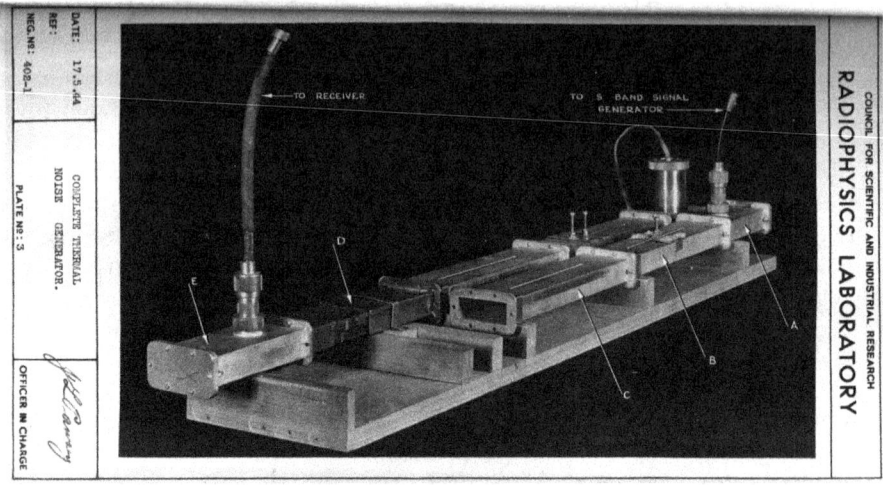

Fig. 5.1 An S band (3 GHz or 10 cm) set up to determine the sensitivity of radar receivers designed by Payne-Scott. The officer in charge was J. L. Pawsey. 'A' is a length of waveguide into which the hot or cold load resistor was coupled. 'B' is a length of slotted waveguide with adjustable tuning screws serving to match the hot or cold resistor to the characteristic impedance of the waveguide. 'C' is a length of slotted waveguide along which a standing wave detector could be moved to check for matching. 'D' is a length of telescopic waveguide set to a length that would keep the crystal current of the receiver the same when either the hot or cold system was connected. 'E' is a standard waveguide to coaxial cable coupling. From RP 211 29 May 1944, a "secret" document: "A Thermal Noise Generator for Absolute Measurement of Receiver Noise Factors at 10 cm" (CSIRO, Radio Astronomy Image Archive)

at the input. Next the load was replaced by a signal generator providing a CW signal within the measurement bandwidth of the receiver. The power being injected was adjusted so that the output level of the receiver doubled (i.e., a 3 dB increase[20]). The power coming from the generator was thus equal to the input noise power. If the actual generator power level being injected as well as the bandwidth of the receiver were known, the noise factor could be calculated using a simple and straightforward formula.

A significant improvement over the signal generator twice-power method was possible if a high level of noise power—much stronger than the 290 K room temperature load—could be used. Payne-Scott developed a noise source (noise tube) based on the Johnson noise produced by a hot resistor of know temperature. The elegant setup (Fig. 5.1 shows the waveguide apparatus, likely using RPL made waveguide components at 2 GHz) utilised a tungsten lamp at 2,100 K embedded within a section of S band rectangular waveguide.

The availability of this high power noise source allowed the RPL to carry out what is now known as the "Y-factor" method of measuring the noise figure of a

---

[20] dB is a decibel, defined as ten times the log of the power ratio. For example 3 dB is a factor of 2, 10 dB is a factor of 10 and 20 dB is a factor of 100.

receiver. By presenting two loads at different temperatures to the receiver, both the gain (level of amplification factor) and the noise factor (sensitivity) of the receiver could be easily determined. In this case, the 2,100 K noise tube acted as the hot load, while a standard room temperature 290 K load was the cold load. This combination resulted in the hot load having nearly ten times more noise power than the cold load. "The increase in noise on switching from the hot to the cold load circuit depends on the ability of the receiver to distinguish small signals above the normal noise, i.e., on its own noise factor."[21] As seen in Fig. 5.1, the two paths (noise generator and the signal generator) were completely interchangeable; the waveguides were simply unscrewed and interchanged.

The cold load was actually a 10 dB (factor of 10) attenuator, as seen in Fig. 5.1. When connected to the input of the receiver, the noise power from the attenuator acted as a 290 K termination. The cold load portion of the setup could also be fed with a CW signal from a signal generator, with its power being reduced by a factor of 10 by the attenuator. The signal generator could be used to line up the receiver in frequency and to determine the bandwidth (although this requirement was not needed for the actual Y-factor measurement). The signal generator could also be used to carry out a twice-power measurement for comparison.

The noise factors based on the noise source and the signal generator methods agreed to better than 25 %. Thus a method of providing a fundamental check on the twice power method, which was commonly used in the field, was available. The cumbersome noise tube method was not suitable for field tests due to its complexity, delicate nature and large size; various receivers could be checked out in the lab using this new and unique noise source. As a procedural step, the more portable signal generator setup could be calibrated against the noise tube back in the lab. Thus, the portable setup could be used at the remote radar stations, having been calibrated at the home base, RPL.

The S band noise tube designed and constructed by Payne-Scott in 1944 had an effective noise power of 2,100 K. Today noise diodes can be bought off-the-shelf that are capable of achieving noise temperatures of 290,000 K or more. The twice power method was commonly used before high power noise sources became available; however, the method is still used for systems with large noise figures where the Y-factor method results can be inaccurate. For modern cryogenic (cooled with helium refrigeration) receivers, which have noise temperatures of much less than 100 K, the Y-factor method is the preferred measurement technique. Currently in the early twenty-first century, the hot load remains at room temperature while the cold load is typically a cryogenic load immersed in liquid nitrogen (i.e., 77 K).

The report, RP 211 was followed within a week by the document TI 121/1, "The Present Position of Low-Power S-Band Measurements in the Radiophysics Laboratory" 6 June 1944 (Payne-Scott 1944b), providing a summary of the full range of power measurements at S band from 0.1 milliwatt to 0.1 picowatt, achieved using amplifiers mounted after the crystal mixers. The calibration and precision of a

---

[21] From RP211 "A Thermal Generator for Absolute Measurement of Receiver Noise Factor at 10 cm", 29 May 1944, author Ruby Payne-Scott.

number of methods were summarised.[22] The noise factors for S band crystals (available in 1944) at the power level of 10^{-13} W were in the range of about 10 dB or a T_{sys} of 2,600 K.

The final technical report, *TI 191/1-* restricted, authored by Payne-Scott (1945) was dated 6 August 1945, 9 days before the end of WWII. This report is the most comprehensive of these documents and is entitled, "Present Position of Fundamental R.F. [radio frequency] Measurements in the Radiophysics Laboratory". This 14-page document provided a description of radio frequency engineering knowledge and practice at the end of World War II. The document provided a primer for the determination of frequency (or wavelength), power at various levels, attenuation, impedance, cable loss, etc. at P band (10–500 MHz), L band (1–2 GHz), S band (2–4 GHz), X band (8–12 GHz), and K band (18–27 GHz). Many of the references were to RPL publications—9 of the 22.

A cryptic and significant paragraph appeared as 3.2.2-c: "Other Methods (for noise factor determinations)". The report stated

> Two other methods for measuring the noise factor of a receiver have been considered in the laboratory; one involves using the radiation from the sun as a standard source, and the other comparing the receiver noise when the aerial points alternatively at "free space" and an enclosure at ambient temperature. Neither effect has yet been sufficiently well investigated to be used to provide a standard noise voltage, but in particular the second, which does not require a directive aerial, has attractive possibilities.

As will be discussed in Chap. 6, Pawsey and Payne-Scott carried out S band observations in March 1944 of the sky and the Milky Way, but did not observe the sun. Sullivan (2009) has pointed out that Pawsey and Payne-Scott were not aware at this time of Southworth's secret reports from 1942 at Bell Labs, describing the microwave detection of thermal radiation from the sun. However, in August 1945, Bowen and Pawsey became aware of the detection of solar microwave radiation based on either rumours of Southworth's observations or from receiving the restricted Bell Lab report. The complex story of the timing of the knowledge of the Bell Lab results in Australia has been described in detail in *Under the Radar* (Goss and McGee 2009). Based on the above quote from Payne-Scott, there is a strong suggestion that she knew about the Southworth results at 10, 3.2 and 1.25 cm from 1942 to 1943 and made the prescient suggestion that the detection of the sun at microwave frequencies might be a useful standard source. As we now know, the quiet sun would never be used as a standard source at microwave frequencies, due to the high intensity of the radiation, as well as its time variability and large angular size, about ½°. The first high frequency observations of the sun at 1.25 cm at RPL were only made in early1948 (see Piddington and Minnett 1949; Pawsey and Yabsley 1949; Pawsey 1950b); lower frequency detections of the quiet sun at

---

[22] Bowen (1947, page 7 of the 1954 edition) has remarked on the remarkable dynamic range (the ratio of the maximum signal to the noise in the system) required by radar systems. The transmitted pulse was typically 10^5 W while the received energy from the target was roughly 10^{-14} W. "The overall operating efficiency is therefore 10^{-19} and it is a great tribute to the pioneers of radar that they persisted in their efforts to attain apparently impossible ends." Note that this footnote in *Under the Radar* had an incorrect exponent of +14 for the received signal.

RPL were carried out by Lehany and Yabsley at 50 and 25 cm in August to November 1947 and published in 1949. Based on the statement by Payne-Scott in 1945 in this technical report, TI 191/1, this delay in the detection of the quiet sun at RPL for a few years remains surprising.

Equality of Pay for Women Ruling: September 1944

Payne-Scott's salary was raised to £A420 per annum, to take effect on 1 January 1944.

In July 1944 there was an exchange of letters from the Commonwealth Public Service Professional Officers' Association and the CEO of the CSIR, Sir David Rivett, about the issue of equal pay for women. Payne-Scott was the only female staff member of CSIR who was a member of this professional organisation, similar to a trade union. The Women's Employment Board (WEB) made a ruling in September 1944 that females were to be paid 100 % of the male rate for the same class of work.[23] This ruling was clearly made to compensate women for carrying out essential services during wartime; otherwise the women's rate of pay was £A97 per annum *less* than the men's pay scale for Research and Technical Officers. However, this optimistic decision was overturned in the postwar era (1949) when the WEB overturned the 1944 regulation which specified that women professional staff would be treated equally. Not surprisingly, Ruby Payne-Scott expressed her opposition in two articles in the CSIRO Officer's Association Bulletin (October and December 1949, Payne-Scott 1949b).[24]

---

[23] In early 1941, the Australian Council of Trade Unions (ACTU) had called a meeting of all Federal Unions with women members for a discussion of policy on the role of women in Australian industry. After adopting a resolution proposing that women be paid an equal wage, the full body of the ACTU adopted this position in June 1941 at a time when the average rate of pay for women was only 54 % of that of males. The resolution had six conditions including details about creating wage parity for women. Connected to Payne-Scott's problems in 1950 about her marriage (Chap. 10), there was the important final provision, "...the removal of all restrictions on the employment of married women in gainful occupations and the recognition of their right to economic independence" (Bayne 1943, page 56).

[24] The story of Payne-Scott's objections to the WEB decision of 1949 is described in Appendix H of *Under the Radar* (2009). This controversial decision was also confusing to CSIRO female employees; the decrease did not apply to individuals appointed before 6 June 1949. In addition the salaries of newly appointed or transferred female personnel were set at the reduced scale. Ruby's suggestion (Payne-Scott 1949b) was simple: "The best advice to the women concerned is to stick like glue to their present fields of work till the situation is defined." In the December article in the Bulletin, Ruby complained in a sarcastic tone about the new WEB and CSIRO ruling. The public meeting in late 1949 to discuss this problem with the new Chairman of CSIRO (Sir Ian Clunies Ross) is described in Chap. 10. In 1951, the professional women of CSIRO lost the case for wage parity in a decision made by the Public Service Arbitrator. Only in 1977 was the gender gap in wages substantially reduced. See Wilde, 1998 (*Unions in CSIRO: Part of the Equation*) and *Under the Radar* (2009).

Fig. 5.2 LW/AWH (Light Weight Aircraft Warning, Height) prototype 25 cm radar with size of antenna 25 by 12 ft. (7.6 by 3.6 m), located at Georges Heights, Middle Harbour, Sydney (note South Head in the distance). The magnetrons were designed at the University of Melbourne and manufactured by AWA. Pulse lengths were either 4 or 1 μs, with a power of 500 kW. Weight was only 7 t compared to 35 t for comparable UK and US radars. This radar was the major technological achievement of RPL during World War II. The end of the war brought a cessation to development activities; only a few prototypes were built. The other antenna to the left is the communications aerial for control of the radar. This radar spearheaded RPL's post-war drive to solve civilian problems for control of aircraft traffic (CSIRO Radio Astronomy Image Archive B1362)

Light Weight Aircraft Warning-Height 25 cm Radar Display

The major project on which Payne-Scott worked from 1944 to the end of the war was the new 25 cm LW/AWH (Light Weight Aircraft Warning, Height) MkII radar. This advanced radar was developed as a prototype in 1945; the first contract for the construction of 47 of these was cancelled in August 1945 as the atomic bomb explosions in Japan brought an end to WWII. One of the radar prototypes at Middle Harbour, Sydney, is shown in Fig. 5.2. In early 1944, the Japanese had developed the ability to "jam" the 200 MHz LW/AW workhorse radars in the South West Pacific Area north of Australia by transmitting a confusing signal at the same frequency as the transmitter. In addition the height determination precision of the 200 MHz LW/AW radars was limited. The absence of low angle coverage at 200 MHz was an additional liability (Briton 1947; Minnett et al. 1999a). The new prototype 1,200 MHz radar sets produced narrower beams and experienced a complete lack of ground—reflected radiation, thus there was no interference pattern due to reflections from the sea in contrast to the 200 MHz radars. The beam widths were determined at the 1.5 dB points or at the 3 dB points for the return echo with values of 1.3° in elevation and 3° in azimuth compared to 10° for the 200 MHz systems. A flat helical scan by the radar beam searched the required volume of space once per minute—up to elevations of 40,000 ft. (12,000 m) at distances of 150 miles (240 km) for heavy bombers. At 25 cm, the peak power of about 500 kW was transmitted with a detection range of 70–100 miles (110–160 km) for small aircraft up to heights of 35,000 ft (11,000 m). Measured heights were accurate to ± 2,500 ft (800 m). at ranges of 30–100 miles (50–160 km), representing a vast improvement in precision compared to the 200 MHz radar sets. This 1 GHz project represented the most challenging and successful technological achievements of RPL in the era 1939–1945.[25]

Both Mills and Payne-Scott had developed skills from the LW/AWH project that were to serve them well in the post WWII rapid growth in radio astronomy. They also played the major roles in planning the scanning method and the design of the advanced displays for this sophisticated radar. A complex of three displays was available for evaluating the system performance, a Plan Position Indicator (PPI) to determine the azimuth and range of the radar echo, and a Range Height Indicator for the determination of the elevation of the incoming aircraft.

Mills has explained the chronology of their collaboration[26]:

> My next work association with Ruby occurred in late 1944 or early 1945 when she began an investigation into the visibility of weak signals on a PPI display. At this time I was interested in developing an automatic signal detector which would wake a sleepy operator when an echo appeared near the detection limit, so we [Payne-Scott and Mills] began together to set up a test system for exploring the effects of varying parameters on signal detectability. However Joe [Pawsey], very apologetically, removed me from the project in order to develop the receiver and display systems for a planned early–warning height–finding radar operating at 25 cm (Fig. 5.2). But as a result of the work with Ruby I was able to see that the proposed scanning method [a Palmer scan, a combination of a circular or raster and conical radar scans], a slowly rotating reflector [1–2 rpm] plus a rapidly vertical scanning feed [4 times a second], was hopelessly inefficient [the signals and noise from all vertical angles were superimposed at any azimuth]. With Ruby I presented an internal report showing that a rapidly rotating reflector [16.5 rpm] combined with a slow vertical scan [entire reflector nodding to cover the 11° of elevation scanning once per min with a return from high to low elevation in only 6 sec; a helical scan was produced] would give about twice the detection range. This was adopted and our predictions of sensitivity were confirmed.

The reports referred to above were: (1) Payne-Scott (1945b), "The Ultimate Visibility of Signals on a PPI Display and the Effect of Electrical Parameters on Visibility", RP 252/1, 20 May 1945 and (2) Mills, "Scanning Considerations in LW/AWH Mk II", RP TI 137/4, 22 June 1945. In addition Payne-Scott had given a

---

[25] This advanced radar system remained the only military project at RPL after the war. (NAA: C3830, D1/2 "Programme of the Division of Radiophysics", dated 8 August 1946) Bowen presented a report of the future activities of RPL. A detailed description of the AWH Mark II was presented: "The Radiophysics Laboratory was formed to develop radar equipment for the Armed Forces and during the war years its programme was determined by their requirements.... This development was started during the war and was so nearly completed at the end of hostilities that under pressure from the Air Force it was decided to complete the construction and perform field tests. [The new radar] ... is approaching the test stage at our field station at Georges Heights [Middle Harbour]... It is expected that the design will form the basis of peacetime radar sets for the Air Force and for various civil purposes."

[26] Letter to Goss from Mills, 14 September 1997.

research colloquium on 30 January 1945, presenting preliminary results of the research.[27] The two reports each had only one author; it is clear that there was a great deal of collaboration between the two. As an example, much of the discussion of the scanning considerations in the Mills report was based on calculations carried out by Payne-Scott.

In late 1945 and early 1946, Payne-Scott was encouraged to convert her research on PPI visibility issues into a publication; at this time there were no restrictions since the war time censorship was no longer in place. The 17 page paper was submitted to the Proceedings of the IRE (Institute of Radio Engineers, USA) on 13 January 1947. The paper entitled, "The Visibility of Small Echoes on Radar PPI Displays", was published in February 1948.

The introduction provided a summary of the goal of the research:

> Detection of an object by radar depends ultimately on the ability of an observer to pick out a small change, in brightness or position, of part of the pattern on the screen of a cathode–ray tube. This ability depends on the one hand on physiological and psychological factors, and on the other, on the parameters of the whole radar system, which determines the nature of the change to be detected. In order to design systems of predictable performance or to compute the effect of any proposed change in a given system, we need to know the laws governing visibility.

She then presented a general theory of visibility with a rigorous derivation of an equation for visibility on a PPI display. The details of the various parameters affecting the response of the total system were analysed in detail. Examples of the parameters that were relevant were: radar pulse duration, pulse repetition rate, rate of the base sweep, bandwidth of the receiver and antenna beam size.

A novel aspect of the PPI evaluation was the creation of an artificial radar station in a small darkened room. In current terminology, this simulation would be called an analogue computer. "The apparatus allowed a wide range of values of system parameters to be arranged in any desired combination." The observations were obtained by three observers, Payne-Scott and two Royal Australian Air Force (RAAF) radar operators. Each observer could be trained in a few hours; various observers were found to have similar responses to the system. The operator was subjected to a simulated attack of enemy aircraft at a random bearing. The next step was a detailed comparison of the measured visibility on the screen as each of the parameters was changed, one at a time; then comparisons were made with the theoretical predictions. In most cases the observations fit the theory, as was shown in the publication with illustrations displaying numerous parameters. The illustrations indicated that the empirically determined observational

---

[27] "Notes on the Research Colloquium held on 30/1/45" by Payne-Scott. (NAA: C3830, D4.) For the first time visibility was defined: "Visibility is in terms of the reciprocal of the power of the minimum visible signal" Thus the smaller the detected signal (for a more distant object) the higher the visibility. The presentation seems to have been controversial based on a number of questions. The advantages of the newer PPI displays compared with the more traditional A type display (simple display of the distance to the target based on the time delay of the echo with no determination of the direction to the target) were discussed by Payne-Scott. She answered questions about the signal to noise of the detection and also the role of the pulse repetition frequency in affecting the observed visibility.

parameters agreed well with the theory developed by Payne-Scott.[28] As Mills has pointed out, this paper remained the definitive publication on PPI displays for many years.[29]

At the end of the paper Payne-Scott acknowledges "many conversations with B. Y. Mills". It is likely that this experience with the detectability of weak signals led to Payne-Scott's interest and expertise in discussions of radio astronomical source "confusion" (the detection of unique radio sources in the presence of nearby sources) in the era 1946–1951. This wartime research led to a perfection of Payne-Scott's research skills; her reputation among her peers was solidified and likely increased the acceptance that she received from her colleagues in the era 1945–1951, as radio astronomy in Australia experienced rapid growth.

Additional Notes

1. Joan Curran was a scientist working on WWII radar research in the UK. R.V. Jones (1978, *Most Secret War*) and Louis Brown (1999, *Technical and Military Imperatives: A Radar History of World War II*) in their comprehensive publications about World War II have described the efforts of Joan Curran (*née* Strothers, later Lady Curran) in the development of Project Window in March 1942; this was called chaff by the Americans. The research was carried out at the Telecommunications Research Establishment (TRE). These aluminium strips used in Project Window, which were dropped from bombers to confuse the enemy, resonated with the 50 cm Wurzburg German radar transmissions; propaganda messages were also written on the strips. The objects were elongated, with a length of about 25 cm and a width of only 1–2 cm. Project Window was first deployed over Hamburg on 23/24 July 1943, causing remarkable confusion among the Germans defensive anti-aircraft radars. Later in the war, the Germans developed counter-measures against Window. In the US the major institution for radar research was at the Radiation Laboratory of the Massachusetts Institute of Technology from 1940 to 1945. Buderi (1996) has written: "The lab employed very few women scientists." From Guerlac's extensive history of the Rad Lab (1987), it is possible to identify at least three women

---

[28] McCready in *The Textbook of Radar,* edited by E.G. Bowen (first edition 1947 and second edition 1954) Chap. 11 "Receivers", described the results of the Payne-Scott research. He pointed out that the signal to noise of a radar receiver is not limited by the noise factor of the receiver but by the sensitivity of the cathode ray tube. He wrote: "Payne-Scott has shown that under these conditions we can detect a signal whose power is 15–18 dB below the noise power, depending upon the type of detector...." He summarised the 1948 publication of the Proc IRE publication by Payne-Scott and concludes "Although many existing radar systems can detect signals whose powers are of the same order as noise in the input circuits, it is preferable to calculate the sensitivity at the cathode ray tube making use of the basic theory and charts in Payne-Scott's paper."

[29] Interview B. Y. Mills with Goss, 1 April 2007.

who carried out independent research during WWII: Dorothy Montgomery, who worked on K band (1.25 cm) research in the group of Edward M. Purcell; Jane Fairbank, who worked on the K band ship to surface vessel radar system; and Louise Buchwalter (later Louise Young), who worked on antenna design for the long range air to surface vessel radar. William Aspray (in the IEEE -Institute of Electrical and Electronics Engineers- Global History Network series (http://www.ieeeghn.org/wiki/index.php/Oral-History:MIT_Radiation_Laboratory) has described an interview with Virginia Powell Strong. She was a chemist who worked on the properties of high burn-out crystals, in order to create components of radars that could withstand combat stress.

2. Marjorie Barnard (1897–1987), a famous Australian author, was the librarian at RPL and National Standards Laboratory from 1942 to 1950. With Flora Eldershaw, Barnard wrote five major works of fiction using the single pseudonym, M. Barnard Eldershaw. Their collaboration included *A House is Built* (1929) and the futuristic novel, *Tomorrow and Tomorrow and Tomorrow*. (This novel was censored on publication in 1947.) Barnard wrote an unpublished manuscript in 1946 entitled *One Single Weapon*, a history of the RPL radar effort. This manuscript of over 300 pages is located in the Basser Library of the Australian Academy of Science in Canberra as well as in the Mitchell Library in Sydney. An accompanying note with the manuscript at the Basser Library states, "This manuscript of the history of radar was commissioned by [RPL] after the Second World War, and owing to some disagreements on factual matters between the author and some members of the Division was never published." In the same note, Bowen also was reported as being uncertain about some of Barnard's conclusions. Like Payne-Scott, Barnard had also attended Sydney Girls High School and the University of Sydney, graduating from university with first class honours in 1918.

3. Kathleen Rachel Makinson was born 15 February 1917 in the U.K. and moved to Australia in 1939 after completing a Physics undergraduate degree at Newnham College at Cambridge. She married Richard E. B. Makinson (1913–1979) in 1939 and worked at the School of Physics, University of Sydney, during WWII with some time off for the birth of her elder son, David, in 1941; her second son Robert was born in 1956. During the War Makinson held positions on an annual, temporary basis as both Research Assistant and Research Scholar. In the early part of the War she worked with Joan Freeman (Freeman 1991) on various defence-related research projects under the direction of V. A. Bailey. (Freeman has described these as "not very satisfying or productive".) As a tutor and demonstrator, Makinson was involved in the "Bailey Boys" (after Professor V.A. Bailey of the School of Physics) courses in electronics and radar techniques for military personnel (mainly the RAAF, Chap. 4). Even though she was working at the University of Sydney with the "Bailey Boys", she was being paid by the CSIR Division of Radiophysics from 1944 to 1945. Makinson's research at the University of Sydney during WWII was described in RP222/1 from October 1944: "Magnetically Controlled Gas Discharges" by R.E.B. Makinson, J.M. Somerville and K. Rachel Makinson.

Chapter 6
1944–1945: Ruby Payne-Scott – The First Woman Radio Astronomer

Ruby Payne-Scott's career as a radio astronomer began in 1944 and extended to her retirement in July 1951. Her remarkable career, which led to many of the early discoveries in solar radio astronomy, can be roughly divided into four phases. The early phase began in March 1944 and extended to late 1945, a period of transition from wartime radar research to early solar noise research. In this period Payne-Scott, under the leadership of J. L. Pawsey, along with others, was laying the groundwork for the beginnings of solar radio astronomy. The following period, from October 1945 to late 1947, marked the groundbreaking solar work at Dover Heights and the publication of their first important research papers on solar physics. Her contributions to Fourier radio astronomy imaging were crucial at this stage. In 1948, Payne-Scott had an interlude working on her own at the Hornsby field station; the detailed properties of Type III bursts were elucidated in this period. Starting in 1949, up to the end of her career at RPL in July 1951, she was mainly involved in the building and use of the high resolution swept-lobe interferometer at Potts Hill Reservoir in Sydney, a collaboration with Alec Little. There was a brief coda in August 1952 during the URSI International Assembly in Sydney.[1] In this chapter we will study the work Ruby did as a proto-radio astronomer in 1944 and 1945.

Propagation Committee at Radiophysics Laboratory: 1944–1954

The minutes of the "Propagation Committee" (PC) of the CSIR, Radiophysics Laboratory (RPL) from 14 September 1944 to 7 March 1949 and the renamed "Radio Astronomy Committee" from 11 April 1949 to 9 April 1954 have provided an invaluable source of information about the research programme of the RPL in

---

[1] Orchiston et al. (2006) have discussed early RPL contributions to solar astronomy in Australia, in addition to activities at the Commonwealth Observatory at Mt. Stromlo and the Physics Department at the University of Western Australia from 1945 to 1948.

these years.² The committee was created during World War II under the leadership of Pawsey, the chair, for discussions once or twice a month. They stopped meeting near the end of World War II for a time, but it was reinstated mid-September 1945 in order to rethink the purpose of the RPL in a post-war Australia. Before or after the war, at these meetings, the discussion of a new topic would often begin by Pawsey asking; "Ruby what do you think?"³

Proto-Radio Astronomer, Ruby Payne-Scott

Ruby Payne-Scott's first radio astronomy experiment was performed in March of 1944. The observations were carried out with J. L. Pawsey at the short wavelength of only 10 cm (3,000 MHz). Payne-Scott played a key role in these observations based on her experience with the calibration of 10 cm receivers and her thorough understanding of the thermodynamics of radio frequency noise. The results were described in a RPL classified memo (RP 209) from 11 April 1944, authors J. L. Pawsey and Ruby Payne-Scott.⁴ The equipment undoubtedly was experimental since most of the radar work being done at this time was in the 1.5 m (200 MHz) range, although another group at RPL was working on the LW/AWH Mk II 25 cm (1,200 MHz) system (Minnett et al. 1999a).⁵ The report, RP 209 had the unspectacular title of "Measurements of the Noise Level Picked Up by an S-Band Aerial", Pawsey and Payne-Scott, 1944. S band—"S" stands for short—was the name given for the 10 cm microwave band during World War II. Today this is the frequency of most microwave ovens.

The short report of five pages, including an appendix with the calculation of the noise factor or noise temperature of the receiver, contained a number of prescient conclusions, including the then surprising fact that the sky 10 cm is cold, an

---

² Copies of the minutes were provided by Don Yabsley (1923–2003) to Goss in 1999 for the period 1945 to mid-1950; Yabsley left the committee membership in 1950. NAA: B2/2 Part1 (144 pages, 1944–1949) and Part 2 (188 pages, 1949–1954).

³ A letter from John D. Murray from 24 January 2004 described the format of the PC meetings. The late W.N. (Chris) Christiansen wrote in October 1977 with a description of the role of Pawsey and Payne-Scott at the PC meetings. At one of the sessions, Christiansen reported that Pawsey initially made a decision, when Payne-Scott was not present. Pawsey said: "Well that is settled." Then suddenly he hesitated when he noticed Ruby's absence. Pawsy continued: "We had better ask Ruby before we proceed."

⁴ W.T. Sullivan III has provided Goss with his annotated version of this report; his notes were written 14 December 1983. A number of his notes have provided valuable background information. In his obituary of Pawsey, Lowell (1964) described the 1944 observations at RPL: "...the work undoubtedly stimulated the solar measurements made immediately after the war". RP 209 was also mentioned in passing by Lovell almost 20 years later (1983): "I had all his documents from [RPL] and this most interesting paper was amongst them. I hope it is still there." In addition, Sullivan (1988, 2009) has described RP 209 in detail. Sir Bernard Lovell, 1913–2012.

⁵ See Chap. 5 for a description of Payne-Scott's role in this 25 cm project.

accurate measurement of the temperature of a clear sky at 20 K, as well as basic calibration techniques still in use today.

The first experiment consisted of determining the response of a 20 by 30 cm microwave horn connected to a receiver with a noise temperature of about 2,900 K as the horn was pointed at various positions within the room and out a window towards the sky. Unfortunately, no detailed description of the exact test set up was provided. The detailed thermodynamics of the noise from the receiver and various sources of this noise were discussed. Various contributing factors were mentioned, including the intrinsic noise of the receiver, the temperature of the room (ground), clouds and "matter in space". The authors were apparently surprised by the low temperature of the sky (between 0 and 140 K) and seemed to detect the increase due to clouds or even rain on 20 March 1944 at 10 am in Sydney. The most precise sky measurement of 20 K was a surprisingly accurate value for the brightness of clear sky conditions. Pawsey and Payne-Scott were surprised that inserting attenuation between the horn and the receiver *increased* the output as the radio horn was positioned looking out of the window, in contrast to pointing the horn within the room. The attenuator reduced the strength of the signal from the sky but added considerably more noise power from the ohmic resistance at room temperature. The result showed that the sky had a low temperature compared to the room temperature of about 300 K. Pawsey and Payne-Scott wrote:

> Despite the theoretical and practical interest attaching to measurements of ultimate noise level the authors are not aware of any reported measurements of received noise powers in the centimetre range of wavelengths. Those described here are of a preliminary nature, and the authors hope to extend them further.[6]

This prediction turned out not to be the case, as no follow up work was reported in subsequent years. The concluding paragraph of the report also suggested a new way to calibrate the noise factors of receivers—at least in the cm bands; the use of cumbersome signal generators could be avoided if it could be established that the temperature of the clear sky were constant. If this were to be the case, the receiver could be calibrated by pointing alternatively at the sky and a room temperature enclosure. This technique is often used by radio engineers today as part of the calibration procedure for cm radio astronomy receivers.

Pawsey and Payne-Scott attempted a single astronomical observation. They used the same receiver and attached it to a 4-foot-diameter paraboloid (the report is filled with mixed units, e.g. cm and inches). They observed a region near the galactic plane in Centaurus (new galactic longitude near 300°) and reported no signals greater than 10 K (the expected value based on current knowledge would be a few K antenna temperature), "very much less than that observed by Jansky and Reber and so small as to have no observable effect.... No attempt was made to observe radiation from the sun". Sullivan (2009) has made the surprising calculation

---

[6] In fact, Southworth and Mueller (Sullivan 2009) had carried out a similar experiment in 1943 at Bell Labs in the US. Pawsey and Payne-Scott were apparently not aware of this in Sydney in 1944 even though there were exchanges of documents between the US and Australia during WWII.

that had they done so, the sun would have been detected easily with a large signal to noise. Pawsey and Payne-Scott were apparently not aware of either the Southworth 3.6 cm solar detection from 1942 (Southworth 1945) or Hey's 5 m solar detection from 1942 (Hey 1946). Both were still classified results in 1944, although RPL was receiving a number of classified documents from the UK and the US during this period. Harry Minnett told Sullivan in 1986 that the March sun may not have been visible from the chosen window at the National Standards Laboratory.[7] The Pawsey and Payne-Scott report contained a number of astronomical references. There were four references to Jansky's papers in the Proceedings of the IRE from 1932 to 1937 (e.g. Jansky 1932), a reference to Reber's paper in Proceedings of the IRE in 1940 (Reber 1940a) (there was no reference to the influential 1940 "Cosmic Static" *Astrophysical Journal* publication, Reber 1940b) and a reference to a paper by Eddington (1926) in the Proceedings of the Royal Society from 1926. Pawsey and Payne-Scott were struck by the divergent predictions of Jansky and Reber concerning the likelihood that the galactic background would have been detectable at 10 cm. Based on Jansky's suggestions that the radio background arose from a large scale distribution of matter at high temperatures, the 10 cm radiation might well have been observable. On the other hand, if Reber's claim that the radiation intensity scaled as wavelength, then there would have been negligible emission at 10 cm. The interpretation given to Eddington's predictions by Pawsey and Payne-Scott was, however, a misinterpretation that others have also made:

> It appears likely that all noise received is of thermal origin and comes from regions of very low temperature. Eddington, working from the measured intensity of starlight, calculates that the radiation from the matter in space is equivalent to that of a black body of 3.2 K. If most of the noise power received by the aerial originates from space, this would account for the very low noise temperature measured.

As described by Sullivan, Eddington had only pointed out that the average energy density of starlight in the interstellar medium near the sun is comparable to that of a blackbody with temperature of 3.2 K. The Pawsey-Payne Scott interpretation, which suggested that the 10 cm radiation field in the interstellar medium would be characterised by a blackbody temperature of 3.2 K, was not correct. The value suggested by Eddington only represents an average value of the radiation field of the optical radiation of the stars and does not represent the radio frequency intensity at this wavelength of 10 cm.[8]

---

[7] In 2007, 2008, and 2013 Goss visited the two inner courtyards at the Madsen Building at the University of Sydney in an attempt to locate the possible room used for the 1944 observation. Several candidate windows were detected with south facing exposures (the sun is always in the north in Sydney); but the exact location remains a mystery.

[8] The approximate agreement with the famous cosmic microwave background of 2.7 K, discovered by Penzias and Wilson (1965, recognised by the Nobel Prize in Physics in 1978), is a coincidence. When Goss was a graduate student at the University of California, Berkeley, in the 1960s, a famous astronomer (in an interstellar medium course) suggested that the Eddington value was consistent with the previously known 2.3 K excitation of the optical interstellar lines of the CN molecule; in fact the excitation of this line is due to the cosmic microwave background radiation.

In summary, there is a strong likelihood that the March 1944 observations at 10 cm by Pawsey and Payne-Scott were the first radio astronomy observations in Australia and even in the southern hemisphere. We agree with Sullivan (1982) that Ruby Payne-Scott is likely the first woman radio astronomer. The 1944 observations described in RP 209 predate the solar observations of Elizabeth Alexander (Orchiston 2005) by about a year. The major importance of RP 209 was the determination of the upper limit on the sky brightness at the short wavelength of 10 cm. This limit was an unexpected result in 1944. The fact that these data were never published is a surprise. The paper was a remarkable contribution, showing an understanding of the thermodynamics of radio receivers that is taken for granted six decades later. The techniques that led to an understanding of the absolute calibration of the instrument would serve the RPL group well in the next few years as the Australian radio engineers participated in the post World War II re-birth of radio astronomy.

During these early years, the people working in the Radiophysics Lab considered themselves engineers or physicists. It was not until they began to realise coherently, as a group of scientists, the far-reaching possibilities of their discoveries that they self-identified as founding members of a new field, "Radio Astronomy". J.L. Pawsey in fact first used the term, "Radio Astronomy" in an otherwise mundane letter in January of 1948.[9]

The incredible situation these pioneering scientists found themselves in is illuminated by Bruce Slee, who worked for some months with Payne-Scott at the Dover Heights station in 1947. Slee states that Ruby, a gifted physicist, had to go to the library to find books on everything from basic to arcane elements of astronomy to better understand what her group was observing. She even lent him a copy of the very dry but information-packed *Text Book of Spherical Astronomy* by W.M. Smart (Smart, 1st edition, 1931) to help him understand elements of classical astronomy such as refraction, time, coordinate systems, orbits, etc. Ruby re-tooled her knowledge set through independent study of astronomy techniques.[10]

Post-war Activities at RPL: 1945- How This Impacted the Role of Payne-Scott

Well before the ending of World War II on 15 August 1945, the leadership of RPL began planning for post-war activities. The nature of their discussions has been treated by numerous authors.[11] The existence of an intact group of scientists was decisive for the emergence of Australia as one of the two major players (along with

---

[9] Sullivan (2009), p. 424, footnote 18.

[10] Slee (1994) and interview with Goss (2007).

[11] Sullivan (2009), Wild (1965, NAA:C3830, D12/1/5. "Origin and Growth of Radio-Astronomy in C.S.I.R.O", delivered at Division of Plant Industry 15 October 1965), Wild (1968, 1987), and Bowen (1984, 1988).

the United Kingdom) in the post-war development of radio astronomy. As Sullivan (1988, 2009) has pointed out, only in certain fields of medical science did the international reputation of Australian scientists rival that of their fellow Australian radio astronomers. As both Sullivan (1988, 2009) and Wild (1968) have emphasised, the fact that the Australian radio astronomers had the Southern sky to themselves played only a minor (but not negligible) role; the experience and expertise of the scientists were the decisive factors.

At the end of his long career, when J. Paul Wild[12] summarised the reasons for the flowering of radio astronomy in Australia, he attributed it to the excellent collaborative guidance of two key individuals (Wild 1987): (1) E. G. "Taffy" Bowen, who was the newly appointed Deputy Chief (Research) of RPL beginning in January of 1944, and was to become Chief of RPL in May of 1946; and (2) Joseph L. Pawsey, the leader of the radio astronomy group and later Assistant Chief of RPL.

Bowen himself wrote in 1984 (Bowen 1984):

> What were the ingredients which led in 1946 to the development of radio astronomy? The first and by far the most important of these was the decision by the Chairman of CSIR, Sir David Rivett [1885–1961], that at the conclusion of World War II, CSIR would be devoted only to peace-time research, and that defence research would be carried out by other agencies. This meant that a highly developed laboratory with a superlative staff became available for a wide range of researches and practical developments in a peace-time environment. ... The next ingredient was that the staff, about 200 strong, was already highly skilled in electronic research and development.[13] They ranged from professors of physics to practical engineers from industry. Many of them had spent months, if not years, at the best overseas laboratories and were saturated with the most recent electronics techniques. In view of later events, it is also rather remarkable that there was not a single astronomer on the staff, nor, for that matter, anyone who had done a university course in astronomy...Next in importance to the people was the store of components of all kinds which had been accumulated during the war years...It is clear that another important factor was morale...Our policy was to try anything that gave promise of useful scientific or practical applications; if successful, we poured in manpower and resources. Radio astronomy was to become one of the most productive of these.

The most far-reaching discussion of the future program of the RPL occurred in an extensive memo written by E.G. Bowen on 2 July 1945, a month before the end of the Pacific War. The memo was an agenda item for the 35th Session of Council of the CSIR to discuss the "Future Programme of the Division of Radiophysics".[14] The future programme consisted of nine possible areas of research: (1) Propagation of radio waves including the ionosphere, radio noise and super-refraction; (2) Vacuum

---

[12] J. Paul Wild (1923–2008) was a future Chief of the Division of Radiophysics from 1971 to 1978 and then Chairman of CSIRO from 1978 to 1985; he was the leader of the solar group at RPL from the mid-1950s to 1971. The Radioheliograph at Culgoora (Chap. 2), which operated 1967–1984, was his creation.

[13] There was some downsizing at the end of the war, as many of the professional staff took up positions in industry, went back to universities or started graduate degrees (for Ph.D.'s the option was to go overseas, in most cases the UK).

[14] NAA:C3830, D1/1 (1945 Programme of the Division of Radiophysics) and D1/2 (1946 Programme of the Division of Radiophysics).

research including the generation of power at millimetre wavelengths and the use of radar techniques to accelerate elementary particles (programme directed by Pulley and Gooden); (3) General radio and radar research including antennas and receiver systems; (4) Radar aids to navigation including long range navigation (programme directed by V.D. Burgmann, a future Chairman of CSIRO; (5) Aids to ground survey; (6) Atmospheric physics research (this became the Cloud Physics research group within RPL); (7) Research and development for the armed forces; (8) Industrial co-operation, and (9) Co-operation with other divisions of CSIR. The future radio astronomy was hidden in a section (1.2) with the obscure title of "Study of Extra-thunderstorm sources of noise (thermal and cosmic)". Bowen wrote: "... a type of noise appears which is thought to originate in the stars or in interstellar space.... Little is known of this noise and a comparatively simple series of observations on radar and short wavelengths might lead to the discovery of new phenomena or the introduction of new techniques". Bowen then suggested that it might be possible to calibrate the radar receivers by pointing the antenna at the sky and then at an object near the antenna which was at room temperature.[15] No mention was made of Jansky's and Reber's results, which in fact were well known at the time. This meagre text was the basis of the formation of one of the major components of the RPL research programme for many years to come. In addition there was no mention of radio radiation from the sun in the Bowen report, possibly due to the fact that the reports of solar detections had not yet reached Sydney.

Indeed, only months before in March 1945, there was a serendipitous observation of the sun at 1.5 m on Norfolk Island, but word of these findings did not reach CSIR in time for the July Session of Council meeting. Once they did, however, within a matter of days in early August 1945, Bowen wrote to F.G.W. White (see Additional Note, No. 1, end of this chapter), who was at this time a member of the CSIR Executive in Melbourne:

> These results [the "Norfolk Island effect" from March 1945, as reported by Orchiston 2005, see below for a discussion of this March 1945 detection of the sun by the Royal New Zealand Air Force] are remarkable in that while one would expect to receive solar noise radiation on S. or X. band [10 cm or 3 cm wavelength], a COL [Chain Overseas Low-Flying] antenna and receiver at 200 Mc/s is quite unlikely to do so. I have heard rumours of the same thing in England, but as far as I am aware, the subject has never been followed up. We are therefore going to attempt to repeat the observations here in Sydney to see if we can track down the anomaly.[16]

Following Sullivan's (2009) conclusion, this quote indicates that the detection of microwaves from the sun by Southworth in Bell Labs in the US in 1942 and reported in 1945 was likely known at RPL, while the Hey detection of solar bursts in the U.K. from 1942 (published in 1946) of solar bursts was the source of the

---

[15] Note the connection between this application of radar techniques to the descriptions by Payne-Scott in the technical report T.I. 191/1 (6 August 1945) and also RP 209 from 11 April 1944 (see above in this chapter).

[16] NAA: C3830, A1//1/1, Part 1.

"rumours". Ironically, J.S. Hey has written in his book *The Evolution of Radio Astronomy* (1973):

> I well remember Sir Edward Appleton's [then Head of the Department of Scientific and Industrial Research in the UK] astonishment at a meeting in 1945 when I remarked that I was contemplating publishing in a scientific journal my 1942 paper on solar radio emission for, by some mischance, no one had informed him of the 1942 episode.

Sullivan (2009) has given a detailed description of how this misunderstanding may have contributed to the conflict between Appleton and Hey over priority for the discovery of solar radio bursts. Thus poor communication of the 1942 results had occurred not only in the UK, but in the far reaches of the Dominions.

The New Zealand results had been sent to Pawsey in August 1945 in a letter from Dr. Elizabeth Alexander of the Radio Development Laboratory of DSIR (Department of Scientific and Industrial Research) in New Zealand; the letter was dated 1 August in Wellington but there was a postscript dated 8 August. The letter summarised the Royal New Zealand Air Force (RNZAF) results and included the report, R.D. 1/518, "Report on the Investigation of the Norfolk Island Effect".[17] Bowen, Kerr, McCready, Payne-Scott and Briton (Chief from Jan. 1945 to May 1946) all received copies of this document (Alexander 1945) from New Zealand. An enigmatic statement in the Alexander letter to Pawsey shows that the Hey results were not yet known: "I think the main differences between Southworth's latest results and ours are first his work in the centimetre band fits more or less with black body theory and ours shows definitely too much energy on 200 Mc/s for theory. Sir Edward Appleton has also taken measurements on 200 Mc/s and confirms our finding. He suggests that at times of increasing sun spot activity there is an increase in energy at both ends of the sun's spectrum, and has encouraged us in our efforts." Orchiston (2005) has suggested that Appleton may well have been disingenuous in this claim (i.e. there is no evidence that Appleton had new solar data of his own in 1945), not substantiated in the Appleton (1945) publication, which in fact included no new solar data of his own. Another possible interpretation is that Alexander may have confused this claim with the "rumour" reported by Bowen above, based on the Hey report from 1942 (Hey 1942). Given the slowness of communication in 1945, it is quite possible that these types of misunderstandings might have arisen.

---

[17] Norfolk Island (an external territory of the Commonwealth of Australia) that lies between Australia and New Zealand; the Royal New Zealand Air Force was responsible for air defence of this Australian territory in World War II. Using a COL radar the RNZAF personnel detected the sun on a number of days starting 28 March 1945 at 200 MHz. Alexander (1946) has given a popular account of the early New Zealand solar radio astronomy. See Orchiston (2005) for the details of this report and a biographical sketch of Alexander.

RPL's First Observations of Solar Noise- Role of Payne-Scott

J.L. Pawsey's leadership in the war time activities of RPL and his experience in antenna design would serve him well as he became a leader in the post-war RPL activities. In an informal manner, Pawsey appointed two key personnel to his group that would have far reaching influence in the next years; Lindsay L. McCready[18] was to head the receiver developments, while Ruby Payne-Scott was to become an overall advisor for scientific issues, engineering planning and mathematics.

A little over a month after the Alexander letter arrived from New Zealand on 14 September 1945, the "Propagation Committee" of RPL was reconstituted; present were Drs. Pawsey (chair), Pulley, Piddington, Messrs. Kerr (secretary), Wood, Iliffe, Yabsley, Parker, Wing Commander Taylor, SLDR Hall and Flying Officer McDonald. The minutes indicate that "the Committee's functions are to review progress in propagation work, and plan new work. The peacetime method of working will be closer to the typical University [sic] research system, one of individual responsibility and we hope to gradually work into this method".

Within the purview of this committee, five topics of research were proposed. The list elaborated on the proposed research topics described by Bowen 2 months earlier. The subjects were: (1) troposphere propagation (super-refraction), (2) ionospheric propagation (Loran—navigation), (3) scattering from clouds, (4) scattering from the middle atmosphere and (5) radio noise levels. In the minutes Don Yabsley described the 2 and 6 MHz receivers which apparently were recording the noise level of the ionosphere. Of noteworthy import was the note that Miss Payne-Scott "is going to look for 200 Mc/s signals from the sun at sunrise and sunset. Such signals, at a level greater than suggested by black body theory, have been reported on COL [Chain Overseas Low- at 200 MHz] sets in New Zealand".

At the 12 November 1945[19] meeting of the Propagation Committee, the following discussion was noted under the rubric of solar noise:

> The programme of future work prepared by Dr. Pawsey was discussed. Present observations will be continued. Mr. Yablsley will shortly take on the development of special equipment for the investigation. He will study the time variation of solar noise, and collect information where possible on the wavelength variation. Miss Payne-Scott will write a report on work to date in T.I. form. Steps will be taken to start observations on the Sun and Milky Way from the Dover Sh.D. [shore defence] station.

This text was followed by a detailed two page report, (B 51/4, in the Propagation Committee file given to us by Yabsley) authored by Pawsey with outlines of possible cosmic and solar programs. The subsections were prescient: study of

---

[18] McCready (1910–1976) joined RPL in 1940 from AWA (as did Payne-Scott). He moved to CSIRO Applied Physics in 1962 and retired in 1971. See the image Fig. 6.1 for McCready (first row, first person on the right).

[19] The minutes of the meeting of 15 October 1945 contained the statement that "reports on this topic [solar noise] will continue to be made to the Research Committee [likely the PC] for the present". Note that the first observations at Collaroy had already begun on 3 October 1945.

solar intensity variations at 200 MHz, and the study of spectra of both solar and cosmic emission at a number of wavelengths above and below 1.5 m, polarization of the radiation at the high frequency of S band (3 GHz or 10 cm), and the precise determination of the direction of arrival of the solar radiation using the shore defence station in a beam "swinging" mode. A number of possible new solar observations at higher frequencies were proposed (wavelengths 50, 25, 10, 3, 1.2 cm). A major new development was proposed to build antennas which were equatorially mounted, so the sun could be observed continuously, not only at sunrise and sunset as was the case with the radars at Collaroy and Dover Heights, which could only be moved in azimuth at a fixed elevation near the horizon. Also the proposal was made to move some of the radio equipment to Mt Stromlo to "carry out observations in close liaison with Stromlo personnel with visual equipment". The remainder of the report described the radio equipment in some detail.

A fascinating detail was mentioned at the conclusion. "For subsequent work it is desirable to obtain a large, say 30 ft. diameter, paraboloid suitably mounted." Had this proposal been implemented, the course of radio astronomy development in Australia in the post-war era would have been vastly accelerated. It would be 7 years before the 36 ft. (11 m) transit telescope was available at Potts Hill at a new RPL field station.

At the next meeting on 10 December 1945, the solar noise report stated: "Observations by the R.A.A.F. station at Collaroy are continuing. Miss Payne-Scott has completed a survey of the subject, to be issued as an internal report..."

As we can see, Payne-Scott was destined to play a major role. Her radio engineering skills acquired during the war, her physics background and especially her mathematical skills were contributing factors to her success in the years 1945–1951. Clearly Pawsey relied on her experience and judgment as he became the force behind the rapid growth of radio astronomy in Australia in this post-war period. She was perhaps so integral to the nascent field of Australian radio astronomy that her marriage to Bill Hall was gladly kept secret from the government by her close-knit group of colleagues. It was known that she wore her wedding band as a necklace, and that she would be stripped of her regular employment status as well as her retirement fund if her status as a married woman were discovered by the rule-enforcing overseers in the CSIR. Her natural talents were given a sheltered place to grow during her 10 years with RPL, and unfortunately we'll never know what more she could have done, had there been a system of maternity leave in Australia then, and had a bureaucratic reshuffling of power in the 1950s not thrown her into conflict with previously ignored laws regarding women's employment in the post War era.

The First Summary Paper in Radio Astronomy: December 1945 Author Payne-Scott

In a wonderfully useful document, cumbersomely named "Solar and Cosmic Radio Frequency Radiation; Survey of Knowledge Available and Measurements Taken at the Radiophysics Laboratory to Dec. 1, 1945" (Payne-Scott 1945c), which we will hereafter refer to as "The December 1945 summary paper", Payne-Scott provided a detailed chronology of events during the course of 1945 that lead to the first Australian solar observations from October, 1945. The December 1945 summary paper provided a valuable source of the details of the research and motivation in these first months after the war. In it, Ruby not only summarised all the proto-radio astronomy work she had done in the previous months, but placed her research in the context of other research carried out in the US, the UK and New Zealand.

She wrote that the initial solar observations: "...were inspired by the almost simultaneous arrival of three reports in the laboratory ..." in mid-1945. Two of these were the classified reports referred to above: (1) the Hey report from 1945 describing the 1942 detections of bursts with the British radars on the south coast of England,[20] (2) the report from New Zealand of a large increase in solar noise due to solar bursts detected on Norfolk Island in late March 1945 using a COL radar. The final decisive report was an article in the *Astrophysical Journal* by G. Reber "Cosmic Static" showing the galactic plane at 160 MHz with a resolution of about 12.5° (Reber 1944). This paper appeared in the November 1944 issue of the *Astrophysical Journal*, having been submitted on 8 May 1944. The detection of the sun was presented in the publication almost as an afterthought. Reber wrote:

> It has been suggested that this long-wave radiation could be set up in the corona of the sun. Until recently no positive evidence was available... In any case the sun had the rather surprising ...intensity of 10×10^{-22} watts.sq.cm,cir, deg., M.C. band.

The galactic centre scans carried out in December 1943 showed the sun and the galactic plane in Sagittarius superimposed, with comparable intensity. Reber did not indicate that the implied brightness temperature of the sun was, in fact, in excess of 10^6 K, if the emission were to arise from a disk the size of the optical sun.[21] Due to the slow speed of sea-mail during wartime, it is not surprising that the *Astrophysical Journal* published in late 1944 in the US did not arrive in Sydney Australia until June or July 1945.

A remarkable, additional, non-solar observation was summarised in the December 1945 summary paper; the North Head antenna had been used. The Milky Way was observable in the daytime in these months and a 200 MHz map of the galactic

---

[20] This was the Army Operational Research Group (UK) report no 275 from 13 June 1945 describing the 27 and February 1942 on G.L. stations at 55–85 MHz (see Sullivan 2009 and above).

[21] An example of an erroneous interpretation of the Reber observation of the sun appeared in the publication by Appleton and Hey (1946a). These authors suggested that the 1.9 m data of Reber implied a brightness temperature of only about 6,000 K, roughly the black body temperature of the optical sun. In fact, Reber made no claims about the intensity of the solar radiation based on his detection of late 1943. As Sullivan (2009) has pointed out, Charles Townes (1947) published a paper: "On the Interpretation of Radio Radiation from the Milky Way" in which an earlier version did include a statement about the large intensity solar intensity (hence non-thermal) implied by the Reber data at 1.9 m. However, the referee of the paper insisted that this conclusion be dropped; the final Townes publication contains no reference to the anomalous solar intensity.

centre at right ascension 17 hours 30 min and declination −33° was presented. (Unfortunately this figure is missing from the copy of the Payne-Scott report located in the early years of 2000 in the CSIRO Division of Radiophysics library by Wayne Orchiston.) With this system the angular resolution was some tens of degrees:

> It will be apparent that, in addition to the radiation from the sun, there appears to be radiation from a more diffuse area centred approximately on the centre of the galaxy.

Also at the end of November, 1945 the Collaroy antenna was used to scan the centre of the Milky Way at sunrise. The observations consisted of scans covering about 20° in azimuth and roughly centred on the galactic centre. The summary of these results followed:

> There are not yet sufficient results to produce a clear picture, and a number of puzzling variations have been observed; it is possible that some of these are due to absorption in the clouds of matter that cause the dark patches observable in the Milky Way.

Likely, Payne-Scott was not aware that the interstellar dust was completely transparent at these radio frequencies. The final figure in the December 1945 report—also not present in the existing copy of the December 1945 summary paper—was a reproduction of the first contour maps of the sky which had been published in 1944 by Reber. This image of the northern Milky Way at 160 MHz with a resolution of about 12.5° most likely would have been far superior to the lower sensitivity and incomplete image presented in the Payne-Scott report. Improved maps of the radio continuum of the southern Milky Way were not available until 1950 when Bolton and Westfold (1950a) and Allen and Gum (1950) presented images of the galaxy at 100–200 MHz, respectively.

The final paragraph of the Payne-Scott report was a plan for the future. Here a number of prescient predictions were made concerning several aspects of the future of radio astronomy in Australia:

> It is hoped to soon begin here a programme of more exact work, in conjunction with the Stromlo Observatory. Among questions to be investigated are the frequency dependence of the radiation, its polarization, further study of the long-term variations and an investigation of the short-period fluctuations. There is also hope to make a survey of the Southern sky; Sydney is almost at the antipodes of Reber's stations (sic), so that we can survey areas inaccessible to him; in particular it will be interesting to see whether radiation can be detected from the Magellan Clouds.

In fact the Magellanic Clouds (the nearest galaxy neighbours of the Milky Way and only readily observable in the southern hemisphere) were detected a few years later in Sydney by RPL scientists in both the HI line at 21 cm and in the radio continuum [HI by Kerr et al. (1954); continuum by Mills and Little (1953)].

Bowen was impressed with the quality of the December 1945 summary paper and commented on this publication in a letter to E.V. Appleton on 23 January 1946[22]: "Miss Payne-Scott, who with Pawsey and McCready has been largely

---

[22] See Chap. 7 for additional details of Bowen's description of the early RPL solar noise research in 1945 in his letter to Appleton. NAA:C3830, A1/1/1 Part 1.

responsible for the work here in Radiophysics, has written an internal report [the December 1945 summary paper] summarising latest ideas on the subject of solar and cosmic noise." Bowen asked Appleton's advice about publication of the Payne-Scott paper:" After adding some further experimental results we propose publishing it in one of the journals. So do you think this is a good thing and would you suggest the *Proceedings of the Physical Society* or the *Astrophysical Journal*?" Unfortunately the paper was never published; this omission remains a serious loss in reconstructing the history of early radio astronomy in Australia since the Payne-Scott report documented so thoroughly early thinking of the Sydney group with regard to both solar noise and cosmic noise. No record has been found that explains the reasons for not publishing. We can only guess that the frenzied pace of research and publications in the following year (and Payne-Scott's likely absence in late 1946 and early 1947, posited to be due to a miscarriage) may have played a role. In 1946 to 1952, Payne-Scott was an author or co-author of 9 publications; during this period RPL had 65 radio astronomy publications.

Symposium on Radar, 5–7 December 1945: Payne-Scott Was Present

A final event in this transition period from war time radar research to post-war radio noise research occurred in late 1945 at the School of Tropical Medicine on the grounds of Sydney University.[23] At this seminar ("Symposium on Radar") the techniques of radar and its military and non-military uses were described to a number of armed service personnel and some representatives of industry as well as numerous RPL personnel. The 3 day symposium was held 5–7 December 1945. The presentations began on Wednesday with a short opening presentation by J.N. Briton (Chief of RPL) followed by "Introduction to Radar" by the assistant chief E. G. Bowen and "Fundamentals of Radar" by J. L. Pawsey. In total there were 19 presentations. The morning of the final day (Friday) was dedicated to demonstrations in the nearby RPL lab concluding with a presentation on "Radar Research" by E. G. Bowen.

Harry Minnett et al. (1999a) has described the development of the Light-Weight Air Warning Radar during WWII at the RPL in the compilation of articles edited by MacLeod (1999). Minnett had a photo of the assembled audience which had been stored for years in a file cabinet and later in a cardboard box in the Minnett home. The photo is shown in the original format in Fig. 6.1, with a number of well known

---

[23] The late Ron Bracewell provided a copy of the programme for this conference. Joan Freeman has described the event in her biography. She wrote that the conference was such a success that: "Bowen decided that our papers should be published in a book... called a *Textbook of Radar* [which] appeared in 1947." The second edition was published by Cambridge University Press in 1954.

Fig. 6.1 Symposium on Radar, 5–7 December 1945, sponsored by RPL at the Lecture Theatre in the School of Tropical Medicine, the University of Sydney. Concluding talk, "Radar Research" by Bowen, and final remarks by the RPL chief, Briton. The *front row* (left to right), is E. G. ("Taffy") Bowen, J. N. Briton, J. Eagles, J. L. Pawsey, H.C. Minnett, L. U. Hibbard, T. Kaiser and L. L. McCready. Note personnel from armed services in the audience. Ron Giovanelli is sitting immediately to the extreme right in the *third row* (Photo used by permission of the late Harry Minnett)

Fig. 6.2 Enlargement of previous figure, showing Payne-Scott and colleagues toward the back of the lecture theatre (pointed out to Goss by Lori Appel). Payne-Scott is in the next to last row (*blond with glasses*). Rachel Makinson is three rows down. Several well known CSIRO colleagues are near Ruby Payne-Scott. To the left of Ruby on the photo is Noel Thorndike. On her other side in the back row is Stuart Dryden of the National Standards Laboratory (NSL) and to the right of Dryden are Gordon Wells and Gordon Stanley, both of RPL (*extreme right of photo, back row*). David Holloway is in front of Dryden (*row in front of Payne-Scott*). Holloway was in charge of microwave standards at NSL. Behind Makinson and to the right is Mel Thompson of NSL. In front of Rachel (*with glasses*) is the well-known expert in optics W. H. ("Beattie") Steel, from National Standards Laboratory of the CSIR. (Identifications provided by Harry Minnett 14 April 2000. Photo used by permission of the late Harry Minnett)

RPL staff who were sitting in the front row. When the original publication arrived in Socorro, New Mexico in early 2000, Goss' colleagues, Loretta Appel, recognised Ruby Payne-Scott in the background. Minnett had an enlargement made at the RPL photo lab and the resultant image is shown in Fig. 6.2. A number of additional RPL and NSL personnel are now evident including W.H. Steel, Rachel Makinson and Gordon Stanley. The names provided by the late Harry Minnett are given in the figure caption. Minnett et al. (1999a) wrote: "It was fitting that RPL's most advanced wartime radar should spearhead the Laboratory's post-war drive to apply radar techniques to civil problems."

Thus in late 1945, Payne-Scott was poised to begin her participation as a major member of the early team of physicists, engineers and technicians who were to revolutionise radio astronomy in the next decade. In the next chapter, we will describe in detail the exciting years from late 1945 to the end of 1947 as the research endeavours were concentrated at Dover Heights, Sydney.

Additional Note

1. Professor (later Sir) F. W. G. White (1905–1994) arrived in Sydney in March 1941 from the University of Canterbury in Christchurch, New Zealand, initially seconded from the New Zealand DSIR for 3 months to provide assistance with the CSIR radar work. This period was extended to 9 months and then indefinitely as the RPL was reorganised. White first became the Acting Head of the Radio Research Board and later in October 1942 became the second Chief of the RPL as the deficient leadership qualities of D. F. Martyn were recognised by the CSIR (Minnett and Robertson 1996, the memoir of White). In January 1945, he became the Assistant Executive Officer of CSIR and then CEO from 1949 to 1956, finishing his career first as Deputy Chairman and then Chairman from 1957 to 1970. White played a major role in the success of the radar effort in World War II with numerous contacts in the UK and the US (Evans 1970). Later on he was a major proponent for the growth of radio astronomy in Australia in the post-war period. In the mid to late 1950s, he was a major backer of the RPL plan to build the Parkes 210 ft. (64 m) radio telescope under the leadership of E. G. Bowen (whom White had recruited during an extensive visit to the US in mid-1943).

Chapter 7
1945–1946: Early Radio Astronomy at Dover Heights

The radio noise group at RPL was part of a fascinating global phenomenon. Throughout World War II, radar researchers all over the world had been encountering, independently, the same anomalies and asking the same questions as they used very similar equipment to try to locate enemy aircraft before it could destroy their own country's pockets of civilization. Once the war ended, these researchers were still pondering just what in the universe they had found. It also quickly became clear to each scientific enclave that they were not alone in their discoveries and that this new form of observing the physical world had the potential to be a whole new branch of science, and to put it simply, a "hot topic". The race, as it were, was on.

Competition over who would publish their findings first was at times even a bit acrimonious, and peace-time careers were being made in a field that would blossom and grow in the years to come.

In early October 1945 the stage was set and the initial participants at RPL were Pawsey, Payne-Scott and McCready. Why did Ruby Payne-Scott play such a major role and why was she accepted as a key member of the team that produced major advances in solar physics? A major factor was that she, unlike many of her colleagues, was a physicist, with a strong background in mathematics. Many of her colleagues were engineers, though both engineers and physicists had made significant contributions to the perfection of radar equipment during World War II. In addition, Payne-Scott was also fascinated by the electrical engineering aspects of radar; this combination of physics and electrical engineering must have insured that J.L. Pawsey, the leader of the nascent radio astronomy group, had complete faith in Payne-Scott. As she became a member of the team, she had the trust of the other members due to her experience and insights. For a few years, she had no rivals in her scientific leadership as the partner of Pawsey in setting the direction of solar radio astronomy research.

The intellectual dominance of Payne-Scott in these first years trumped the difficulties that she experienced as a woman in the CSIR/CSIRO; her colleagues accepted her and likely covered up for her for a few months in late 1946 or early 1947 when she likely had a miscarriage. Certainly admiration for her remained until her resignation in 1951 and continued throughout 1952, during the time of the URSI international conference in Sydney. In this chapter, the genesis of solar radio astronomy in Australia, with particular emphasis on Payne-Scott's role will be described. Only with the arrival of J. Paul Wild at RPL in 1947 was Payne-Scott's dominant role at RPL diminished; the day to day leadership of the solar group was taken over by Wild in the 1950s after Payne-Scott's departure.

The nature of RPL in 1945 was also critical for the success of Payne-Scott; the leadership of Bowen and Pawsey and the strong support of White within the CSIR/CSIRO administration stimulated the research environment. Thus a strong support for fundamental astronomical research was fostered during a period when there was little astronomical research at Australian universities, outside of mainly optical investigations being done at the Mount Stromlo Observatory in Canberra.[1] In the years 1945–1951, Payne-Scott played a major role in the foundation of solar radio astronomy. Her prominent contributions were the introduction of "Fourier Synthesis" into radio astronomy, the first use of interferometry in radio astronomy, and the discovery of Type I, II and III solar bursts. It is vital to understand that in this time period, the obstacles these nascent Australian radio astronomers faced came down to four major factors.

Where could they find a place to set up their equipment and have regular access for consistent observations when all the locations they could use were in fact already in use by the Australian armed forces. They were, after all still government employees, who until recently had been part of the war effort themselves.

Even more constantly pressing, how could they redesign and improve the receivers, antennas, calibration systems, etc. that they were using in order to obtain more reliable data? In essence, they were vigorously modifying the wartime transmitter/receiver radars into what would become radio telescopes.

They also needed to continuously rethink the physics, mathematics, and astronomy they thought they understood in order to explain all the new knowledge and subsequent questions their observations garnered, even though *none* of them were trained as astronomers.

Lastly, and not unimportantly, they needed to justify to the world and more directly to their employer—the Australian government—that they really were onto something new and interesting by publishing their findings, in order to stay funded. Since, in Australia these scientists were working for the government rather than at universities, they had the additional burden of needing to show at least some inkling of practical applications of their work in their regular reports to the higher ups. One can read in early reports that the solar noise group would regularly attempt to show that solar research was in fact a practical avenue of study due to the influence of the sun on weather, radio communications and possibly agriculture.

---

[1] Originally known as the Commonwealth Solar Observatory (CSO), the observatory at Mt. Stromlo became associated with the nearby Australian National University (ANU) in the era 1946. It became commonly known as the Mt. Stromlo Observatory (MSO), and the title became official when it formally joined ANU in 1957. For the purposes of this book, it shall be called Mt. Stromlo Observatory, even though it was still officially named the CSO in the post—war era. Frame and Faulker (2003) have discussed the history of MSO.

The Collaroy Campaign: October 1945: First Solar Radio Astronomy in Australia

On 17 October 1945, Bowen wrote Fred White, assistant executive officer of CSIR in Melbourne, with an enthusiastic assessment of the first solar noise campaign: "... Pawsey and Miss Payne-Scott's noise measurements are bearing considerable fruits".[2] He summarised the results, in particular the correlation of noise level with sunspot numbers. A letter to *Nature* was planned in a few days. The entire operation happened quickly. The first observations from the Collaroy site were obtained on 3 October 1945 and the experimental work continued to 23 October. The publication to *Nature* was submitted on 23 October—including data from that date—with the final publication on 9 February 1946[3]: "Radio—Frequency Energy from the Sun" with authors Pawsey, Payne-Scott and McCready.

The initial Australian detections were obtained at a Royal Australian Air Force location at Collaroy, a northern seaside suburb of Sydney, about 24 km north of the city centre. The site was located about 120 m above sea level with the coastline running NNE to SSW. These initial observations were obtained on a COL (Chain Overseas Low-flying) 200 MHz radar antenna. A COL is a special version of aircraft warning radar made to detect low-flying, incoming aircraft. The one at Collaroy was similar to the British built COL sets that had been used by the Royal New Zealand Air Force with which the sun had been detected earlier in March and April of 1945 on Norfolk Island (an Australian territory defended by the New Zealand military during WWII). The sun was observed at sunrise and sunset since the COL antenna could only be moved laterally, or "in azimuth". The December 1945 summary paper gave the characteristics of Radar Station Number 54—the observations at Collaroy would continue until 15 February 1946[4]:

> The aerial is a broadside array of four horizontal rows each of ten half-wave dipoles with a reflector, having a gain of 80 relative to a half-wave dipole (i.e. G = 130) and a horizontal beamwidth [to half power] of 10 deg.

The first detection was on 3 October 1945 at 0531 EAST (Eastern Australian Standard Time), with a signal increase of about 27 % over the general receiver level

---

[2] National Archives of Australia-NAA: C3830, A1/1/1/, Part 1.

[3] This paper was RPP 1 (Radiophysics Publication No. 1); this series ran from 1945 to 1997 and produced 3,934 publications. Copies of the first notebook from the publications office, which ran from RPP 1 to RPP 667, ending in May 1961 are available. The notebook consisted of a series of handwritten records providing the publication history of the publication, e.g. dates of submission, publication, etc. There were two additional RPL series of internal reports. The RPR series was a series of preprints. The series RPL was a collection of lab reports; some of these were later converted into polished reports or publications.

[4] All references for the December 1945 summary paper are taken from "Solar and Cosmic Radio Frequency Radiation; Survey of Knowledge Available and Measurements taken at Radiophysics Laboratory to December 1, 1945" by R. Payne-Scott. SRP 501/27, unpublished manuscript from CSIRO Division of Radiophysics archive, Payne-Scott 1945c.

at a bearing (azimuth) of 94°, essentially in the direction of the eastern horizon. Then 9 min later the noise power on a bearing of 93° was 4.5 times the normal noise power. The level corresponded to an equivalent temperature of the sun (size 30 arcmin) of 15×10^6 K or a flux density of order 10^6 Jy (Jansky). The publication of these data in 1946 also summarised short term fluctuations:

> We observed, from the direction of the sun, a considerable amount of radiation having the apparent characteristics of fluctuation "noise" when observed on a cathode-ray oscillograph or head-phones.[5] However, the output meter reading fluctuated considerably, a characteristic which is not typical of normal thermal agitation "noise". The variation of apparent azimuth of arrival and of intensity with horizontal rotation of the aerial and the sun's elevation was qualitatively consistent with the assumption of radiation from the body of the sun modified by the known directional characteristics of the aerial.

The December 1945 summary paper had given addition details about these sudden increases in solar noise, or "kicks":

> One feature...is the short period fluctuations; the noise from the sun causes a fairly steady meter deflection on which are superimposed at intervals of perhaps a few seconds kicks which may be of the same order as the steady deflection; the relative magnitude and frequency of occurrence of these kicks seems to be independent of the elevation of the sun over the hour or so during which it is observed.

The 1946 *Nature* publication did describe the short term fluctuations, named for the first time as "bursts". There was no mention, however, that the burst time scale was a few seconds; the time scale of the bursts was discussed in subsequent observations in 1946–1947. The paper concluded:

> Furthermore, because of the very high levels relative to expected thermal radiation...and the observed short-period meter fluctuations, it seems improbable that the radiation should originate in atomic or molecular processes, but suggests an origin in gross electrical disturbances analogous to our thunderstorms.

With great foresight, the RPL solar noise group cultivated a working relationship with the astronomer, Clabon W. Allen,[6] who worked at the Mt. Stromlo

---

[5] There has been some criticism by astronomy colleagues of the headphones worn by the Ruby Payne-Scott figure in the 28 May 2012 Google Doodle celebrating her 100th birthday (see Fig. 1.4). It is encouraging to see that she must have used headphones to ascertain the nature of the characteristics of the solar "noise". For example did the audio version of the radio frequency signal sound like receiver noise or lightning? The Google Doodle web page lists a number of "Tags": Radar, Headphones, Sun, Cool Gal, Glasses, Receiver and Soundwaves. Of course, soundwaves from the sun were not detected by Payne-Scott.

[6] Clabon ("Cla") W. Allen (1904–1987) was on the scientific staff of the Commonwealth Solar Observatory (CSO) near Canberra, in subsequent years the Mt. Stromlo Observatory of the Australian National University. The RPL group's appreciation of the importance of optical collaboration had already been mentioned at the PC meeting of 12 November 1946 where it was proposed for the first time to install a 200 MHz at the CSO in order to insure: "...close liaison with Stromlo personnel with visual equipment." The impressive solar events of late July–August 1946 were observed at CSO; the publication was prepared by Allen (1947). The 200 MHz data shared with Payne-Scott et al. from the March 1947 event also was obtained at Mt. Stromlo. See also Allen 1957.

Observatory. He would telephone the RPL to let them know when unusual solar activity (e.g., sun spots) was visually documented via the optical telescope there; he was also a key figure in helping these radio physicists and engineers to re-educate themselves as astronomers.

The December 1945 summary paper contained three plots that illustrated the detection of the sun in October 1945. All of the measured sizes were consistent with a point source—less than about 10° in diameter—of radiation arising from the sun. In the 1946 publication in *Nature* by the RPL group, there was no indication that the sea-cliff interferometer technique to determine either the location or the size of the emitting regions had been employed.[7]

These short term events, "kicks", or "bursts"—which were likely Type I bursts, though the term Type I burst actually was not used until sometime in the course of 1949–1950—were not mentioned in the 1946 *Nature* article. Possibly the reason that these features were not mentioned was the observers' uncertainty, which was emphasised in the December 1945 summary paper:

> The meter fluctuations observed over a period of a few seconds ... may be due either to absorption or scattering of the radiation in the earth's atmosphere or to genuine fluctuations in the solar radiation. There is so far little evidence one way or the other, but this will be one of the first points to be investigated in future work, as it is critical in deciding the origin of the radiation.

As we will see below, the future work carried out in 1946 did elucidate the later known nature of these "kicks", "noise storms", or Type I bursts. Of course, these did arise from the sun.

In Fig. 7.1 we see the main results of the October 1945 Collaroy observations as shown in the solar scan plots from the 1946 publication; the sunspot data were provided by C.W. Allen of Mt Stromlo:

> It is apparent that the peaks of 1.5-metre radiation coincide with peaks of the sunspot area curve and with the passage of large sunspot groups across the meridian (1946 *Nature* publication).

The December 1945 summary paper contained additional conclusions:

> It will be seen that there is good correlation between the two curves [the radio intensity and the sunspot area], particularly between their peaks, the peaks of the radiation curve being sharper than those of the curve for sunspot area. It is possible that the radiation may emerge from say, a *crater* [emphasis added] on the sun's surface, and so be highly directional. This suggestion is borne out by the sketches of the sun's disc, which show that the peaks coincide with the passage of large optically visible spots across the meridian.

The meaning of the word *crater* is puzzling; likely this concept was related to the *holes* discussed below. Bowen (1945) wrote at about this time in the *Australian Journal of Science*[8]:

---

[7] The McCready et al. publication of 1947 stated that no interference pattern was observed in these October 1945 data; this absence was probably due to a wide distribution of sunspots over the sun at this date, in contrast to the period in early 1946, when a compact sunspot group formed.

[8] Bowen, E.G. (1945).

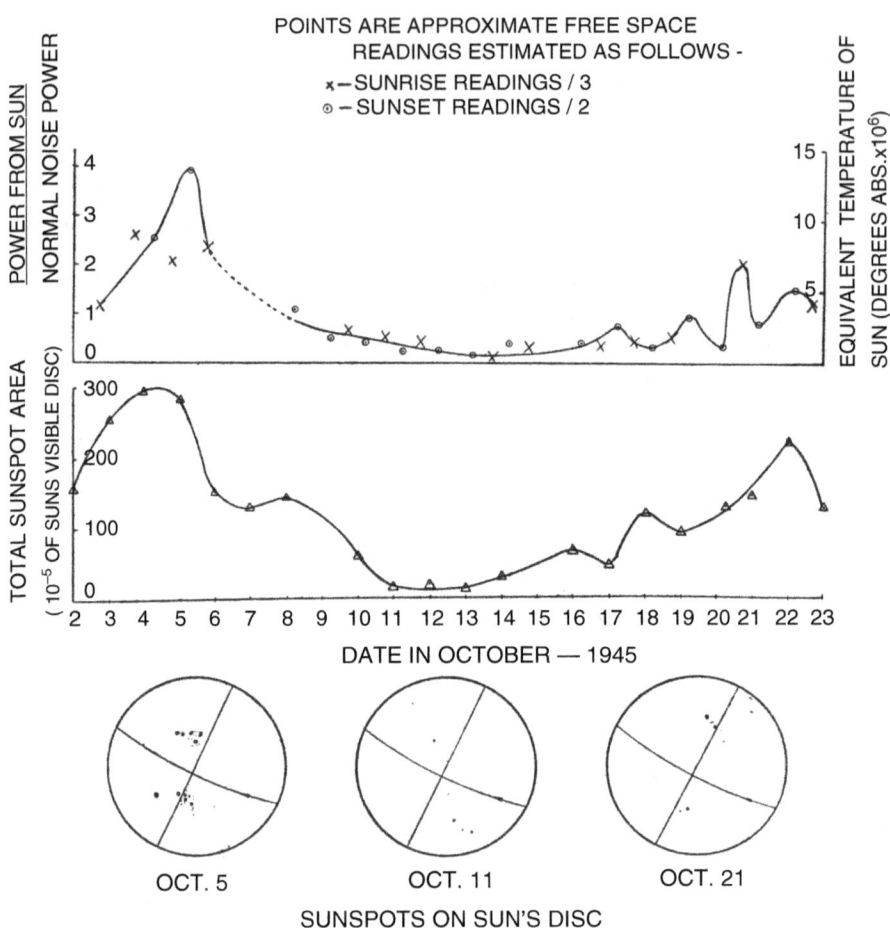

Fig. 7.1 The main result, showing a correlation between intensity of radio noise from the sun and visible size of sunspots from the same dates in October 1945, from the first publication in radio astronomy from Australia, submitted to *Nature* on 23 October 1945 and published 9 February 1946. The two graphs at *top* are plotted along the same dates in October 1945 along the x-axis. Of the two, the topmost plots the level of radio noise on the y-axis, indicated on the *left* as power from the sun and the sun's temperature on the *right*. The lower of the correlated graphs plots the optical size of sunspots as observed at the Commonwealth Solar Observatory at Mt. Sromlo. Major sunspots on the solar surface are shown in the *bottom* of the figure from three dates in October 1945 (Reprinted by permission from Macmillan Publishers Ltd: *Nature*, vol. 157, p158, Figure 1, "Radio-Frequency Energy from the Sun", copyright 1946 (Pawsey et al. 1946))

> Within a few weeks of the end of the war, the phenomenon [solar noise detected by Hey in 1942] was investigated afresh in Australia, and a few weeks' continuous observation was sufficient to show a very close correlation between excess noise from the sun and sunspot activity. It was shown that the existence of active sunspots corresponded to an average noise temperature at the sun of two or three million degrees, and on occasions as much as twenty-five million degrees absolute. One theory of the production of this noise is that

sunspots correspond to holes in the surface layers of the sun through which energy corresponding to the internal temperature can escape in the direction of the earth. The fact that little or no energy comes out in the visible spectrum is readily explained by the heavy absorption of these frequencies which is bound to occur. Alternatively, someone is letting off atom bombs in the sun and has been doing so for some considerable time.

Perhaps the *holes* referred to above are related to the *crater* described in the December 1945 summary paper; the likely connection was being made between the hot internal temperatures in the solar interior and the cooler solar surface at 6,000 K. No explanation for the bizarre and likely humorous statement concerning atom bombs was provided. As far as we can ascertain, this novel explanation was never discussed again. Within a short period, there was no discussion of craters on the sun being involved in the emission of the excess solar radio emission; the RPL group later realised that solar bursts, in fact, arose higher in the solar corona. They would later learn that the lower the frequency, the higher in the corona the phenomena were taking place.

The December 1945 summary paper and the *Nature* publication of early 1946 also mentioned additional solar observations that were noteworthy. Likely using the higher frequency antenna at Georges Heights at Middle Harbour (see Orchiston et al. 2006), a marginal detection of the sun was reported at 25 cm on 4 and 5 October 1945, with an equivalent temperature—the average values of brightness over the 30 arcmin diameter solar image—in the range of 6,000–20,000 K. Based on current knowledge about the properties of the sun at the 20 cm wavelength, the low value of 6,000 K is implausible and must indicate the large uncertainty in these first observations in the cm range. At 50 cm, the sun was not detected with an upper limit of about 60,000 K. In other words, their equipment was only able to detect temperatures at or above 60,000 K during observations at 50 cm. At that wavelength, any real signals would have been less than this value.

Both the December 1945 summary paper and the 1946 publication contained an attempt to compare the solar radiation with the previously known properties of "cosmic static" reported by Jansky (1933), Reber (1940a, b), as well as Fränz (1942). The conclusions from the two sources are somewhat different. From the 1946 *Nature* publication:

> In view of observations of such intense bursts of radiation from the sun at the wave-length at which "cosmic static" is known, it appears desirable to question the suggestion that the latter originates in the interstellar space. It seems more reasonable to attribute it to similar bursts of radiation from stars which, because of their large number, could yield an approximate constant value for any one area in the sky.

The December 1945 summary paper contained three relevant conclusions concerning the comparison of the radio radiation from the Milky Way and the sun:

(8) Similar radiation [i.e., the solar radiation] is obtained from the collection of stars in the galaxy; it is most intense in the direction of the centre of the galaxy, and has a distribution corresponding approximately to that of the stars in the galaxy.

(9) The effective temperature of the Milky Way increases with decreasing frequency [has a non-thermal spectrum in modern terminology].

(10) Because of their similar nature and behaviour, it seems likely that the radiation from the sun and the Milky Way have similar origin.

As is now well known, these parallels between the solar emission and the galactic emission did not stand the test of time. Within a few years the realization that the filling factor or dilution factor (i.e., the fraction of the sky covered by the stellar disks, a value of about 10^{-14} for stars similar to the sun) was tiny and the luminosity of the sun was very low, implied that a collection of "suns"—stars similar to our sun—in the Galaxy could not explain the non-thermal emission of the Milky Way.[9] By the late 1950s astronomers accepted that the extended emission of the Galaxy did arise in the interstellar matter via synchrotron emission from relativistic electrons distributed throughout the galaxy (see Alfvén and Herlofson 1950; Kipenheurer 1950; Ginzburg 1951).

The final conclusions in the December 1945 summary paper concerned the physical nature of the solar radiation. There is a prescient discussion of the importance of the solar corona, influenced by conversations with D.F. Martyn:

> The radiation [the metre-wave non-variable emission] may come from the corona, which has been recently shown to have a much higher temperature than the photosphere, and which, although transparent to visible light, may well be opaque to long radio waves. Dr. D.F. Martyn, of the Mt. Stromlo Observatory, Canberra has suggested this origin.

Payne-Scott pointed out that Martyn's (1946b) theory did not seem to account for the greatly increased solar radiation associated with sunspots. "It would [also] not give a continual increase in temperature with decreasing frequency, as appears to occur for the Milky Way radiation. There is not yet enough evidence to determine how the solar radiation varies at frequencies below 160 Mc/s."

Other possibilities for the origin of the solar radiation were discussed such as non-thermal emission similar to atmospherics in the earth's atmosphere, as well as plasma oscillations in the ionised atmosphere of the sun. Payne-Scott then pointed out that that the previous theory of the Jansky-Reber observations of the Milky Way, which had postulated that the origin for this was emission from the ionised matter between the stars in the galaxy, was simply not applicable to solar data. She wrote, "This would, of course, not account for the observed solar radiation", presumably due to the vastly different physical scales and densities in the interstellar medium of the Milky Way as compared to the solar corona.

---

[9] Sullivan (2009) has emphasised that Payne-Scott did not point out a problem mentioned by Reber and Jansky: if the solar radio radiation and the radiation of the Galaxy had the same origin, then the fact that the ratio of sun to Galaxy for the optical regime was much higher than for the radio was unexplained. The optical sun is much brighter than the optical Galaxy as compared to the radio; at intermediate frequencies of 200 MHz the sun and the Galaxy are, in fact, comparable in intensity. Greenstein et al. (1946) also pointed out the serious discrepancies due to the small dilution factors of stars in the Milky Way which would "vitiate this interesting new suggestion" (from Pawsey et al. in the 1946 *Nature* paper) in order to explain the galactic radiation as arising from the collective effect of emission from stars in the galaxy.

Bowen, then deputy chief of RPL, to become chief of RPL in May 1946 at age 35, was clearly proud of this first publication. He wrote E.V. Appleton on 23 January 1946, after receiving the E.V. Appleton paper from November 1945[10]:

> This laboratory obtained what I think is the first direct experimental verification of this effect [solar bursts] during October 1945 when over a period of three weeks we measured solar noise on 200 Mc/s and obtained a close correlation with sunspot activity. A letter to *Nature* was immediately concocted and you may have seen the published version by now. ...We are now speculating on the source of the noise, and have departed somewhat from the suggestion in the letter [the February 1946 publication] that it was of electromagnetic origin. Personally I feel it must be of thermal origin either from the depths of the sun or in some way from the corona.[11]

A number of authors have commented (e.g., Sullivan 2009) on the long publication delays that the early Australian radio astronomy articles experienced after submission to British publications, in particular the first two submissions from Sydney (see Additional Notes, No. 1, end of this chapter). Two publications from the UK predated the Australian paper of 1946: the Appleton paper "Departure of Long-Wave Solar Radiation from Black-Body Intensity" (submitted 24 September 1945 and published 3 November 1945 in *Nature*) and the Hey publication, "Solar Radiation in the 4–6 m Radio Wave-Length Band" (submitted 17 October 1945—a week before the Sydney submission—and published in *Nature* 12 January 1946). The references in the RPL paper of 1946 were striking in that there was no reference to the Southworth paper from April 1945; the references that appeared were Reber (1944), Jansky (1933), and the restricted reports from Hey and the New Zealand group.[12]

The Pawsey, Payne-Scott and McCready paper published in the February 1946 issue of *Nature* was definitely noticed by the higher-ups in Australia. In mid – 1946, Payne-Scott was given a promotion from Assistant Research Officer to Research

---

[10] NAA: C3830, A1/1/1/, Part 1.

[11] Bowen's claim that the RPL detection was the first observation of the excess radiation associated with sunspots was not correct. As Pawsey, Payne-Scott and McCready pointed out in their 9 February 1946 *Nature* paper, earlier reports were cited from the British AORG (Army Operations Research Group, Hey, 1942) and the New Zealand group. In addition the discussion from Bowen about the timing of the Nature paper is ironic, since the prevailing view held in Australia was that Appleton (as the likely referee for *Nature*) may well have held up this first RPL solar publication (Pawsey, Payne-Scott and McCready) in favour of the Hey paper (Orchiston et al. 2006). The publication times were roughly 40 days for the Appleton paper, 80 days for the Hey publication and 114 days for the Australian solar publication.

[12] The use of these still classified reports caused some trouble back in the UK; although the results of both reports were then widely known in the ionospheric and radio noise field, the reports had not been cleared with various committees in the UK, responsible for the declassification of World War II documents. Sir Edward Appleton wrote White with a complaint about the citation of two classified reports. Both Bowen and White wrote letters of apology to the UK. Both did not really see that major damage had been done. White wrote Bowen on 30 April 1946 that "...I do not think it matters a great deal- it is only a formal point". Bowen had a more biting response in a letter to White on 26 April 1946: "I am sorry that Appleton is making a song and dance about our letter [*Nature* paper of 1946] ... but I suppose he is just expressing his well-known ownership of all radio and ionospheric work." (NAA: C3830, A1/1/1 Part 1).

Officer. The memo from the Chief of the RPL, John N. Briton, dated 25 February 1946, stated:

> During the past 12 months Miss Payne-Scott has demonstrated her real ability as a research officer. You will recall that earlier we were in some doubt as to her suitability for continuing with the Division. Since then, however, her work, particularly in connection with investigations concerning the visibility of signals and noise measurements, has been quite outstanding and I wish to recommend that she be reclassified as Research Officer, with effect as from 1st July, 1946.

No descriptions of the doubts referred to by Briton have been found in the personnel files. A few months later (May 1946), Taffy Bowen[13] became Chief of RPL. Payne-Scott's salary was raised to £A458 per annum at this time; however, a series of events began in July 1946 that were to have far reaching consequences. Bowen realised that many of the more senior RPL staff still did not have "with-commitment" appointments and were thus not eligible for superannuation or a CSIR sponsored pension. Perhaps the exigencies of the War had led to this neglect. In any case, Bowen wrote to G. A. Cook (the new Secretary of the CSIR) insisting that due to these long delays, 19 individuals be given "with-commitment" status, i.e., a permanent appointment with pension rights. Within weeks the ball was rolling; these individuals were called to a medical examination by the Commonwealth Medical Officer at Circular Quay in Sydney to investigate whether the staff member was healthy enough to be expected to "discharge his or her duties until 60 years of age without more than average sick leave". After some minor delays, Payne-Scott appeared for her medical examination on 17 September 1946 and was then referred to an eye specialist, Dr. St. Vincent-Welch. Apparently, Payne-Scott's poor eyesight produced some doubts about her long-term health prognosis. On 4 November 1946 Cook wrote to the Minister for Scientific and Industrial Research, J. J. Dedman, with the recommendation that 15 CSIR staff be "gazetted" as "employees within the meaning of Section 4 of the Superannuation Act". Payne-Scott was in this list but recommended for the "Provident Fund" instead of the "Superannuation Fund" (see Additional Notes, No. 2, end of this chapter). The official list was published on 21 November 1946, the date of "gazettal".

A few days later Payne-Scott's classification within the CSIR was changed to Research Officer, Grade III, with a substantial salary increase from £A458 to £A588 per annum. Clearly, her professional star was on the rise.

As 1946 began, plans were being made to extend the observations of the sun and the Milky Way. At the 15 January 1946 meeting of the Propagation Committee a "cosmic and solar noise section" had been established under the leadership of

---

[13] E.G. ("Taffy") Bowen (1911–1991) completed a PhD degree with E.V. Appleton at King's College, London in 1934. He worked with Sir Robert Watson-Watt starting in 1935, playing an important part in the early development of radar in the UK. He went to the US as a member of the Tizard Mission in 1940 and worked at the Radiation Laboratory in Cambridge, Massachusetts (USA) until late 1943, when he joined the CSIR RPL in early 1944. In 1946, Taffy Bowen became the Chief of RPL, where he provided the leadership in the post—war research programs of RPL, mainly in cloud physics and radio astronomy. See the obituary by R. Hanbury Brown et al. (1992).

Pawsey and McCready.[14] The members of the group included Payne-Scott, Yabsley, a technical officer (Don Urquhart) and a technical assistant. The observations were still being carried out at Collaroy; however the sun was then in the centre of the Milky Way and observations were difficult due to confusion with the intense radio noise from the centre of the Milky Way in December and January. The solar and cosmic noise section planned to concentrate on equipment development in the interim.

Ground-Breaking Developments in Solar Noise Research and Techniques of Interferometry: February 1946, Dover Heights

After the October 1945 campaign was complete and the *Nature* paper submitted at the end of the month, the daily monitoring at the Collaroy site with the COL antenna continued up until 15 February 1946. The air force controlled the Collaroy site and due to high level of use by the air force, it became a less desirable site for RPL research. In contrast, a new site at the ShD site (Shore Defence radar) at Dover Heights RPL field station at the Army Reserve (see Fig. 7.2), which was controlled by the army and much more accessible for regular use by the RPL group, was inaugurated in early February 1946. This was situated in the eastern suburbs of Sydney (closer to the RPL, see Fig. 1.4) about 17 km south of the Collaroy site.[15] The cliff site was slightly lower in elevation with an eastern horizon 85 m above the sea.

The antenna has been described by Bird (1993) as having an original design from World War II with the broadside array designed by Minnett in 1943, using individual elements designed by Pawsey, Minnett and Dobbie in 1940. The 200 MHz antenna consisted of 36 half—wave elements, arranged in a plane, 33 cm in front of a wire—mesh reflector. A number of improvements in the receiving apparatus were made, such as the use of continuous paper chart recorders—instead of visual reading of meters. In addition, several 200 MHz Yagi tracking antennas were installed at North Head, Mt. Stromlo, and even Dover Heights, which could follow the sun throughout the day. These simple Yagi antennas had a lower gain and thus a weaker signal than the radar aerial (ShD), but the ShD and COL antennas could only observe for about an hour at sunrise (and sunset at Collaroy) due to the lack of elevation mobility. Calibration was a major issue with a claimed precision of the intensity scale of only 40 %; the calibration of the systems would improve in subsequent years. The fringe spacings—the effective angular resolution of the

---

[14] NAA: C3830, B2/2, Part1.

[15] This site is now Rodney Reserve, Dover Heights. A memorial plaque was unveiled by the Governor of New South Wales, Her Excellency Prof. Marie Bashir, on 20 July 2003; in addition, a model of the 100 MHz sea-cliff interferometer used in the early 1950s by Bolton, Slee and Stanley was put on display.

Fig. 7.2 RPL field station, Army Reserve, Dover Heights; the Fortress Fire Command Post was in this reserve, 2.2 km south of the Macquarie Lighthouse and 2 km north of Bondi Beach. (Also 7 km east of the centre of Sydney.) Photo, 18 February 1943. Possibly the two men near the Shore Defence Radar (ShD- 200 MHz) are John Worledge (head of the NSW Railways radar structures group) and Fred White (then Chief of RPL; see Minnett et al. 1999b). This 200 MHz antenna was used for the solar radio astronomy experiments in 1946. The view is to the north, North Head is clearly visible. The antenna to the north (on top of the smaller building) is the experimental array (CSIRO Radio Astronomy Image Archive B81-1)

sea-cliff interferometer—were about 30 arcmin at Dover and 21 arcmin at Collaroy. The primary beam was about 9–10° (1945 December report) for the Collaroy aerial and about 25° for the ShD aerial (Pawsey 1942).

The sun rotates, completing a full turn at its equator in about 27 days. This means that an observer could see a particular area of the sun for about 2 weeks as it makes its way around the side facing the earth before disappearing from view for about 2 weeks as it travels around the side of the sun facing away from the earth. Large sun spots, such as the ones from 1946, are typically active for many weeks, and though we lose "sight" of them as they rotate around the sun, they might even be observable again about 2 weeks later as they come back around from the far side of the sun.

The sun was detected on practically every day on which observations were carried out in early 1946. As a number of authors have pointed out, an event of great good fortune occurred in early February 1946. The largest sunspot the group ever detected up to this time was observed to cover 0.0052 of the solar area.[16] (Figure 7.3 shows an optical photograph from the Royal Greenwich Observatory

---

[16] Newton (1955) has summarised major sunspots from 1874 to 1952. The largest sunspot in this period, with a maximum area 6,100 millionths (0.0061) of the sun's hemisphere was observed on 7 April 1947, followed by a 6 February 1946 sun spot measured at 5,202 millionths. The list of major sunspots can be extended to at least 2001 using the NASA web site www.science.nasa.gov/spaceweather/sunspots/history.html. Four of the top ten sunspots since 1874 occurred in the years 1946–1947. Newton gave little credit to RPL for the later development of solar radio astronomy, with only references to the work of Appleton and Hey carried out in the UK.

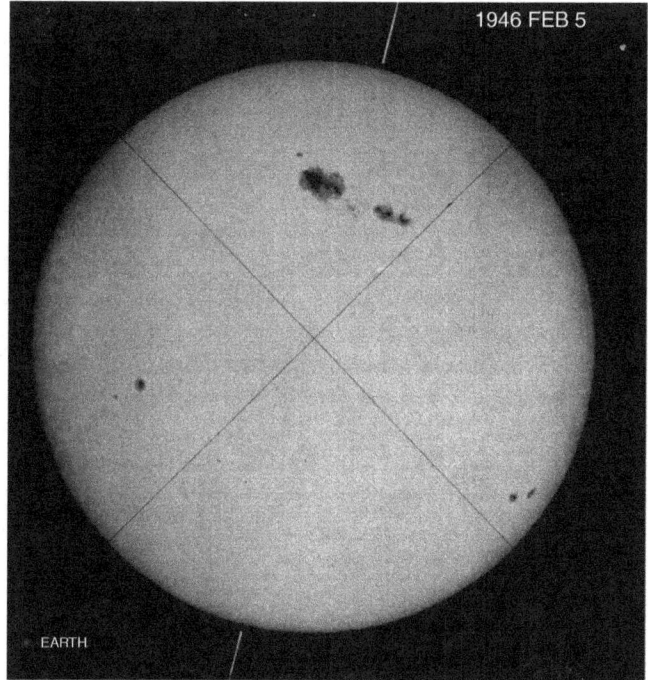

Fig. 7.3 An optical image of the great sunspot of February 1946. Note the orientation with north to the *top* (see Fig. 7.7 for the orientation with respect to the eastern horizon as observed from Dover Heights). This was one of the larger sunspot groups ever recorded. Major radio solar events were associated with this sunspot group. Observed at the Royal Greenwich Observatory 5 February 1946. The radio emission from this sunspot group was observed in the UK and Australia (Adapted from Appleton and Hey 1946a, Fig. 1 in their publication, *Philosophical Magazine*, (*Ser 7*), vol. 37, p. 73,1946, "Solar Radio Noise.-I". Used by permission of Taylor and Francis Ltd., http://www.tandfonline.com)

from 5 February 1946) C.W. Allen from Mt. Stromlo contacted the RPL group by telephone with the news of the new sunspots, and a series of intense observations began.[17]

The rapid bursts observed were the "kicks" referred to in the Payne-Scott report of 1945. The intensities ranged up to 12×10^6 Jy (North Head 7 February 1946 at about 1600 EAST-Eastern Australian Standard Time). There were two types of variation, which in modern terminology would both fall under the category of Type I bursts: (1) a relatively slowly varying type with time scales of many days; these

---

[17] At the 14 February 1946 meeting of the PC, there was a lengthy discussion about the major solar event of the previous week. The report stated: "...a series of records obtained at sunrise on the 200 Mc/s Dover and Collaroy sets showed a strong correlation with the actual visible sunspot area.... These records indicated that the activity originated from an area small compared with the total area of the sun although the analysis had not yet located the source with respect to the spot. The effective temperature at 200 Mc/s of the active area was of the order of 10^{10} K".

were likely what we now call Type I "enhanced radiation" or "noise storms", with intensities in the range 0.05–10×10^6 Jy (a variation of a factor of 200); and (2) intense bursts with time scales of seconds ("the kicks", likely Type I bursts) "of widely varying frequency of occurrence". The intensities of the bursts could be an order of magnitude or greater than the mean level. Later on in the 1950s, the classification of these Type I events was described: "thousands of short-lived spikes [seconds duration] superimposed on a slowly varying continuum", continuing for periods from a few hours to a few days (Kai et al. 1985).

Two important pieces of data indicated that the rapid variations were intrinsic to the sun and did not arise in the atmosphere of the earth; a comparison was made to the twinkling of stars in contrast to the more steady appearance of planets in the night sky. There were no systematic changes in the character of the rapid variations as the sun rose from the horizon to the zenith; also the variations always consisted of increases above a mean level with no decreases. The major reason that the RPL group claimed that the rapid variations were due to intrinsic processes on the sun was the close similarity of the intensities as observed between Sydney and Mt. Stromlo at a distance of 260 km; the modern terminology for this technique is "spaced receivers". They wrote (McCready et al. 1947):

> It is highly improbable that variations having such a high degree of correlation at widely separated sites should be due to any effect in the atmosphere, and it seems certain that most of them are extraterrestrial, and presumably solar, in origin.[18]

In addition to the 200 MHz data, a few additional observations were obtained at 3,000 MHz, 1,200 MHz and 75 MHz. The latter frequency represented an extension of solar metre wave observations to a considerably longer wavelength by the RPL group. During disturbed periods the 75 MHz (4 m) intensity was much higher than at 200 MHz (1.5 m). The two high frequency observations agreed roughly with the Southworth (1945) data at 3,000 MHz.

The correlation of the flux density of the solar 200 MHz emission and sunspots that had been found from the October 1945 data was now extended using the data from 3 October 1945 to 15 February 1946; the Collaroy data were still being provided by Air Force personnel. The data were extended further using the Dover Heights data—the first data having been obtained on 7 February and continuing to about 6 May 1946.

---

[18] At the meeting of the Propagation Committee on 16 April 1946, a report was given summarising the coordinated Mt. Stromlo and Dover Heights data at 200 MHz. The main result was that the bursts occurred simultaneously at the two places, suggesting that the bursts were not due to the atmosphere or the ionosphere of the earth. Publication of the results was being prepared. At the subsequent meeting on 13 May 1946, the announcement was made that the paper was almost complete and was to be submitted "in a few weeks". This prediction was in fact repeated in subsequent meetings in June and early July (in fact submitted on 22 July 1946). The topics that were to be discussed in this paper were: correlation of solar noise with sunspots, the short burst phenomena and their simultaneous nature at the two observing sites, and the location of the bursts on the solar surface. A new development described was that the Mt. Stromlo scientists were attempting to carry out a correlation of the radio observations with visual data on sunspots.

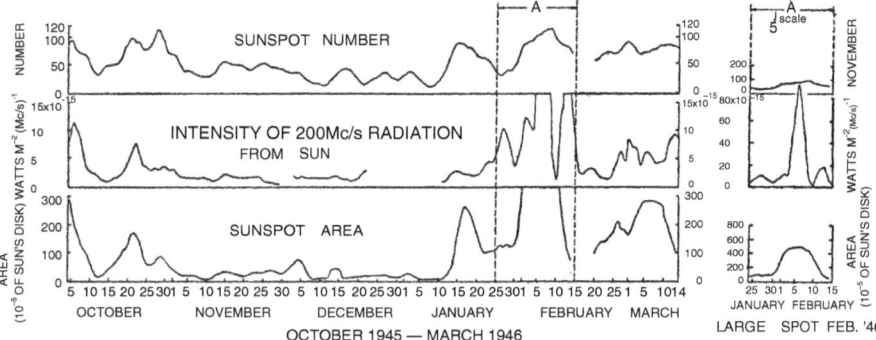

Fig. 7.4 Day to day variation of the radio noise at 200 MHz compared to sunspot number and sunspot area; the connection of radio noise from the sun and sunspot activity was established (Figure 3 of McCready et al. 1947. *Proceedings of the Royal Society, Series A*, vol.190, p. 357, 1947, "Solar Radiation at Radio Frequencies and its Relation with Sunspots". Used by permission of the Royal Society)

A partial summary for a portion of this period is shown in Fig. 7.4; during the great sunspot of early February an expanded scale is used on the right hand side of the figure. The correlation earlier suggested between sunspot area and radio intensity was once again confirmed. During the period the mean flux density of the 200 MHz sun was 13×10^5 Jy on 5 October 1945, 1.2×10^5 Jy on 11 November 1945, 10×10^5 Jy on 26 January 1946, and a whopping 106×10^5 Jy on 7 February 1946. Data from a simple Yagi at North Head on a few days in early February were utilised while the Mt. Stromlo 200 MHz system was available after early April.

The major result from the 1946 February campaign was the interferometry which began on 26 January 1946. This was an historic occasion as it represents the first use of interferometry in radio astronomy. Sullivan (2009) has provided a thorough historical summary of this event. As he points out the first use of interferometry in radio astronomy used only a single radio telescope! In a moment of extreme cleverness, Payne-Scott's group used a technique they had developed while detecting aircraft during the war, where both the direct ray from the sun to the receiver, as well as the rays reflected—in a slight delay—from the sea to the receiver were combined in order to effectively detect the radiation at two distinct directions, thus simulating a virtual radio antenna located exactly twice the height of the cliff below the true location.

The method made use of the sea-cliff interferometer,[19] a radio analogue of a "Lloyd's mirror". The unpublished RPL 9 report by Payne-Scott (August 1947) provided a succinct description of the method:

---

[19] We follow Sullivan (1991) in using the term "sea-cliff" interferometer; in 1946 the interferometer was called either a "sea interferometer" or a "cliff interferometer". The paper did refer also to "Lloyd's mirror" via an optics analogy and also "lobes", as with a radar set. At the PC meeting of

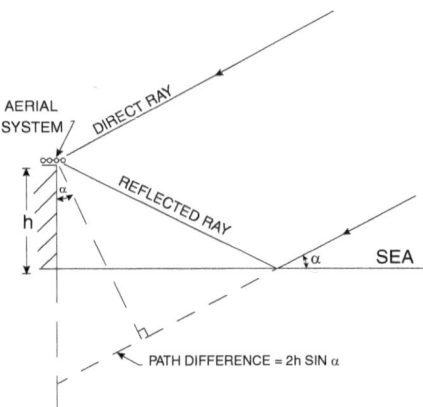

Fig. 7.5 A schematic diagram of a sea-cliff interferometer. The effective baseline of the virtual interferometer is twice the cliff height. The direct ray from the radio source and the reflected ray interfere to form the interferometer. From Dover Heights Sydney, the sun and "radio stars" could only be observed as the source rose in the east over the sea (CSIRO Radio Astronomy Image Archive B1639-4)

> Briefly, at dawn both direct radiation from the sun and that reflected from the surface of the sea are received, and as the sun rises the receiver output varies sinusoidally as the phase difference between the direct and reflected rays increases; from the times of minima relative to the known time of sunrise, the elevation of the radiating source relative to, say, the centre of the sun can be calculated, while the ratio of the minimum to maximum power gives a measure of the width of the source.

The interference arose from the portion of the incoming wavefront that arrived directly from the source to the antenna and from the reflected radiation from the sea; the latter wavefront would travel an additional distance, twice the cliff height times the sine of the angle the source is above the sea (elevation). This is equivalent to having another virtual antenna located at a cliff height below the base of the cliff, Fig. 7.5.

With the Dover Heights antenna fringe or lobe separation of about 30 arcmin at 200 MHz (compared to the primary beam size of 25°), the solar sources could be located with a precision of a few arcmin. (The multiple lobes or fringes can be seen in Fig. 7.6 at sunrise; the sun was detected for only about an hour since the antenna could not be moved in elevation. The fringes disappeared after the sun moved out of the primary beam of the aerial.) The upper limits for the angular sizes of the widths of the sources on the sun were typically 8–13 arcmin. The use of this type of interferometer was well known from radar experience during the war as the technique was used to determine the height of aircraft.

---

12 November 1945 an impressive list of research projects for the fledgling radio astronomers was presented by Pawsey. The use of interferometry was anticipated as a method to study small scale structure if the "noise originates in single small area". The method was described as the use of precision directive finding techniques, obviously influenced by World War II radar practices.

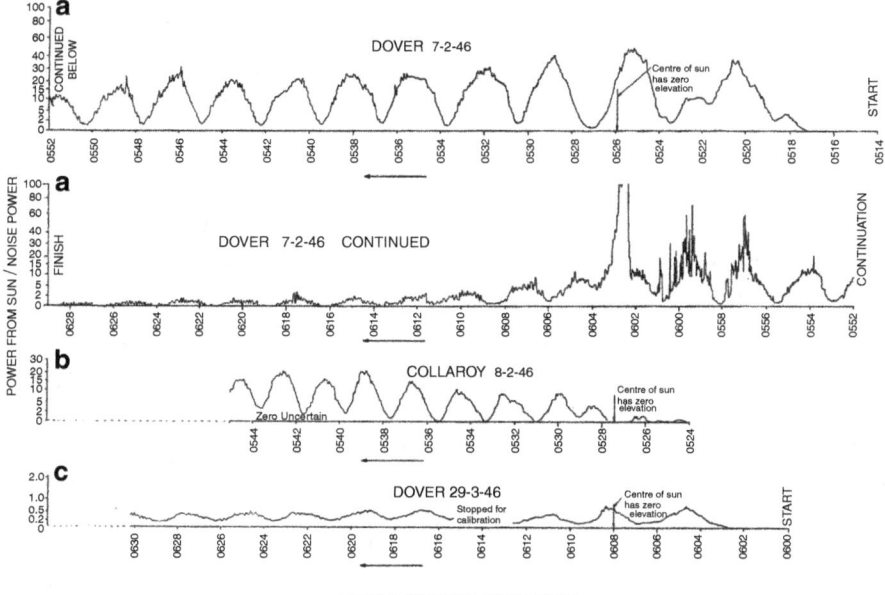

Fig. 7.6 Interference patterns at 200 MHz at sunrise for the great sunspot of February 1946; note the radio emission is observed before optical sunrise (centre of sun has zero elevation) due to increased radio refraction compared to optical refraction Panel 'a' from Dover shows deep minima indicating a small source size. Panel 'b' is the next day at Collaroy; where a higher cliff height makes the period of the interference fringes shorter. Panel 'c' shows a 'normal' day when the shallow minima indicates a larger angular size. See Fig. 7.7 for the appearance of the radio and optical sun on these days. Time increases to the left. Optical sunrise is about 0524 EAST (Figure 5 of McCready et al. 1947. *Proceedings of the Royal Society, Series A*, vol.190, p. 357, 1947, "Solar Radiation at Radio Frequencies and its Relation with Sunspots". Used by permission of the Royal Society)

As Kellermann and Moran (2001) and Wild (1955) have pointed out, the sea-cliff interferometer had a number of advantages: it had twice the sensitivity as compared to an interferometer consisting of two separate but similar antennas that are connected as an interferometer; no interconnecting cables or preamplifiers were required; and the sharp horizon could be used to eliminate confusion due to sources which had not risen and there were no lobe ambiguities in identifying the maxima on the records with their corresponding lobe numbers. The obvious disadvantages were the limited observing time as the source was rising or setting and the low elevation of the observations with resulting large (also uncertain) ionospheric and tropospheric refraction corrections. In addition the roughness of the sea due to wave action caused a loss of coherence, especially at shorter wavelengths.[20] In addition no delay compensation was possible as in a conventional interferometer of the

---

[20] At 20 cm (1,400 MHz) no fringes were detected at Dover Heights since variations in the height of the waves were comparable to the wave length.

Michelson type (which is much more flexible since the beam can be positioned at any position if the source is above the horizon, not just for a source that is rising or setting).

The first fringes on the active sun were observed on 26 January 1946, Australia Day, with deep minima; the modulation of the 200 MHz intensity was almost complete implying small scale structure in the distribution of the solar radio radiation. Fringes were obtained from the "sunspot radio radiation" (the enhanced radiation or noise storms) and even from the quiet sun,[21] with angular size of about 40 arcmin at 200 MHz. It was possible to determine the absolute elevation of this radiation while no positional information was obtained for the short time scale bursts. Immediately the observers realised that a large fraction of the 200 MHz solar radiation did arise from a source(s) much smaller than the sun itself. During the great sunspot activity of the first days of February, data were obtained at Collaroy on 6–9 February as well as 7 and 8 February at Dover Heights. The data are shown in Fig. 7.6 which shows the deep minima on 2 days in February from Dover and Collaroy; fortunately interference data existed at both antennas on both 7 and 8 February. The comparison was encouraging as the positions agreed from both sites. "In each case the radiating strip has a width considerably less than that of the sun's disk, being of the order of the size of the sunspot group, and passes through the group." (McCready et al. 1947) The fringe periods at the Collaroy site were shorter, due to the increased height of the cliff at this site. Due to the short duration of the intense Type I bursts observed from 0556 to 0604 EAST (compared to the fringe period of about 2 min), it was impossible to determine the angular size of these events; the fact that the maxima occurred at the same time as the steady radiation did imply that the two originated in the same vicinity—as would be expected for Type I bursts. The inferred positions and widths of the equivalent radiating strips are show in Fig. 7.7. Thus for the first time there was direct evidence that the bursts and the (Type I) noise storms originated from a region close to a sunspot group.

In summary, the new data extended the results of the previous campaign from October 1945: the radiation at 200 MHz consisted of possibly two types of processes: (1) a slowly varying type of emission, and (2) bursts of duration with time scales of about a second. The emission originated from different positions on the sun and could usually be associated with a compact radio source near a sunspot. Finally, there was no evidence to make a detailed association with particular optical phenomena (e.g., flares) and radio emission. For the first time a derivation of an approximate brightness temperature for the storm radiation became possible. On 7 February 1946, the radio flux density was about 10^7 Jy with an angular size less than 6.5 arcmin (the optical disk of the sun is about 30 arcmin). The equivalent

---

[21] The radio quiet sun represents the thermal bremstrahlung (free-free emission arising from free electrons as they interact and are accelerated by the nearby protons) radiation from the million degree corona. The concept of a quiet sun arose simply due to the fact that the sun is a hot body, with an expected thermal emission. The intensity depends on the electron temperature and the total number of electrons along the line of sight.

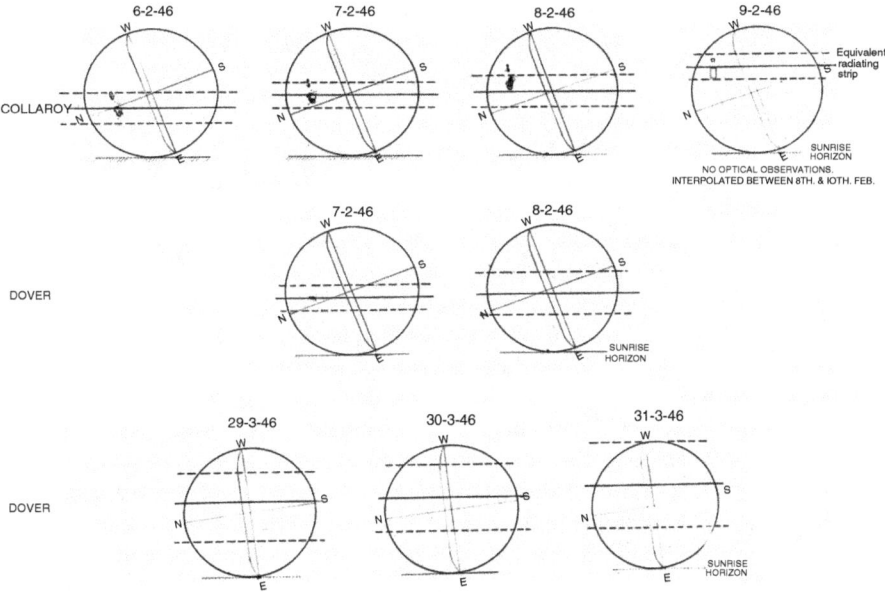

Fig. 7.7 Sketches of the sun with the position and the width of the radio noise source for a number of days in February and March 1946; the sea-cliff interferometer was used to derive the radio noise position and equivalent size. The positions were uncertain to a few arcmin (solar diameter is about 30 arcmin). The widths were upper limits. The *top panel* refers to Collaroy data while the *bottom two* refer to Dover observations. The sun is shown with the horizon to the east shown as a hatched area. The equator of the sun is shown (east-west) as well as the solar poles (north-south). The great sunspot of February 1946 is sketched. For the *top two panels* the radio emission was localised, arising from a region near the sunspot. In the *bottom row*, the sources of radio emission are scattered, although the emission arises mainly from one sunspot group. The intense radio emission arose from an area much smaller than the optical sun, close to the optical sunspot (Figure 7 of McCready et al. 1947. CSIRO Radio Astronomy Image Archive B779-1)

temperature was thus greater than 3×10^9 K, much larger than any conceivable temperature on the sun. As was the case for the 1945 data, gross electrical disturbances (similar to electrical storms in the atmosphere) were suggested as possible causes. A parallel was made with the generation of noise in radio valves (tubes). However, there was some doubt expressed by the RPL group about the applicability of the electrical storm analogy.

There was discussion about the uncertainty concerning the association of the two types of emission:

> The appearance of bursts superposed on a more slowly varying background suggests that there may be two separate mechanisms, though it is possible that the background may consist of a large number of overlapping bursts (McCready et al. 1947).

Even today the nature of Type I bursts remains uncertain; some groups favour the "distinct emission mechanism" while others favour the "superposition hypothesis".

This Type I storm radiation has a background component upon which Type I bursts are superimposed. A few years after the initial Dover Heights observations, Wild (1951) summarised the properties of Type I bursts. The bursts have a short duration (about 1 s) with a bandwidth of only some 10's of MHz and can be associated with noise storms which last hours to days. This emission originates from fundamental frequency plasma radiation with high degrees of circular polarization. The polarization is left handed when the magnetic polarity of the dominant sunspot is positive and right handed when the field is negative. This pattern indicates that the Type I emission bursts are polarized in the ordinary wave mode (Kai et al. 1985).

The major impact of the two research campaigns of 1946 at Dover Heights was due to a number of factors. The team of Pawsey, Payne-Scott and McCready was building on the experience gained from the results from October. Relevant scientific questions were posed and revolutionary instrumental techniques were applied: "directional observations, based on interference phenomenon", which enabled both the location and size of the radio source on the sun to be determined, and yielded an association of the radio bursts with sunspots. This experience led directly to the concept of radio Fourier synthesis, a technique applied in the construction of numerous radio telescopes during the remainder of the twentieth century. Each member of the team made major contributions to the ultimate success of the endeavours. Finally, there was an amazing piece of luck: this period was close to the time of a major sunspot maximum, with major sunspot groups occurring in February 1946 and July 1946. The RPL team had the relevant expertise and equipment and they were present at the right time.

A major achievement of the early 1946 research was the introduction of Fourier synthesis in radio astronomy. McCready, Pawsey and Payne-Scott suggested at this time that measurements of the amplitude and phase of observations taken at a large number of antenna spacings would provide the complete Fourier spectrum from which the true distribution could be derived. This result was based on the result of their analysis of the response of the sea-cliff interferometer as the source rose over the horizon. They wrote (McCready et al. 1947):

> Since an indefinite number of distributions have identical Fourier components at one frequency, measurement of the phase and amplitude of the variation of intensity at one place at dawn cannot in general be used to determine the distribution over the sun without further information. It is possible in principle to determine the actual form of the distribution in a complex case by Fourier synthesis using information derived from a large number of components. In the interference method suggested here Δ [delta, the phase difference between the direct and the reflected ray from a distant source] is a function of h [height of the cliff] and λ [lambda, the wavelength of the radiation], and different Fourier components may be obtained by varying h or λ. Variation of λ is inadvisable, as over the necessary wide range the distribution of radiation may be a function of wave-length. Variation of h would be feasible but clumsy.[22] A different interference method may be more practicable.

---

[22] Examples would be sea-cliff interferometers at locations with different cliff heights or even observations from an aircraft at different heights as suggested by Ryle in an unpublished memo from August 1955 discussed by Sullivan (2009). Sullivan wrote: "... in essence a sea-cliff

Of course, the more practical method that has been adopted in the last 60 years has been interferometry with separable and movable antennas. Payne-Scott and collaborators also pointed out that for an arbitrary distribution the response of the interferometer consisted of a steady term and another term that had the form of a Fourier cosine series. For the first time Fourier synthesis was described in radio astronomical context, a milestone acknowledged by Ryle in 1952.[23]

We have been able to determine the origin of the concept of Fourier synthesis within the RPL group.[24] There remains little doubt that Pawsey was the originator of the concept, with major mathematical contributions from Payne-Scott. The association of Fourier synthesis and interferometry can be ascribed to Pawsey and Payne-Scott, a major highlight in the development of twentieth century astronomy. Since Pawsey was averse to mathematics (Sullivan 2009, quote from Steve Smerd), the equations were certainly formulated by Payne-Scott. Kevin Westfold was a newly arrived staff scientist in 1946. Westfold wrote[25]:

> I joined the Lab just at the time the McCready, Pawsey and Payne-Scott paper had been submitted to the Royal Society. In order to familiarize me with the subject Joe [Pawsey] had put me on to reading all about the sun, particularly the chromosphere and corona. I was given the MS [manuscript] of their paper to read and, as a mathematician, I thought I should check the derivation of equation (4) [to determine R, the ratio of the minimum to maximum power of the interference fringes] and the approximation (5) for a spherical earth. Their derivation had assumed a plane earth. When I put the earth's radius into the calculation I came to the conclusion that there was a significant error in the formulae they had used. I reported this to Joe, who immediately referred me to Ruby. I think Joe's action established that Ruby had been responsible for these derivations, but I am pretty sure that it was Joe's idea to use the sea as a Lloyd's mirror interferometer. We can assume Lindsay had been responsible for the design of the equipment. The sequel: Ruby was, of course, correct. What I had done, being all clued up with my research on the sun, was to estimate orders of magnitude by inserting the radius of the sun into my formulae instead of the radius of the earth. The factor of ~100 was not insignificant! This was a salutary lesson for a tyro [novice] theoretical radio astronomer. Joe, of course, never forgot the fright I had given them, though I am not sure that he was aware of exactly how this had come about. Ever after he would speak of me thus: "Westfold is a good mathematician, but he makes mistakes."

For cases for which the minima were quite deep, the radio power was concentrated in a compact source; it was then possible to derive both a position and width of the equivalent emitting strip. The width (in elevation) was derived

---

interferometer with variable fringe spacing- simply observe a low-elevation source from an aircraft flying at a different heights above the sea!" As far as is known, this novel experiment was never attempted.

[23] In his 1952 paper, "A New Radio Interferometer and its Application to the Observation of Weak Radio Stars", Ryle wrote, "The relation between the magnitude of the varying component of power intercepted by an interferometer and the Fourier transform of the distribution across the source was first pointed out by McCready et al. (1947)."

[24] Correspondence with Christiansen, Mills, Wild and Bracewell 1999–2005.

[25] Correspondence via email 5 November 1997. Westfold (1921–2001) was a professor of mathematics and later astronomy at Monash University in Melbourne, Australia, as well as Deputy Vice-Chancellor, 1982–1986.

from the value R and the elevation from the timing of the minima compared to the known time when the sun had zero elevation. For a number of days the emission did arrive from small regions and some of these are shown in Fig. 7.7. For the days in February 1946, the radio emitting strip was much smaller (about 1/5th) than the size of the sun. The radio emitting region was quite compact and roughly corresponded with the major sunspot. The strip moved across the sun (whose rotation rate is 27 earth days) with the sunspot; of course, the sea-cliff interferometer only provided information about the elevation (up-down) of the sunspot radiation, with no information of the azimuth of the radio signal. The detailed structure of the radio emission was, of course, unknown with this crude interferometer; however the position determination and size information were decisive in indentifying the compact radio sources with sunspots. Thus the inferred association of these two that had been suggested by Appleton, Hey and the earlier RPL paper was now clarified. In addition the bursts of 1 s duration had the same times of maxima as the steady noise storms; thus both arose from the same location on the sun.

The bottom row of images in Fig. 7.7 shows the inferred radio properties in late March 1946, when the intensity of the radio emission was much weaker, with fewer and less prominent sunspots. The radio minima from the dawn records were then less pronounced and the inferred width of the radio emitting strip was larger. The radio storm emission was at this time a few times 10^5 Jy while the quiet sun emission (angular size of about 40 arcmin) had a flux density of about 10^5 Jy. Thus the centroid of emission was pulled to the centre of the sun as the storm radiation weakened, because about half of the intensity then arose from the disk of the quiet sun.[26]

Publication of Dover Heights Research and Interferometry Techniques: Proceedings of the Royal Society, 12 August 1947

A number of major questions had been raised by the RPL group and it was at this time that the publication of these results from early 1946 occurred, though the group had apparently been talking about publishing their findings for some time. The novice

---

[26] During the exciting observing campaign of early 1946, the weekly progress can be followed in the minutes of the PC. A new development (PC 16 April 1946) was described by the Mt. Stromlo group of Allen et al. as they attempted a correlation of the radio data with "visual" data. A month later (13 May 1946), Pawsey reported that "visual observations are very difficult. No correlation had yet been obtained at Stromlo between radio and visual phenomena". In the midst of these reports, a very intriguing report appeared from the 19 March 1946 meeting provided by McCready and Payne-Scott: "Some explanatory investigations of stellar noise have been conducted with inconclusive results." No additional details were provided. The first detection of Cygnus A in Australia was only made by Bolton and Stanley in June 1947 from Dover Heights. Bolton (1982) claimed that initial futile attempts had in fact been carried out by Pawsey in September 1946. Hey, Parsons and Phillips reported the discovery of Cygnus A in the issue of *Nature* from 17 August 1946 (submitted 4 July 1946); the observations were carried out in late May and early June 1946.

radio noise group had learned many of the techniques and was becoming familiar with astronomical concepts. "Solar Radiation at Radio Frequencies and its Relation to Sunspots", co-authored by McCready, Pawsey and Payne-Scott[27] appeared in the *Proceedings of the Royal Society* on 12 August 1947, after being submitted by Dr. David Rivett FRS almost 13 months earlier on 22 July 1946 (see Additional Notes, No. 3, end of this chapter). This publication was to have far reaching impact on the development of solar physics and radio astronomy techniques.

In addition to the solar noise discussion based on the data of late 1945 and early 1946, the *Royal Society* publication concluded with an attempt to unify the known properties of solar noise and cosmic noise—the radio emission observed since 1933 by Jansky and later by Reber, associated with the Milky Way. The problem with the comparison Payne-Scott had made in the December 1945 summary paper was now partially recognised (McCready et al. 1947):

> Cosmic noise was originally attributed to radiation from interstellar matter, rather than from stars, at a time when similar radiation from the sun had not been detected. The discovery of solar noise raises the question as to whether the cosmic noise is due to similar processes in stars. The basic difficulty remains that the intensity of cosmic noise is vastly greater than it should be if the stars emitted the same ratio of radio-frequency energy to light as does the sun. Nevertheless, the great variability of solar noise suggests the possibility of vastly greater output from stars differing from the sun and it seems that data at present available leave the question completely open.

The major problem was the low radio luminosity of the sun. One parsec is equal to 3.26 light years, and the sun is 0.000005 pc from the earth. If the sun were placed at the typical distance of a nearby star—some tens of parsecs—the flux density (dependent on the intrinsic luminosity and distance) would be about 10^{-8} Jy. This value would provide a meagre contribution to the background of radio emission in the galaxy. As discussed earlier in this chapter, the problem was resolved later in the 1950s when astronomers recognised that the non-thermal galactic background radio emission arises from synchrotron emission from high energy electrons radiating in the magnetic field of the Milky Way Galaxy.

The 1947 *Royal Society* paper also contains a number of technical discussions about the subtleties of the sea-cliff interferometer. Likely these appendices were written by Payne-Scott. An example was the calibration of the intensity scale. In addition a major problem was the effect of *radio refraction*, the bending of the radiation due to an atmosphere of variable density above the telescope. The radiation is bent towards the zenith. As an example, at an elevation of 45° the refraction is about 1 arcmin. Near the horizon, the radio refraction is about 1° compared to the optical refraction of about 0.6°. This comparison is the explanation of why the radio sun rises in the east about 8 min of time earlier than optical sunrise (Fig. 7.6). Payne-Scott and colleagues carried out a clever scheme using the solar data itself to calibrate the effect (i.e., the refraction angle as a function of elevation) at Dover

---

[27] Most of the astronomical interpretation in the paper was provided by Pawsey and Payne-Scott.

Heights (see *Under the Radar*, appendix L). In 1982, Bolton[28] criticised McCready, Pawsey and Payne-Scott's refraction correction process, claiming that the positions derived by him for Cygnus A and other sources in 1948–1952 were in error by 0.2–1° due a faulty theory used by Pawsey and colleagues. In *Under the Radar* (Chap. 8 and appendix L which contains a quantitative treatment of the procedure), Goss and McGee have shown that there is no foundation for this claim; the accuracy of the *Royal Society* procedure was about 3 arcmin at elevations above 2°. However, the major disadvantage of the sea-cliff interferometer was the restricted range of low elevations; not only was the observing time limited to an hour or less per day, the low elevations implied that the positions had large uncertainties due to the imperfect calibration of the refraction corrections. Within a period of less than 8 years, in the 1950s the RPL group abandoned the sea-cliff interferometer technique in favour of the conventional Michelson interferometer.

Soon after the publication was submitted to the Royal Society in late July 1946, Pawsey and Payne-Scott attended the ANZAAS (Australia and New Zealand Association for the Advancement of Science) meeting, 21–28 August 1946 in Adelaide. Both gave 15 min presentations: Payne-Scott, "Discovery of Cosmic and Static [sic, solar?] Noise" and Pawsey, "Interpretation of Observations". The presentations were summaries of the exciting new data in the *Royal Society* paper. Pawsey wrote:

> This work is a new branch of astronomy. . . . the discovery of this [radio] radiation will come to be recognized as one of the fundamental advances of astrophysics. The first stages [after this discovery] are those of general exploration of the phenomenon. To these our work belongs.[29]

Clearly, the RPL solar group was justifiably proud of their contributions after less than a year spent on solar noise research.

Additional Notes

1. The authors themselves felt that the lapse between submitting the paper and publication caused problems. Five days before the publication on 4 February 1946, Pawsey sent a telegram (costing 19 shillings 4 pence) to the editor of *Nature* complaining about the delay. A concern was "due to unfortunate and incorrect reports of our work originating in the US would appreciate early publication". Bowen even characterised this as: ". . . irresponsible sources in

---

[28] Bolton and colleagues Stanley and Slee also considered a number of technical issues regarding the use of the sea-cliff interferometer in this era (Stanley and Slee 1950; Bolton and Slee 1953). They considered limitations due to the curvature of the earth (the sea is not a flat surface as observed from a cliff) and sea waves, leading to imperfect reflection. Payne-Scott had also considered this latter problem in her unpublished report of 1947, RPL 9.

[29] NAA: C3830, A1/1/1, Part 1.

the US". The Australian High Commission in London had sent a telegram to RPL on 31 January 1946, with a confused summary of press reports claiming that Piddington had made *radar* contact with the sun and moon. "We are being pestered for details..." Pawsey also tried to kill this story in a return telegram to the High Commissioner on 6 February, pointing out that they had only detected "noise from the sun...apparent temperature over one million degrees...and shown close correlation with sunspot activity". The assertion that the blame lay with the US press is, however, inconsistent with a letter from Pawsey to Southworth 10 days later; the latter colleague had sent the US press newspaper clippings to Pawsey in Australia. Pawsey wrote, "The Australian press got it thoroughly mixed up, they were even claiming echoes from the sun...The New York Post seems to have copied a false Australian press report." The two press reports (NAA:A1/1/15 Part 1 and D9/4H Part 1) indicate that at least the New York Post of 29 January 1946 only reported the false claim that the Australian group were the first to detect radio [not RADAR] emission from the sun. However the Sydney Sun newspaper article of 30 January 1946 did assert incorrectly that radar echoes had been detected from the sun. The hyperbole was extreme: "It is likely that many... new discoveries will follow which will undoubtedly affect the lives of everyone." Bowen wrote in the margin, "What is this about?"
2. Since Payne-Scott was judged to have poorer health prospects, she was forced to join a 'Provident Fund' instead of the superannuation fund. With the superannuation fund, the employer made contributions to supplement the employee contributions. At retirement (after working for over 10 years), the employee received an annual benefit that was a fixed fraction of the last yearly salary. Peter Davidson of the Australian Council of Social Service wrote to Goss on 13 December 2006, "... There was [most likely a] CSIRO Provident Fund and the level of benefit reflected the level of contributions made and the length of membership ... CSIRO Staff Provident Fund ... would have been owned by the contributing members and the employer (CSIRO) may well have made small contributions but [these] would have been considerably less than the contributions to staff super[annuation]. I assume that the limitations on membership were much less restrictive than those pertaining to the super[annuation] fund. Hence temporary staff and casual staff and staff with some medical conditions would have been eligible for membership. Payments would have been less than the super [annuation] entitlements, even ... comparing equal periods of employment". Thus the Provident Fund was a reduced benefit for those individuals not thought likely to attain a retirement age of 60 or 65. Payne-Scott lost all her accumulated benefits in 1950, after a refund of her own contributions with no interest as a result of the discovery of her marriage (Chap. 10).
3. The 1947 *Royal Society* paper has a complex history, summarised in *Under the Radar* (Chap. 8). The order of authors originally represented the relative contributions of the three authors: Pawsey, Payne-Scott and McCready. The editors of the *Proceedings of the Royal Society* alphabetised author lists in this

era as pointed out by B.Y. Mills and John E. Baldwin to Goss. In fact a preprint is available in the RPL archive with the "Pawsey, Payne-Scott and McCready" author ordering. The claim has been made (Orchiston 2005) that the *Royal Society* paper "suffered an inordinate and inexplicable delay". In *Under the Radar*, the authors did discuss this issue, concluding that there was "no definite proof". In 2010, John Baldwin (1931–2010) and Goss both investigated this claim. Both of us found no evidence at all of an unfair delay; the publication time of almost 13 months is quite in line with the median delay of 10 ± 2 months of all papers in volume 190 (1947). The paper might have been delayed by one or at most 2 months compared to other papers in the *Proceedings of the Royal Society* at this time. Likely in this post-war era, there was a substantial backlog of research output that had been delayed by the fact that most scientists had been involved in war time research.

Chapter 8
1946–1947: Personal Tragedy and Professional Triumph

Payne-Scott as Solo Scientist at Dover Heights, Mid-1946

In the period after the major solar events of February 1946, the solar emission at 200 MHz had decreased to modest levels. The typical 200 MHz intensities were less than 10^6 Jy—less intense by a factor of 10—by the next solar rotation in early March. The RPL solar group began at this time to concentrate on multi-frequency observations of the solar emission. Already at the PC (Propagation Committee) meeting of the previous November, elucidation of the metre wave spectra of the short term bursts had been proposed (i.e., the determination of the radio energy as a function of frequency or wavelength). At the PC meeting of 11 June 1946, Pawsey and Payne-Scott wrote: "Future work will concentrate on an exploration of the spectrum, both of the mean level and of the bursts."[1] A month later on 8 July 1946, it was clear work had begun in earnest. At 60–200 MHz, Payne-Scott had found that bursts did not correlate and simultaneous recording at 60–75 MHz had begun (see Additional Notes, No. 1, end of this chapter). In other words, detecting a burst in one frequency did not mean the burst would be identically or simultaneously active at other frequencies.

By mid-1946, Payne-Scott was planning a new campaign at Dover Heights to follow up the successes associated with the prominent solar activity of February 1946. With incredible luck, amidst the increased intensity of observation, a new and prominent sun spot appeared in July, with an associated large increase in solar noise. Newton (1955) has classified this sunspot as the fourth largest single spot in the period 1874–1952, a rank still maintained in 2008. The maximum area was 4,720 millionths (0.00472) of the solar hemisphere. A number of solar groups observed the radio emission from this event in addition to Payne-Scott: Allen (1947) from CSO (Mt. Stromlo); Hey et al. (1948) from AORG the Army Operational Research Group in the UK; Lovell and Banewell (1946) from Jodrell Bank; Martyn (1946a) from Mt. Stromlo (CSO) and Ryle and Vonberg (1946) from Cambridge (see Additional Notes, No. 2, end of this chapter).

---

[1] National Archives of Australia-NAA: C3830, B2/2, Part 1.

In addition to the importance of determining the spectra of solar bursts, the realisation of the existence of various types of bursts was becoming apparent in this period 1946–1948. The elucidation of the temporal, frequency and polarization state of the solar bursts and outbursts was a major challenge; several years were required before the taxonomy of the radio sun was sorted out. This period has been described by Wild (1985):

> The situation ... was one characterized by mystery, incredulity and intense interest. A whole new field of research lay ahead with obvious objectives: to disentangle the confused conglomeration of phenomena; to interpret and understand them; and to put results to use in the mainstream of research for solar physics, astronomy and physics.

The classification of the various types of bursts was necessary in order to understand the physics of the distinct emission mechanisms. While Ruby was active in the field, these problems were just beginning to get sorted out. The terminology of Type I, II, and III Bursts was first invented in 1950, close to the period when Ruby left RPL in mid 1951. In addition, the association of the varieties of radio bursts with well studied optical phenomena (sunspots, flares, prominences, etc.) was required in order to compare the radio properties with the well known optical sun.

As a test bed for the multi-frequency approach to this problem, Payne-Scott observed at Dover Heights for an intensive 2-week period during the transit of the large sunspot in July–August 1946. A total of about 110 hours of observing was accumulated. These pilot observations—to be the basis for future multi-frequency programmes—were carried out in order to lead to "several specific phenomena that seemed particularly worth[y of] study".[2] The observations were carried out at 60, 75, and 200 MHz using tiltable Yagis and dawn observations at 30–200 MHz using the ShD (shore defence) broadside array. Some limited data were obtained on the low frequency of 30 MHz (10 m). Security was a major issue. A problem for the recording of the data was the occurrence of a burglary at the beginning of the campaign in which a number of recorder pens were stolen! The result was that the timing and recorder pens were not optimal, ("the timing and recorder pens were slightly out of line"), causing additional errors in the timing of the various frequencies. Later in 1946, vandals attacked the ShD antenna; likely Payne-Scott's observations were the last use of this broadside array.

Compared to the earlier observations in the February campaign, the method of reduction of the dawn records in the determination of the positions and widths of the sources was substantially improved in Payne-Scott's July 1946 campaign—with a precision of a few arcmin. For the first time, a detailed treatment of the issue of the interferometer response in the presence of multiple sources on the sun was included. With this simple interferometer, the properties of weak burst sources could not be determined with any certainty. The confusion over the determination of the properties of weak "enhanced radiation" sources (noise storms) in the presence of the larger scale quiet sun—about 40 arcmin—was discussed.

---

[2] NAA: C3830, B2/2, Part 1.

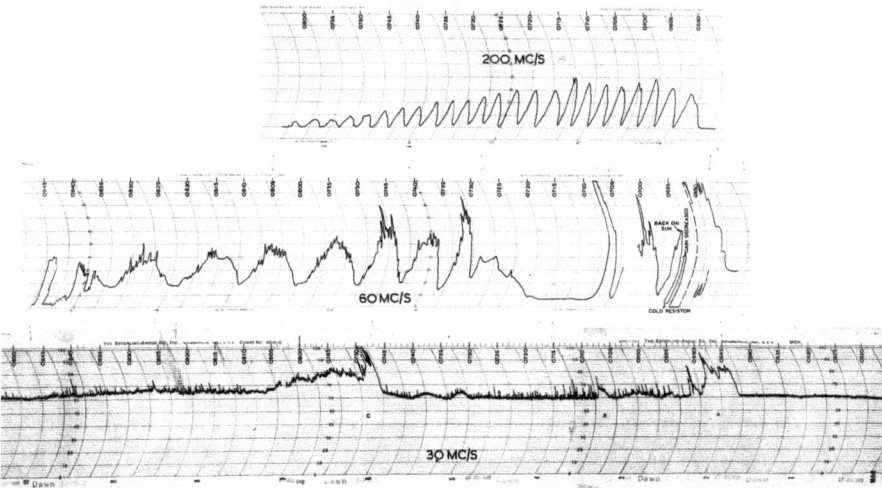

Fig. 8.1 Radio noise data from the major sunspot group of July–August 1946. For the first time solar radio data were observed at a number of different frequencies: 30, 60 and 200 MHz. The ShD (shore defence) radar antenna was used (perhaps the last data obtained with this instrument) as well as a number of simple Yagi aerials. A 200 MHz broadside array was also used with a polar mount for continual observations throughout the day, not just at sunrise. The dawn records from 27 July 1946 showed severe fading (intensity of the solar noise attenuated) at 30 and even 60 MHz, likely due to ionospheric effects at these elevations close to the horizon. Time increases to the left; optical sunrise is about 0655 EAST. The 200 MHz fringes are attenuated about 45 min after sunrise as the sun moves out of the primary beam of the ShD antenna (Figure 4 from the unpublished internal report from 1947, RPL 9 by Payne-Scott. CSIRO Radio Astronomy Image Archive B1285-10)

With the multi-frequency observations, detailed comparisons indicated that in general, there was an absence of correlated behaviour at different frequencies. Bursts tended to be more intense and variable at the lower frequencies. For much of the period when the intensity was most intense (27–29 July during the meridian transit of the great sunspot), the intensities at 75 MHz were almost ten times greater than at 200 MHz. During a period when the total emission level was much lower, however, the 200 MHz intensity was usually comparable or even in excess of the lower frequency data.[3] For many of the bursts, the bandwidth of the radiation was inferred to be small since the data were not well correlated at the different frequencies (as now expected for Type I bursts). The sunrise observations at the three lower frequencies (Fig. 8.1) are impressively variable. The observed solar radiation fadeouts at 30–60 MHz (partial disappearance of the signal) suggested that the ionosphere was the likely cause. The RPL group also reported the detection of solar emission at the lowest frequency (30 MHz).

---

[3] The quiet sun provided a minimum for the 200 MHz intensity ($6-8 \times 10^4$ Jy); at the lower frequencies, the sensitivity was not sufficient to detect the quiet sun, a few times 10^4 Jy.

The simple radar antennas used as a passive radio telescope employed the sea-cliff interferometer technique. The technique produced a one-dimension image of the sun (Fig. 7.7) with sensitivity along narrow elongated strips parallel to the horizon i.e. the resolution of the system was only along the axis perpendicular to the horizon. If more than one burst was located at a comparable elevation above the horizon, the determination of the identification of the radio source with an optical sunspot was ambiguous. Modern solar radio telescopes produce a two-dimensional image of the radio sun. The determination of the positions of the 200 MHz emitting strips showed results comparable to the *Proceedings of the Royal Society* publication of 1947; the strips included the majority of the sunspots suggesting that, as previously, the radio emission arose from regions near the centres of solar activity. On 27 July 1946, the strip was quite narrow and agreed with the optical position of this prominent sunspot. On the other 16 days, the strips included more than one spot; towards the end of the period, the strips were quite broad (20 arcmin), arising from the thermal radiation from the entire corona. Luckily sun spots tend not to cover the surface of the sun, occurring rather in isolated areas, which meant these early observations made with this simple one-dimensional interferometer were still effective.

Some of the bursts, which had a typical increase of a factor of a thousand and in a few cases a factor of a million, were isolated with double-humped shapes—probably what would later be called Type III bursts. Type III bursts are now known to be broadband fast drift bursts that can be associated with electron streams moving at velocities of a fraction of the speed of light. At metre wave lengths, the drift rate is about -100 MHz/s, meaning there is a declining shift from higher to lower frequency. This drift rate is about 100 times faster than the slower Type II bursts ("outbursts"). These events are often associated with the impulsive phase of solar flares. Payne-Scott spent much of her time in the next years working on these Type III bursts, called by her "unpolarized" bursts.

For these bursts, the time delays were of the order of seconds, as we would expect for Type III bursts. The delay histogram is shown in Fig. 8.2. A major result from the July–August 1946 campaign was that some bursts showed some level of correlation at 60–75 MHz, with a definite delay of about 2 s as the higher frequency preceded the lower. For some bursts, the 200 MHz radiation could be associated with the lower frequency events and this frequency also arrived earlier by a few seconds. However, this result was not published until the events of early March 1947. A few possible Type II bursts (called "outbursts" by Allen) may have also been observed; the intensities were even more impressive than the Type III events. However, the recognition of the delays (from 200 to 60 MHz) at the level of 5–6 min was not all certain at this time in mid-1946. The major result was, of course, the recognition of the seconds of time delays for the Type III bursts, a controversial result that was not generally accepted for a few years.

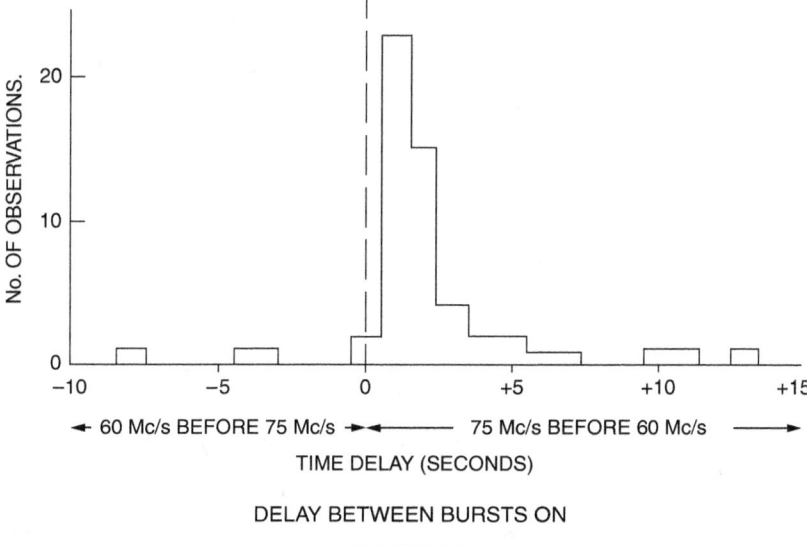

Fig. 8.2 Time delays (x-axis) in seconds for a number of Type III bursts observed at Hornsby. Data were recorded simultaneously at three frequencies, twice in 15 min; delays were observed for some events. The 200 MHz bursts arrived first followed by bursts at 75–60 MHz. The time delays were about 1 s. The distribution of these delays is shown here where the delays between peaks as observed at 75–60 MHz are plotted. Clearly the data suggest that the 75 MHz isolated bursts precede the 60 MHz events by about 1.5 s. This result formed the basis for the many month campaign at Hornsby in 1948 (Figure 11 from RPL 9. CSIRO Radio Astronomy Image Archive)

Miscarriage and Missed Work: Late 1946–June 1947

It is here in the story of Ruby Payne-Scott's life that it becomes necessary to paste together scraps of information from multiple sources in order to get an imprecise yet coherent picture of what happened in the year following the July–August 1946 observation campaign at Dover Heights.

Ruby was pregnant around this time; her pregnancy ended dramatically and publicly in a miscarriage during a meeting at RPL, from which she was likely transported to hospital. The exact date of the miscarriage or the state of the pregnancy is unknown.

Payne-Scott's colleagues at the RPL were very much aware of the legal and financial consequences that would befall her should their employer, CSIR, become aware that she was married. A married woman, by law, could not be hired as a permanent employee, and thus would be ineligible to participate in a superannuation retirement fund (Chap. 10), and would also need to be re-evaluated every year to see if she were worthy of being rehired for the coming year.[4] For Payne-Scott, who was a respected and integral part of the "radiophysics" group, such a change would not only have been personally humiliating, it would have destabilised group projects to which she contributed, and severely cut into the economic viability of her working life. It seems clear that most if not all of Ruby's immediate colleagues knew of her 1944 marriage to Bill Hall. No-one spoke of it; she wore her wedding band on a chain around her neck rather than on her finger. By simply not mentioning her married state, Payne-Scott and the rest of the group were able to keep her status secret for 6 years. B. Y. Mills, her colleague, wrote in a letter to Goss dated 14 September 1997:

> An interesting event occurred early in this work when, quite uncharacteristically, Ruby left early one afternoon without explanation. She later[in fact about a year later] told me that this departure was to marry Bill Hall, something which she found necessary to keep secret.

It is clear that if B. Y. Mills eventually knew of her marriage to Bill Hall, yet acknowledged her need for keeping it secret, then likely most of her colleagues knew as well.

It is not surprising then that no official documentation is found in Payne-Scott's personnel file regarding time away from work or medical leave for the miscarriage that was well known. What we do have to go on are two personal accounts and noteworthy gaps in the documents of regular meetings of the PC.

K. Rachel Makinson, a physicist and colleague of Payne-Scott's at CSIR, who knew both Ruby and Bill quite well,[5] recounted a tragedy that happened at a meeting of the OA (Officer's Association) sometime in 1945–1947 when Ruby suddenly was taken ill and then taken to hospital as a miscarriage began. Makinson did not witness this event but heard about it within a day. Joan Murray, who worked in the CSIRO administration during these years, also remembers the miscarriage which began during a meeting in the NSL Library.[6] Neither source is sure of the date when this occurred, but it is clear that Ruby's miscarriage began quite publicly in the company of her colleagues.

---

[4] Joan Freeman was sympathetic when she wrote in her autobiography (1991): "There was only one case involving sex discrimination, beyond the powers of RP to control, which Ruby went to extraordinary lengths to circumvent. It was in 1944 that she let it be known that she was living with a man (Bill Hall) to whom she was not married. Nowadays little would be thought of such a situation, but in the 1940s 'living in sin', as it was called, was looked on askance ... [Ruby] carried on as usual, unperturbed ... Ruby had hoped, by her deception, to evade what she considered to be an outrageous and discriminatory law. All her RP fiends, having developed a strong affection for Ruby as well as respect for her scientific abilities, greeted the story with hilarity, and sympathized with her attitude." Payne-Scott did not see the humour in the consequences of the CSIRO reaction in 1950 concerning her marriage (Chap. 10).

[5] Makinson told Goss about a number of interactions she had with Payne-Scott during the years 1941–1950 which included letting the Hall family live in their house for a year while Bill and Ruby Hall were building their own house in Oatley. In late August 1951, only a few months before Peter was born, Bill and Ruby Hall moved into their partially completed Oatley house which had as yet no water or electricity. NAA: A1/1/1, Part 6.

[6] Letter from John D. Murray and Joan Murray, 26 January 2004.

Ruby was away from work for some time after this terrible loss. In trying to pinpoint when this all happened we can reference the complete set of PC minutes from 11 June 1946 to 6 June 1947. She was not present at any of the meetings starting 8 July 1946—the beginning of the major observations going on at Dover Heights—and she does not show up in the minutes again for another eight meetings, or about 11 months,[7] at which time she is recorded as attending a conference called, "Meeting of Solar Noise Group" on 6 June 1947. Of note, we see that none of the reported activities for the eight meetings in the intervening period concerned her work, save only at the 14 October 1946 meeting, in which McCready reported that observations at Dover Heights would begin again when Slee arrived at the end of October, implying that Payne-Scott was possibly not available. From her personnel file we know that during the process of hiring Ruby as an employee "with commitment" she was sent for a special exam with an eye doctor in October 1946. The entire process of permanently appointing Payne-Scott and eighteen other RPL employees was begun in July 1946, and yet Ruby does not show up for her medical exam until September, with the subsequent eye exam in October. What is more, the recorded observations in (Fig. 8.1) only go until August 12, 1946, even though the sunspot activity may have continued through September 1946.[8]

What does this mean? The observations ended abruptly in mid-August; her medical exam was delayed until September. Could Ruby have passed a medical examination for her permanent appointment if she'd been pregnant? Could she have gone to an exam while still recovering from a miscarriage? How long would it have taken her to recover her full working capacity after such a physically and mentally draining event, one that doubtless became known in varying levels of accuracy around the lab?

Unfortunately we don't know the answers to these questions.

At the extensive PC meeting on 6 June 1947, descriptions of projects lead by Yablsley, Payne-Scott, Bolton, McCready, Piddington, Minnett, Pawsey, Smerd and a joint project with Bolton and Payne-Scott, as well as plans for the next few years were presented. The main purpose of this meeting was to prepare for the absence of Pawsey as he and his wife were about to leave for a 13 month visit to the US and Europe. At this meeting, a proposal for the Payne-Scott report (RPL 9, August 1947) was outlined to summarise the solar work of the previous year

---

[7] The frequency of meetings was once per month from mid-1946 to mid-1947 1946; Payne-Scott attended about 65 % of the 63 PC (plus Solar Noise Group) meetings from September 1945 to the date of her retirement in July 1951. She was, however, active in mid-January 1947; on 14 January 1947, she wrote a short summary of the July–August 1946 observations at 60, 75 and 200 MHz. The details of the bursts are shown, similar to the report RPL 9 from August 1947. NAA: C3830, A1/1/5 Part2.

[8] The choice of end date on 12 August 1946 is puzzling. The solar noise activity associated with this prominent sunspot was far from over. Payne-Scott in RPL 9 mentions a burst that was "off scale" on 12 August at both 60 and 75 MHz. Allen (1947) indicated large 200 MHz bursts up to early September 1946. Could this abrupt cessation of the campaign in mid-August 1946 be related to her absence later on in 1946, perhaps be due to her miscarriage?

July–August 1946. "This report is hoped to provide a basis for a paper for publication jointly with other observers."

By June 1947 she had begun a number of projects. The long delay in writing up RPL 9 (a year after the observations) and the fact that no publication was ever completed may also be related to this possible loss of work of a few months. There is a strong likelihood that her colleagues, especially Pawsey, helped in "covering" for her absence. There is no discussion of any problems (such as medical leave) in her personnel file. By early to mid -1947, she was active and ready for a new phase of her research career.

Clearly, Payne-Scott's professional status was not endangered by this period of lower intensity in her work. On 1 July 1947, her salary was raised from £A588 to £A675 per annum; at this point Payne-Scott's salary was comparable with or higher than those of many of her colleagues, e.g., Minnett at £A725 and both Bolton and Wild at £A500. Both Payne-Scott and Freeman—also now a Research Officer, Grade III, with an annual salary of £A650—were still being paid at the Women's Employment Board (WEB) rates.[9] In addition, in mid-1947 Payne-Scott also received a substantial windfall; she was paid £A129 in arrears for the WEB award from the period 14 September 1944 to 1 January 1947. This back pay represented an amount of over 2 months' salary.

Although the RPL 9: "A Study of Solar Radio Frequency Radiation on Several Frequencies During the Sunspot of July–August 1946" (laboratory internal report) was never published in a journal, it does contain much valuable material about the spectral indices of the bursts and especially the short time scale delays for the Type III bursts. The August 1947 report has 17 pages of main text with 15 pages of appendices dealing with various technical issues (e.g., appendix II in RPL 9 provided a detailed description of the interpretation of the dawn records from the sea-cliff interferometer, providing more detail than the *Proceedings of the Royal Society* paper published August 1947) and details of each day's observations from 22 July to 12 August 1946. There were 15 figures in the report. This report laid the basis for the multi-frequency campaigns to come in March 1947 and especially for her long range campaign of 1948 at Hornsby. The introduction of RPL 9 provided a valuable summary of existing solar noise knowledge in mid-1947 while the "Conclusion and Future Programme" section of the report provided clues to the issues that were to be tackled in the near future, including the intriguing problem of the reality of time delays between bursts. Some of the data from RPL 9 do find their way, however, into Payne-Scott's next publication with Yabsley and Bolton in the August 1947 issue of *Nature*.

---

[9] In September 1944, the Women's Employment Board (WEB) ruled that for wartime employment, women should receive equal pay to men for equal work. This state of equality was terminated by the WEB on 6 June 1949 when the ruling was rescinded.

Fig. 8.3 John Bolton in front of the blockhouse at Dover on 1 May 1947, about 2 months after the discovery of the giant Type II burst on 8 March 1947 by Payne-Scott, Yabsley and Bolton. The 200 MHz ShD antenna is no longer present. The 100 MHz Yagis are to the left while the 60 MHz Yagi is to the right (note the orthogonal placement of the dipoles on the Yagi antennas, used for the determination of circular polarization of the solar radiation). The 200 MHz Yagi was on the far corner and not visible; for "radio star" observations the Yagis were used in a parallel configuration (CSIRO Radio Astronomy Image Archive B1031-6)

The Behemoth Type II Burst of March 1947: Payne-Scott, Yabsley and Bolton Observe an Amazing Event

In the meantime towards the end of 1946, a twin 100 MHz Yagi replaced the 75 MHz Yagi at Dover Heights (see Fig. 8.3, the blockhouse at Dover Heights with John Bolton, a newly arrived- September 1946—colleague at RPL). The discovery of the behemoth solar outburst on 7 March 1947 has been described in a dramatic fashion by John Bolton (1982). The largest sunspot of the modern era appeared in March and April 1947 (Newton 1955). The March 10 spot is number five in the Newton ranking (area 0.004554), followed by its next appearance 27 days later on April 7 with an area of 0.006132. This latter iteration of the same sunspot remains the record breaking sunspot since modern records have been maintained.

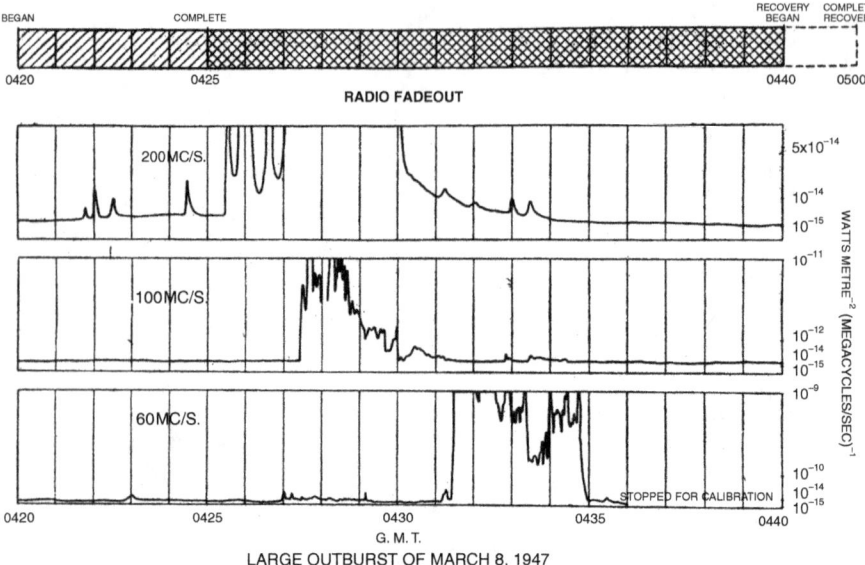

Fig. 8.4 The giant Type II solar burst of 8 March 1947 as observed at Mt. Stromlo and Dover Heights. The 200 MHz data were obtained at Mt. Stromlo and the 100–60 MHz data were observed at Dover Heights. The 60 MHz peak may have been a hundred billion Jy, but is uncertain. The minutes of time delay between the various frequencies is obvious. The commercial broadcast radio fadeout at a short wave frequency close to 20 MHz was used as a proxy for the time of the major flare. About a day later, a major aurora was observed in Australia on 9 March. In the publication the delays of several seconds for Type III bursts are also described (Reprinted by permission from Macmillan Publishers Ltd: *Nature*, vol. 160, p256, Fig. 7.1, "Relative Time of Arrival of Bursts of Solar Noise on Different Radio Frequencies", copyright 1947 (Payne-Scott et al. 1947))

At this time the spectacular outburst of 8 March 1947, which became the prototypical Type II burst,[10] was detected. The data are shown in Fig. 8.4. The claim was made that the 60 MHz peak was about 10^{11} Jy, the strongest extragalactic signal ever received.[11] However, Paul Wild told Sullivan (2009) in 1987 that both he and Payne-Scott regarded the intensity calibration of this event to be quite uncertain.

"Relative Times of Arrival of Bursts of Solar Noise on Different Radio Frequencies", the paper to *Nature*, by Payne-Scott, Yabsley, and Bolton describing

---

[10] Type II bursts are the strongest of the solar events, occasionally reaching 10^9 Jy at 100 MHz. These are rare slow drift events (mean drift rate of -0.25 MHz/s at about 100 MHz with a range of drift rates in the frequency range 30 to 100 MHz of -0.1 to -0.5 MHz/s, Roberts, 1959) that occur in association with flares. They have lifetimes of a few minutes up to 15 min. Narrow band emission often is the occasion for harmonic lanes, i.e., radiation is observed at the fundamental plasma frequency and twice this value (first detected by Wild et al. 1953, 1954). The emission mechanism is plasma oscillation, stimulated by outward moving shocks.

[11] Dale Gary and collaborators at the Owens Valley Solar Array (OVSA) did report a decimetric (30 cm) burst of about 10^{10} Jy on 6 December 2006. It is possible that at metre wavelengths this burst exceeded the March 1947 event (www.physorg.com, 15 December 2006).

this event was submitted within 2 months. The paper was likely written by Payne-Scott and Pawsey and submitted on 21 May 1947 with publication on 23 August 1947. This publication had an important impact on solar research for many years. The delay in publication was about 12 weeks, a few weeks less than the delayed publication of the February 1946 paper McCready et al. describing the initial Sydney detections.

Three main results were described in this publication: (1). Most of the bursts were not correlated at the three frequencies, suggesting that the emission at the various frequencies arose from widely separated levels in the solar corona. These bursts would be mainly described as Type I storm events in the future. (2). For events that would be called Type III bursts in the modern epoch, there was a good correlation in shape and ordering of the arrival times of the emission with 200 MHz first, then 75 and finally 60 MHz. This data were, of course, just the data from the July-August campaign of 1946, obtained by Payne-Scott as reported in RPL 9. Of 60 cases of associated bursts at 75 and 60 MHz, the most common delay was found to be 2 s (Fig. 8.2). The determination of the delays between 200 and 75 MHz was comparable, but quite uncertain since the form of the bursts was found to be disparate. In a few cases, delays were obtained with the 30 MHz emission arising a few seconds after similar events at 60 MHz.

The *Nature* paper of 1947 indicated that some of the Type II bursts (with delays of minutes in the progression from high frequency to low) had probably been detected in the 1946 data; however the 8 March 1947 event was the most prominent event observed up to that time. The 200 MHz data were provided by the Mt. Stromlo instrument. There was a delay of 2 min between the onset of the burst at 200 MHz and the 100 MHz event; then after a delay of a further 4 min the 60 MHz outburst occurred. The shortwave fadeout (at an unspecified shortwave frequency, perhaps 20 MHz due to excess ionisation of the D layer of the ionosphere at a height of 90 km from the earth's surface) could be used as a proxy for the onset of a major flare on the surface of the sun; at 20 MHz, Radio Australia disappeared and then reappeared about 20 min later, as reported in the *Nature* paper. When optical observations began at Mt. Stromlo on this day at 0440 UT (1440 EAST in Australia), a major flare was already in progress and the optical flare ceased at 0450 UT. The interpretation was that: "...the outburst was related to some physical agency passing from high-frequency to lower-frequency levels. If we assume, following ideas suggested by Martyn, that radiation at any frequency originates near the level where the coronal density reduces the refractive index to zero, we can derive a rough estimate of these heights from electron-density data..." (Payne-Scott et al. 1947). The derived velocities were in the range 500–750 km/s, as compared to the 1,600 km/s transit speed of auroral particles from the sun to the earth's magnetosphere. The displacement would correspond to about 0.3 solar radii in 6 min or an angular displacement of 5 arcmin in this interval. In fact a prominent display of the Corona Borealis was observed on 9 March in Australia, over a day later, indeed a rare event in eastern Australia. The inferred velocity of the particles causing the aurora was consistent with this time delay.

Payne-Scott, et al. did not attempt the same calculation for the Type III burst data. The implied relativistic speeds of 25,000 km/s (almost 0.1 the speed of light) seemed too unreasonable at the time. It was only after Wild's 1950 paper (Wild 1950b) that this speed was proposed; it took some time before the implied high relativistic speeds of the outward moving electrons giving rise to Type III bursts were generally accepted.

Payne-Scott did in fact attempt to calculate these speeds, though the results were not published. Indeed, about a year after the huge solar event, on 9 February 1948, Payne-Scott gave a talk to the Propagation Committee on her progress with the Hornsby observations of time delays for the "unpolarized bursts" (Type III). (Note that Payne-Scott's spelling "polarized is used".) Her presentation was "Measurements of time delays between solar noise bursts on 65, 60 and 18.3 Mc/s."[12] These observations are described below. Delays had been determined from the 18.3, 60, 65, and 85 MHz observations and she reported:

> Using Smerd's computations on radiating levels, based on Martyn's theory, the measured time differences would give a velocity of about 0.5×10^{10} cm/sec. for 60–18.3 Mc/s and 0.7×10^{10} cm/sec for 65–60 Mc/s- in very good agreement, considering the errors and scatter in the measurements and the empirical values used in the calculations, but an apparently unlikely velocity. However, if bursts do originate at the borders of prominence materials, it is quite possible that the much higher density gradient there, may lead to a crowding of levels by as much as 100 times, reducing the above velocities to the order of prominence velocities.

The inferred velocities were about 0.2 times the velocity of light; but these values were dismissed as being unlikely at this time and no mention of this revolutionary explanation was included in the *Nature* paper of 1947. This presentation is likely the first time the possibility of Type III bursts being caused by electron streams moving at a substantial fraction of the speed of light was discussed.[13]

One of the major frustrations building in the RPL group was the lack of international attention and respect for their work, particularly in Great Britain. It was not until Ron Bracewell in 1950 began a series of five short articles summarising the Australians' work in the UK journal, *Observatory*, that word began to spread. In the meantime, the frustrated, nascent astronomers at times resorted to minor tricks of semantics in order to make it appear as though their work was being cited abroad.

This *Nature* paper began with just such an unusual citation. While on an overseas trip to Europe and the US, Bowen gave a colloquium on 9 December 1946 at the RCA Laboratory in Princeton, N.J. The talk was briefly written up in the February 1947 issue of *Observatory* (Reber and Greenstein 1947):

---

[12] NAA: C3830, A1/1/5, Part 2.

[13] As Suzuki and Dulk (1985) have discussed "these streams move out through the corona along open field lines at a speed of about c/3, and their passage sets up plasma oscillations – Langmuir waves- which then radiate at their characteristic frequency". The energy range of the burst of electrons is typically in the range 10–100 kev. The electrons may be generated by instabilities located where opposing lines of [magnetic] force are in close proximity (Wild 1974).

If the coronal absorption depends on frequency, time lags of several seconds might exist between radio outbursts as observed at different frequencies. Unpublished work at the Radio Physics (sic) Laboratory at Sydney, Australia was reported on by E.G. Bowen in a talk.... He stated that the time of arrival at the Earth of bursts of solar energy is a function of frequency. Bowen also reported that the direction of circular polarization behaves rather erratically with time during solar outbursts and is not directly correlated with meridian passage of a sunspot group.

Thus the results from Payne-Scott's campaign of July–August 1946 were shown to have been cited in a British publication. Quoted in this rather circuitous manner, it is in some sense a self-citation.

There are at least two versions of the events that lead to the *Nature* paper of August 1947.

The Bolton version describes a sudden directive in late February 1947 by Pawsey to find a new use for equipment that had been destined for a cancelled eclipse observing expedition to Brazil in 1947. He and Gordon Stanley took all the equipment and tools to Dover Heights, and got to work. Amusingly, they first constructed a broadcast band receiver to listen to the cricket Test Match between England and Australia, and then they installed the solar receivers. Bolton (1982) reported[14]:

Finally on a Saturday afternoon [8 March 1947], as I unlocked the door of the blockhouse on my return from lunch, I heard the pen of one of the recorders hit the stop at the end of its travel. It was the 200 MHz recorder. I switched all three recorders form inches-per-hour to inches-per-minute and reduced the gain settings on all receivers to a minimum. Shortly afterwards the 100 MHz recorder hit its stop as the activity at 200 MHz decreased and three minutes later the 60 MHz recorder went off scale. Activity at all three frequencies ceased after about 15 min.

Bolton gave this account of the 8 March 1947 observations in a lecture at the annual meeting of the Astronomical Society of Australia at Noosa Heads, Queensland in 1982 (Bolton 1982). The story was written up and published in the proceedings of the meeting. Don Yabsley, who had not been present, took Richard X. McGee, then the editor of the *Proceedings of the Astronomical Society of Australia*, to task as he had not discussed the story with Yabsley before publication and Yabsley thought that the story had been misrepresented. Yabsley wrote[15]:

This session [the July–August 1946 Dover Heights campaign described above] provided confirmation of something that I had already noticed, namely that when isolated, intense bursts were observed on two or more radio frequencies, the burst at the higher frequency generally preceded that at the lower frequency by a few seconds. Hence my inclusion as one of the authors of the "Relative Times of Arrival..." letter to *Nature* in 1947. In actual fact I don't remember playing any active part in the preparation of this letter - by early 1947 I had ceased participation in observations at Dover Heights, and was preparing new receivers for simultaneous observations at 200 Mc/s, 600 Mc/s and 1200 Mc/s at Georges Heights using a 16 × 18 ft. paraboloid antenna. [See Orchiston et al. 2006 for a description of this research and for a photo of Yabsley and Pawsey in front of this antenna.] I think it is probable that Ruby and Joe Pawsey were responsible for the greater part of the text. John

---

[14] This account contains a number of misrepresentations of the solar work carried out in 1946–1947.

[15] Letter to Goss, 22 September 1977.

Bolton was the observer (at 100 Mc/s and 60 Mc/s) on March 8, 1947. The 200 Mc/s receiver at Dover was out of action, and the 200 Mc/s record was provided from Mt. Stromlo [as stated in the *Nature* publication]. My personal feeling is that Pawsey's name should have been on the letter as an author - perhaps he felt that three names on a Letter to *Nature* were quite enough.

Sullivan (2009) also had a similar interview with Yabsley in 1986:

> With regard to the paper by Payne-Scott et al. (1947), Pawsey contributed much to the project and to the paper and that originally he was intending to be a co-author. But in the end he withdrew his name because he felt that three authors were quite enough for a *Nature* letter.

As we have seen, though Bolton claimed to have used the 200 MHz receiver at Dover Heights, it was not working on this day and the time interval for this remarkable Type II burst was in fact about 9 min, somewhat less than the 15 min he describes. The 200 MHz data were provided by Allen and his colleagues at Mt. Stromlo. The 200 MHz recorder went off the edge of the recorder paper with a peak intensity far in excess 3×10^7 Jy. On the next day, during an outburst, Allen states "a rush of noise was heard in the loud speaker attached to the noise recorder" (Allen 1947).

The initial order of authors found in the RPL publications notebook is surprisingly *Bolton, Payne-Scott and Yabsley* in contrast to the published version of Payne-Scott, Yabsley and Bolton (certainly not alphabetical)! We have no idea of the reason but as we shall see in Chap. 10, a lingering irritation may have continued, involving Bolton and Payne-Scott. Pawsey was thanked for "his interest and helpful discussions" as well as Woolley, the Commonwealth Astronomer for the 200 MHz data.

Plans for Pawsey's Departure Abroad: September 1947

As has been mentioned before, it was around this time that Joe Pawsey, the leader of the radio astronomy group, began gearing up for an entire year spent abroad in the US and Europe. The fact that the RPL scientists were not astronomers by training, but physicists and engineers, meant that their acceptance in the world of astronomy was a complex process. That their acceptance did occur within a decade is partially due to the extensive contacts Pawsey made both within Australia as well as abroad during this long overseas tour, which began with his departure from Sydney on 25 September 1947 and ended with his return on 29 October 1948.[16]

---

[16] Leaving their three young children behind with the two grandmothers, Pawsey and his wife arrived in San Francisco on 28 October 1947 and visited the University of California and Stanford. He left the US on 27 March 1948 after numerous visits to US institutes in California, the Midwest and on the east coast. He met Reber during this period, as well as Struve, Minkowski and many other well known scientists. After arrival in the UK on 1 April 1948, he was based at his old college, Sidney Sussex, at Cambridge with his host, J.A. Ratcliffe (his former Ph.D. thesis

Pawsey had visited wartime radio researchers in the US in 1941, and he had contacts in Britain from his time as a postgraduate fellow at the Cavendish Laboratory of the University of Cambridge as well as his work on an early television system at EMI (Electrical and Musical Instruments Ltd.). He had been planning this 1947–1948 trip since the end of the war as a way to re-establish those connections and to make new contact with astronomers in the US, Britain, and other European countries. In short, he was networking, trying to build the reputation of the RPL group and bring much needed attention to their work. Pawsey impressed the overseas astronomers with his understanding of the technical aspects of radio astronomy and his newly acquired astronomical experience. As we know, the group had taught themselves astronomy in just the previous few years.

As Pawsey left Australia at the end of September 1947, he was likely quite proud of the remarkable achievements of the RPL solar group in the previous 2 years. A major goal of this trip was to publicise their successes. Already in this short period, this group had begun to make a name for themselves. An example of the early recognition of the group can be gleaned from the following exchange with J.A. Ratcliffe (1902–1987, Pawsey's Ph.D. advisor from 1934). About a year earlier, 2 August 1946, Pawsey wrote[17]:

> I have been principally interested over the last six months in the problem of radio frequency noise from the sun.... At the moment [indeed Payne-Scott was observing the major solar emission associated with the July 1946 sunspot] we are doing a bit of exploring, taking measurements of intensities at a number of different frequencies, some during the day and others at dawn [the latter is the sea-cliff interferometer]. We have found that the variation of solar noise on different frequencies is dissimilar and that that the dawn effect on 60 Mcs. is much more complicated that it is on 200 Mcs.

On 17 September 1946, Ratcliffe wrote back a letter of praise:

> First let me say how admirable I thought your paper on solar noise was [the *Nature* letter of February 1946]. There have been so many people scratching at this subject and making a few half-hearted measurements, that it is nice to see someone who has done it so thoroughly as you...I think you will agree with me that this question of the emission of radio wavelengths from the sun is such a big one that it needs several workers on it. I do not view with any more dismay the possibility of overlap here than I do in the case of ionospheric research, for example. Now that that the Air Mail works so quickly (Ryle showed me your reply yesterday to his letter) we will make a special attempt to keep you fully in touch with what we are doing.

The implication of this letter is that Pawsey was concerned about duplicating work between the two groups, in the UK and Australia. Ratcliffe assures him that there is no need to worry, and indeed it was lucky that some overlap in observations occurred in these early years, as the different research groups were able to check each others' work for errors as they developed their skills, knowledge and equipment.

---

advisor). They arrived back in Sydney on 29 October 1948 after multiple stops in Australia— Aden, Perth, Adelaide, and Melbourne. In Melbourne, Pawsey met Bolton as his young colleague was travelling to the 1 November 1948 solar eclipse in Tasmania, shortly after Bolton and Stanley returned from "The Cosmic Noise Expedition to New Zealand" in August. NAA:C3830, A1/3/18.

[17] NAA: C3830, A1/1/1, Part 1.

Since Pawsey had been Ratcliffe's graduate student, although he was only 6 years younger, it was natural that he was a guest of his former advisor when he visited the Cavendish Laboratory in May 1948. Pawsey spent quite some time discussing solar radio noise with his colleagues in Cambridge; these discussions led to numerous letters to and from staff at RPL. Since Pawsey had his foot in both camps, his role was decisive in guiding the nature of the interactions.

Pawsey was clearly adept at leading the members of the radio astronomy group, and as we will see in the next chapter, the absence of his calming presence lead to increased personality conflicts after his departure. Most notably, Ruby Payne-Scott and John Bolton would end up at loggerheads on more than one occasion.

A wide ranging discussion of the "Solar Noise Group" was held on 6 June 1947 to discuss the impending departure of Pawsey; 2 days before his departure there was a more thorough discussion to plan the work allocations for 1947–1948 at a "Meeting of Dr. Pawsey's Solar Noise Group". Payne-Scott's major project, described as Dover Heights Programme No.2, was to finish the analysis of the July–August 1946 Dover Heights campaign, leading to a joint publication with other members of the solar group. As discussed previously, this paper was never published, although the internal report RPL 9 was completed in August 1947.[18] The other objective outlined for Ruby was to determine the validity of her data on the delays in radio frequency activity during bursts from the March 1947 campaign.

Payne-Scott Winds Down at Dover Heights: Late 1947

In the period towards the end of 1947, Payne-Scott's activities at Dover Heights started to wind down as the preparations were being made for the transition to the Hornsby Valley site. At the PC meeting of 16 October (Pawsey was by then overseas), she reported on her "Dover Heights No. 2" progress[19]: "timing equipment is now in operation and the few bursts [presumably Type III's] that have occurred recently have been timed. Waiting for further solar activity and meanwhile planning photographic equipment [to film the bursts from the CRT display]". At the next PC meeting on 14 November, 1947, a number of reports of solar work at

---

[18] Surprisingly the plan called Dover Heights Programme 1 was an extensive set of solar observations by Bolton and his group of Stanley and Slee. Included in this programme was the continuation of simultaneous observations of the solar enhanced radiation (Type I bursts) with total and polarized intensity at 60, 85, 100 and 200 MHz, a programme that was in fact carried out in 1948 by Payne-Scott at Hornsby (Chap. 9). In addition, Bolton's group was to also continue quiet sun observations with the sea-cliff interferometer during periods of low solar activity. "It is particularly hoped to see the beginning of a large spot; also another meridian transit." In the end little came out of these solar plans for Bolton's group; most of their effort in 1948 was concentrated on the 1948 expedition to New Zealand. Under the rubric "Dover Heights Programme no. 1", the minutes of the 23 September 1947 meeting stated: "More accurate location and size estimate of Cygnus source. (New Zealand?)."

[19] NAA: C3830, A1/1/7.

Dover Heights were presented. Bolton, Stanley and Slee were busy with sunrise observations at 100 and 200 MHz and the surprising report was given that the [weak] thermal level of the quiet sun was being detected at 60 MHz (about 1.5×10^4 Jy). Payne-Scott reported that no useful data had been taken at Dover Heights on the relative times of arrival of bursts since the sun had ceased being active. She was anticipating new data from Hornsby; apparently the Dover Heights site was becoming less useful for solar work due to 100 MHz daytime interference from the increasingly built up surroundings in Sydney. None of this late 1947 data from Dover Heights was apparently published; likely the improvement of the data quality obtained in early 1948 at Hornsby rendered the poorer quality Dover Heights data from late 1947 less valuable.

Payne-Scott would end up observing at the Hornsby Valley site for 9 months, doing much to illuminate the solar activity that would later become known as Type III bursts.

Additional Notes

1. At the March to June monthly 1946 PC meetings, discussions of the determination of spectra of solar bursts continued. At the 8 July 1946 meeting:" General opinion however favoured a spectrum analyser style of approach, with all frequencies being studied at one place." This was the first mention of the swept frequency type of instrument that was to become so decisive in the classification of Type I, II and III bursts by Wild and McCready (1950). Finally at this meeting there was the surprising statement that a letter had arrived from Appleton describing the results of solar noise coinciding with a sudden ionospheric disturbance, the controversial paper of 3 November 1945 (Sullivan 2009). Bowen knew about this paper when he wrote Appleton in early 1946. Is it possible that this publication was not known to the RPL scientific staff as late as July 1946? If so, the internal communication at RPL was indeed poor. Also at the 13 May 1946 PC meeting, a discussion was held about a letter that Trevor Pearcey had received from a colleague in Cambridge describing Hey's spectra of solar bursts from 1.5 to 12 m wavelength (25–200 MHz). The result was published in July with almost no publication delay in the Philosophical Magazine (Appleton and Hey 1946); large increases in the radio emission were correlated with short-wave radio communication fadeouts (frequencies of 12–20 MHz) suggesting enhanced ionisation in the D layer of the ionosphere. The instantaneous radio spectra of a few bursts were determined, with some evidence of a low frequency cut-off due to the earth's ionosphere. Based on the observed sharp maximum of radio noise within 1 or 2 days of central meridian passage, the authors suggested that the sunspot radio noise was beamed in a sharp beam perpendicular to the sunspot. This letter with the report of the new Appleton-Hey data may well have spurred on the RPL group. Appleton and Hey proposed that a comparison with optical data suggested that the most intense

optical flares were associated with radio bursts with delays of several minutes (after the flare onset).
2. The data represented the first interferometer observation of Ryle and his group at Cambridge; the spacing of the interferometer ranged from 17 to 200 m. Also circular polarization was determined by crossing the polarity of the two elements of the interferometer. The 100 % circular polarization indicated that strong magnetic fields in the sunspot group (already known from optical data) could impact the radio emission process. Two additional groups published results on the circular polarization of these Type I bursts at about the same time: Martyn (1946a) the previous week (31 August 1946) based on CSO data at 200 MHz from 26 July 1946 (using the Yagi provided by RPL) and Appleton and Hey (1946b) in an adjacent publication (just preceding the Ryle and Vonberg publication 1946) from data at 60 MHz obtained on 27–28 July 1946 with the AORG antenna. In October 1946, Bowen wrote Pawsey from London a letter reporting on radio astronomy activities at Cambridge; he reported that Ryle was quite critical of the Martyn claim about the change of polarization state at meridian transit of the sunspot (from right hand to left hand), in disagreement with the Cambridge data. Their data changed polarization state at random times depending on whether the source of radiation arose on one side or the other of the sunspot.

Chapter 9
1948: Hornsby Field Station: Daily Observations

Move from Dover Heights to Hornsby

The year of Joe Pawsey's absence from RPL—25 September 1947 to 29 October 1948—was both frustrating and encouraging for Ruby Payne-Scott. There is evidence that without Pawsey there to manage the group's personality conflicts as well as guide their scientific research, a low level of needless chaos ensued. Even so, Payne-Scott used the time productively.

As discussed in Chap. 8, her plans for the 1948 campaign were formulated on 23 September 1947 in a decisive meeting of "Dr. Pawsey's Solar Noise Group".[1] Seven major research campaigns were discussed and for Payne-Scott there were both the Dover Heights Programme No.2 and a campaign to confirm the delays between frequencies during solar bursts from the March 1947 data.

The compelling rationale was to follow up the controversial claims of the *Nature* paper of August 1947. Many colleagues were not convinced of the reality of the "seconds" delay possibly associated with Type III bursts. In addition Payne-Scott had become very uncertain about the reality of the "minutes" delays for the Type II outbursts associated with flares.[2]

Just before Pawsey's departure from Sydney in late September 1947, there had already been an exchange of letters with Martin Ryle, one of the leaders at the Cavendish Laboratory in Britain. Ryle expressed doubts about the frequency time dependence of the Type III bursts at the level of some seconds described in the *Nature* paper concerning the 8 March 1947 event. Pawsey wrote back to Ryle on 3 July 1947[3]:

> With regard to the "seconds delay" cases, I am not entirely happy about the evidence. I believe that the tendency exists but do not know how often relative to zero delay, which is common [Type I], or to unco-ordinated [sic] effects.

---

[1] This group (Solar Noise Group) met seven times from June 1947 to April 1948.
[2] Correspondence with Christiansen, Mills, Wild and Bracewell 1999–2005.
[3] NAA: C3830,.A1/1/1, Part 1. The letter from Ryle to Pawsey has not been found.

Payne-Scott, with an assistant, Marie Clark,[4] set out to confirm the reality of the delayed bursts as a function of frequency. The goal was "To make sure [underlined in original] if there is a systematic delay and sequence for bursts on different frequencies. Delay due either to time of travel ... or to selective wave retardation. Result to be lined up with magneto-ionic theory to decide later. This is the most important detail to fit into any explanation." [5]

Before Pawsey's departure, the leadership issues in his absence were also discussed. No deputy was to be appointed, with the overall direction left in the hands of the Chief of RPL Bowen. The details of the research programs were to be supervised by Lehany and McCready, neither of whom it seems were prepared to mediate conflicts between big personalities.[6]

It must be pointed out here that though Ruby Payne-Scott was an integral part of the group and a brilliant mathematician, she had a tendency to become impatient with colleagues whose work did not measure up to her standards. Payne-Scott was well known as a decisive individual with strong opinions. By many accounts she could be abrupt to a point bordering on rudeness when in debate. On top of that, there were definite instances of sexist behaviour in the workplace[7] that doubtless affected her attitude toward some of her colleagues.

---

[4] Marie Couts Clark's career has been summarised by Hooker (2004); a contemporary of Payne-Scott, she graduated in physics at the University of Sydney in 1932. As Payne-Scott had done, Clark worked for a short period at AWA after teaching at a school. Clark functioned as an assistant, helping Payne-Scott at Dover Heights and later at Hornsby. Clark left RPL in July 1950. John Murray (letter, 26 January 2004) has provided additional details of Clark's role at RPL.

[5] NAA: C3830, A1/1/5.

[6] The text read: "Dr. Pawsey told members that during his absence he believed and expected that the research could carry on without appointing a deputy. Dr. Bowen would be in charge and would keep in fairly close touch with the work. To avoid burdening him with unnecessary details, Messrs. Lehany and McCready, as senior officers of the Group, were requested to keep Dr. Bowen informed of events by meeting at suitable intervals and co-opting other Group members whenever considered necessary." The project to investigate the spectral analysis of solar noise was discussed. The personnel mentioned were McCready and Medhurst (part-time) "until he leaves, then new man". This new man was probably Paul Wild, who joined this project in 1948. There is a note that this new instrument represented a "common interest with Ruby Payne-Scott's time delay experiments... Should be available to extend Ruby's results".

[7] Ruby was strong-minded, forthright and made her opinions known, on both scientific topics and daily discussions of current news. Slee reported that she "did not suffer fools lightly" and the rivalries with Stanley, as well as with John Bolton, were intense. She did not give in but held her own. An unfortunate incident took place in this period which is reported by Kellermann et al. (2005) in their description of the life and career of Gordon Stanley. Stanley painted "Men Only" on the only toilet door. Payne-Scott was not intimidated, however; she just ignored the sign and went into the toilet, laughing. In later years Payne-Scott's colleagues recognised her brilliance. In a letter to Goss (9 October 1997), Stanley wrote, "She was part of my early education on women's issues, and despite early insensitivities on my part, I grew to have a great respect and liking for her".

Harry Minnett, an RPL colleague wrote to Goss in 1998:

> She was a very pleasant but determined girl, especially when debating her ideas. I remember well how in a discussion, if a contrary point of view started an argument, her voice could rise in pitch and intensity to over-ride and then silence the opposition. Although such discussions were very wearing, she was always responsive to the logic of an opposing point of view and, if convinced, would readily concede.[8] Her ideas were always worth listening to and I learned much from her.

B.Y. Mills admitted that some found Payne-Scott abrasive[9]:

> Her most obvious characteristic was that she was very forthright and outspoken. She held opinions firmly, both scientific and social, and was always ready to defend them and to expect the same from people who disagreed. This did not endear her to some who regarded her as very aggressive! However those who knew her well respected her integrity and honesty, particularly in scientific matters. In this she was similar to Joe Pawsey although they were so different in other respects. As a scientist I think of her as extremely competent and knowledgeable, always ready to embrace new ideas if they appeared valid but not, I think, outstanding as an originator.

Lyn Brown, a librarian at RPL as well as a neighbour in Oatley and fellow member of the Sydney Bush Walkers had this assessment of Ruby's character[10]:

> I remember Ruby as a clever, forthright, honest, generous and outspoken individual. She could come across as forbidding due to her strong opinions with no fear to express them. But she was a warm-hearted individual who seemed to have little sense of humour. We had quite different philosophies; I am a Christian while she was as an honest agnostic. We never discussed politics, except in so far as it affected our children in schooling ... Ruby had a continual concern for the status of women in the work force ...

Strikingly, another female colleague of Ruby, Joan Freeman, also did not seem to be as bothered by her outspoken nature as some of the men. In her 1991 book, *A Passion for Physics,* Joan Freeman wrote:

> [She made] -a strong impact with her distinctive, very positive personality.... She was tall, solidly built, with straight fair hair, a strong-minded, no nonsense disposition, and a shrill voice which she could use very effectively in an argument...she had a sincere, kindly, and generous nature, to which I instinctively warmed.[11]

---

[8] Her colleague Robert Coulson did not share this opinion about an open-minded nature (email 21 February 2007 to Goss), see below.

[9] Based on an interview with Bernard and Crys Mills in Roseville (Sydney) on 1 April 2007 and letter to Goss, 14 September 1997.

[10] Based on a recorded statement of 17 min' duration on 17 February 1999, an interview in August 2003, and letters from Lyn Brown and Fred Brown (27 November 2007 and 4 December 2007).

[11] A dissenting point of view has been expressed by Robert Coulson, a War time colleague of Payne-Scott, and a friend of Joan Freeman and Ron Bracewell. He was a managing director of the English Electric Valve Company for some years, responsible for establishing the traveling wave tube section of EEV. Coulson has written that Payne-Scott was a "uniquely difficult person to get close to and rather daunting". He was impressed by her low tolerance for "non adherents to her faith [communism]" (email, 21 February 2007). Bracewell and Freeman did not share this negative assessment. B. Y. Mills (email, 23 August 2008) has pointed out that most conservatives at RPL shared Coulson's view, but those who worked with Payne-Scott or knew of her achievements were quite positive in their assessment.

Of more direct import to her working life in this period as she transitioned from working at the Dover Heights facility to her stint at Hornsby was the irritation she and Bolton had with each other while sharing equipment at Dover Heights. Slee has described the complex interactions at Dover Heights during this period.

For daytime observing, there were two solar groups, Payne-Scott's and Bolton's. During the course of 1946, Bolton began his survey of circular polarization of solar bursts, but the sun showed no activity at the end of 1946. Bolton continued his solar observations in early 1947, but by early 1948 he had lost interest in working on radio emission from the sun.[12] He and his group stopped their solar projects, while Payne-Scott continued. In October 1947 his team began the "radio star" observations in earnest, which led to the discovery of Taurus A on 6 November 1947.[13] This was probably the source of much of the conflict as Bolton's group wanted to test the equipment in the day time (the Yagi antennas at 100–200 MHz), at the same time as Payne-Scott's team wanted to carry out solar observations. Testing and calibrating equipment while others actually were trying to use it would clearly cause problems. Slee has used the terms "competing", "tension" and "confusion" to describe the relations between Payne-Scott and the others during this period. The nature of these tense conflicts was likely one of the reasons for Payne-Scott's transfer to Hornsby in late 1947 and early 1948.

These conflicts became more acute in the course of late 1947. Within 2 months of Pawsey's departure, on 18 November 1947, Lindsay McCready wrote by hand to Pawsey, describing many of the events of the preceding 2 months (see Additional Notes, No. 1, end of this chapter):

> The letter would be incomplete without some gossip on personalities. To cut a long story short, Bolton and Ruby have had a "bust up" at Dover [Heights] partly due to technicalities (e.g., Ruby's local oscillator and his 100 mc/s) and partly due to, I fear, her personality and, last but not least, both parties wanting to use the same gear for different experiments at the same time! Anyway, after careful examinations of the rights of all and of [the] facts we decided it be [sic] better if Ruby moved to Hornsby. No-one objects and both ok and Frank Kerr ... now are quite happy about it she says she can get to Hornsby [by public transport] and arrive at the site before 9:45 am and that she can't do that except on rare occasions at Dover. We are fitting her up in a separate trailer. I did not tell Taffy the full details—I mainly concentrated on technical difficulties they were naturally experiencing.

It is notable that McCready laid blame for this antagonism upon Ruby's personality rather than on Bolton's or even the way in which the two strong personalities mixed. From a feminist perspective, it is possible to see that Ruby's unwillingness to compromise her research for the benefit of John Bolton, who was her junior both in years and seniority at RPL, would possibly have seemed acceptable to McCready had she been a senior male researcher instead.

---

[12] At least some fraction of Bolton's solar noise work on the circular polarization of bursts and outbursts was later published by Pawsey in his review paper of 1950.

[13] The Cygnus A 'radio star' had been detected in Australia on 17 June 1947 and additional observations were made in July, August and September. During this period, Virgo A (although then called Coma Berenices A) and Centaurus A were discovered.

Pawsey's reaction to McCready's message is not known. However, he did write to Bowen on 8 December 1947 from the Embassy in Washington, making a surprisingly forthright choice on behalf of Bolton[14]:

> The second point concerns solar noise in R.P. I had a letter from Lindsay [McCready] in which he mentioned that there had been some sort of a showdown between Ruby and John Bolton. This is not unexpected to me as Ruby seemed to me to be getting in the way at Dover. As I understand it, Lindsay has the situation well under control, at any rate when he wrote the letter, having arranged a transfer of Ruby's work to Hornsby. I don't think Lindsay had mentioned this to you [In fact from other correspondence we know that Bowen knew all about the conflict and assumed that McCready had "smoothed it over very effectively".], and I am only doing so because there might be some future complications in which Lindsay might require your backing. My feeling on the matter is that Lindsay's actions are likely to be above reproach. I also think that Bolton has, through his hard work and effective results, earned the right to take control of Dover, so that anyone working there, shall be doing so at his invitation.

The move to Hornsby removed the day-to-day conflicts, although the clash of personalities continued.[15] Also the sunrise solar observations (with the sea-cliff interferometer) started to become less important at this time as the new swept-lobe interferometer, which would provide high angular resolution throughout the day, was already under construction initially at Bankstown and later at Potts Hill.

The first definite suggestion that Payne-Scott would transfer to Hornsby was made at the 14 November 1947 solar group meeting. 100 MHz observations had become almost impossible in the daytime at Dover Heights, not surprising given its location close to the centre of Sydney. Plans apparently were more developed, since 10 days later Payne-Scott wrote to Pawsey[16] who was by then in Washington, DC:

> I am in the process of moving myself to Hornsby taking a trailer and the 65 MHz gear. The sun has been quiet lately, but I suppose may spring a surprise on us anytime.

At Hornsby, Payne-Scott was free of the inhibiting atmosphere of Dover Heights and she could concentrate on her own work. During the 9 month period that Payne-Scott spent at the newly established Hornsby Valley field station, she laid the foundation for unravelling the mysteries of the "unpolarized" bursts or "isolated bursts", which would later be called Type III bursts. These bursts are probably the most studied burst phenomenon in astronomy; during periods of activity on the sun

---

[14] NAA: C3830, F1/4/PAW/1, Part1.

[15] An additional fact that may have reduced the day-to-day tension was that Bolton and Bowen had worked out a deal in early 1948 in which Bolton was to stop future solar work: "It does not appear practical to continue with any observatory work [probably implying that the group would occasionally observe the sun for testing purposes] on solar radiation." On 1 February 1948 Bolton wrote a long letter to Pawsey at the Embassy in Washington: "Bowen has given me permission to give up on the sun—at any rate just regard it as another source—and concentrate on the galactic source problem. I intend to go into solar evolution and atmospherics and see if I can find a home for my 100 Mc stars." (NAA: C4659, 8). Pawsey wrote back on 18 February 1948 and agreed with the *fait accompli*: "... am entirely in favour of your concentrating on the galactic work [radio stars]. The astronomers in the U.S. are waiting in a body on your results—so go to it".

[16] NAA:C4659, 8.

about three of these events per hour can be observed, with a drift rate to lower frequencies of about −100 MHz per sec. The theory that the bursts are plasma emission from electron streams in the corona is the accepted explanation.

Payne-Scott's assistant, Marie Clark, contributed to the day-to-day observational chores in this period of continued solar activity. Clark did not participate in the scientific research; her name never appeared in an acknowledgement in any of the Payne-Scott publications.[17] It is interesting to note that Payne-Scott was given a female assistant at Hornsby after her "bust up" with Bolton at Dover Heights. Is it possible that McCready decided that a female assistant would somehow make for a quieter working set up with Ruby? One can imagine that for a scientist of strong professional standards such as Payne-Scott, the gender of an associate with whom she worked closely would not matter so long as their abilities were at a high enough level.

McCready wrote to Pawsey[18] in the UK on 6 June 1948 that there were conflicts between Clark and Payne-Scott. McCready had heard Payne-Scott speaking to Clark on the telephone in an impatient manner; he also thought that Clark was given too much responsibility "with observing work and maintenance at Hornsby". McCready seemed to side with Clark: "I am satisfied that the girl is willing and doing her best. I think Ruby's main concern at the moment is to get [well?] ahead of John Bolton in papers (my opinion only but not without circumstantial evidence)..." Later in the letter McCready says, "I will do my best and of course watch relations and the group's welfare and have a chat with Miss Clarke [sic]." Since the observations were carried out successfully through September 1948, we can assume that the personal friction was to some extent reduced.

The Hornsby Valley field station was an important RPL outpost from 1946 to 1955 (see Orchiston and Slee 2005); the focus of the science at this site was *low frequency* radio astronomy. The location was a northerly suburb, Old Mans Valley; about 25 km north from the centre of Sydney near Hornsby (see Fig. 1.4). Kerr et al. (1949) performed some of the early lunar radar experiments at 18 and 22 MHz using Radio Australia as a transmitter; as a by-product, it was possible to study the earth's ionosphere (see Figs. 9.1a, b). Later on Shain and Higgins developed low frequency arrays at 9 and 18 MHz (Shain and Higgins 1954; Higgins and Shain 1954). Early continuum images of the Galaxy with absorption due to intervening HII regions were important results from this instrument (see Additional Note, No. 2, end of this chapter).

---

[17] In early August 1948, a report was written by McCready to send to Pawsey in the UK "Proposed Publications of Radio Astronomy Group (Dr. Pawsey's Group)". The report listed 17 publications under way. Paper No. 11, "Correlation of Solar Noise Intensity with Sun's Rotation", by Payne-Scott and Clark, remains a mystery. This proposed publication for *Nature* never appeared; the claim was made that "analysis should be complete in 6 weeks and 1st draft ready in 8 weeks". Perhaps the friction between the two colleagues prevented the completion of this project. No other information about this project has been found.

[18] NAA: C3830, F1/4/PAW/1, Part1.

Fig. 9.1 (a) The low frequency arrays used by Payne-Scott during 1948 to determine the properties of solar bursts. The field station at Hornsby was used initially for Moon bounce radar experiments carried out by Kerr et al. (1949). The transmission was at 17.84–21.54 MHz from the Radio Australia station in Shepparton, Victoria; the reception of the Moon bounce radar signals was at the Hornsby site. During 1948, Payne-Scott observed solar bursts at 18.4 MHz (plane polarized broadside array), 19.8 MHz (fixed rhombic with a bearing 60° east of north, elevation 15°), 60 MHz, 65 MHz and 85 MHz (CSIRO Radio Astronomy Image Archive B1266-15). (**b**) Close up view of the broadside array, transmission lines and instrument huts where the receivers were located. From 11 February 1948 (CSIRO Radio Astronomy Image Archive B1266-5)

The late Merle Watman[19] of the SBW (Sydney Bush Walkers) provided a humorous comment concerning Payne-Scott at this time. "[We]...heard that Ruby was one of the most brilliant science students to come out of Adelaide (sic) University." Ruby often left the train at Hornsby station after a weekend bushwalk to "take some readings somewhere" and the party would joke that "Ruby had gone to take her echoes off the moon". In fact as stated above, Kerr, Shain and Higgins were engaged in lunar radar experiments at 18 and 22 MHz but Ruby was not involved in these observations.

Payne-Scott made excellent use of two important properties of the new Hornsby Valley site—improved radio frequency isolation from man-made interference (in 1948 Hornsby was at an outer suburb of Sydney) and a variety of the low frequency instruments at her disposal. She used an 18.3 MHz broadside array (Fig. 9.1a, b) and a 19.8 MHz rhombic antenna for these 15 m wavelength solar observations; these data were to be compared with 60, 65 and 85 MHz Yagi antennas being installed at Hornsby. The low frequency instruments near 19 MHz added a new dimension in the determination of the delays compared to the 60 MHz data; the low frequency 19 MHz time of arrival was, clearly and consistently, many seconds later than arrival time for the frequencies near 60–75 MHz (Fig. 8.2).

Work at Hornsby began at a rapid pace towards the end of 1947 and early 1948. The publication, submitted exactly a year later in early 1949, stated that the total observing campaign ran from January 1948 to 23 September 1948. The solar group meeting of 6 January 1948 implied, however, that the observations may have begun before Christmas 1947. At this meeting, Payne-Scott already announced that data

---

[19] From an interview with Dr. Elizabeth Hall, 1999.

Fig. 9.2 The two Yagis used by Payne-Scott for solar work at the frequencies of 60 and 65 MHz at Hornsby Valley in 1948. The additional Yagi at 85 MHz is not shown; this consisted of a pair of crossed Yagis for reception of circular polarization. From 11 February 1948 (CSIRO Radio Astronomy Image Archive B1266-2)

had been collected at 60–75 [in fact, probably 65] MHz with generally little activity; however, several events had been observed with fade-outs (caused by increased ionisation due to the radiation from the solar flare) at 14.4 MHz (Radio Australia, VLQ3). The exciting news was that large delays in the unpolarized bursts of 5–10 s between 60 and later 18 MHz had already been discovered. The 85 MHz system was not yet operating.

As the campaign got underway in early 1948, the variable component of solar radiation at 85, 65, 60 was observed (separate simple Yagis with a single polarization were available at all frequencies; at 85 MHz crossed Yagis were available and thus circular polarization could be determined). In Fig. 9.2, the Yagis at 60–65 MHz are shown, amidst the 19 MHz broadside array. These three frequencies and 19 MHz were used on an almost daily basis. In contrast to the earlier observations at Dover Heights, these systematic observations enabled Payne-Scott to characterise both the detailed time behaviour and polarization properties of these metre wave bursts and outbursts. The continuous nature of the observations in 1948 was a major component of the success; Payne-Scott wrote an internal memo (1948b) in December 1948—RPL 30—illustrating the coverage of all observations made at RPL from 20 MHz to 24 GHz from 1947 to the end of 1948. The only long-term continuous coverage before this had been provided by Yabsley at 200, 600 and 1,200 MHz from August to November 1947 (Lehany and Yabsley 1948, 1949). Then there were the observations by Payne-Scott for the whole of 1948 at 85 and 60 MHz. She noted that all other data "are very scrappy" for this 2 year period. The low frequency antennas could not track and thus considerable corrections had to be made to the intensities as the sun was observed off-axis.

Clark and Payne-Scott were certainly busy in this early period of 1948. The 20 February 1948 solar group minutes state that they had already succeeded in creating histograms of burst timing between 65 and 60 MHz (only 0.3 s) and 85–60 MHz (0.7 s). The more reliably determined difference between 60 and 19 MHz was about 9 s. These results had been presented by Payne-Scott at the controversial talk she gave a week earlier at the Propagation Committee meeting of 9 February 1948 at which the high velocities (on the order 0.2 times the speed of light) had been discussed and rejected as being implausible. As has been pointed out, Payne-Scott's theory and calculations on these velocities were simply ahead of their time, and possibly unacceptable to the group of scientists when presented by Ruby, without Pawsey there to give support to her surprising conclusion. As we know, it was not until Wild's publication (1950b) that such a theory was given real credence.

On 12 April 1948 the Propagation Committee met with Bowen in the Chair; Pawsey was to return six and a half months later. Payne-Scott reported that fast cathode ray tube photos were obtained to determine the rise times of bursts at two frequencies with the Hornsby antennas. Three weeks later, on 30 April 1948, at a meeting of the solar group, the circular polarized antennas at 85 MHz were reported to be in operation; the bursts showing the delays showed no polarization in contrast to the storm bursts associated with the enhanced emission (storm bursts) (Type I continuum). This is the origin of the name favoured by Payne-Scott—*unpolarized bursts* (at this time, Pawsey preferred the term **isolated bursts**, also a characteristic of this type of event, and later they became known as Type III). At the end of April, Payne-Scott suggested the Hornsby work should cease within a month or two and then continue in a limited fashion at Potts Hill; the data collection in fact continued through September, but we know from Propagation Committee minutes that Ruby was in charge of the Potts Hill site planning by July 1948. The implication is that her assistant, Clark, took over the basically routine, daily observations at Hornsby, while Payne-Scott spent most of her time overseeing the development of Potts Hill. Doubtless, she had begun to find the Hornsby work stultifying and needed encouragement to finish the campaign.

Payne-Scott Has Doubts at Hornsby: Advice from Pawsey

After a few months of observing at Hornsby towards the end of 1947 and early 1948, Payne-Scott must have become quite uncertain concerning the value of the new multi-frequency data, with doubts about the reality of the frequency delays of the unpolarized bursts (Type III). At this time, her fellow RPL colleagues, Wild and McCready (1950), were within a year of using the swept-frequency spectrograph at Penrith (70–130 MHz); this data produced the ground-breaking classification of the Type I, II and III solar events, based on observations from March to June 1949. Also, as reported above, a number of colleagues in both Australia and the UK had already expressed doubts about the reliability of the seconds delays in the unpolarized bursts. She and Bowen both were becoming convinced in early 1948 that the Hornsby research was of little value. McCready wrote to Pawsey[20] on 6 June 1948: "... it appeared that her [Payne-Scott's] work at Hornsby was getting nowhere. Taffy [Bowen] himself suggested she fold it up and get into something more profitable ... I got her to give a talk at the Propagation Committee [the 9 February 1948 colloquium where she summarised her findings and put forth the revolutionary calculations on the inferred but unlikely extreme speeds associated with the bursts] mainly for Taffy's benefit ..." Another major factor contributing to Payne-Scott's anxiety was the planning and construction of the 100 MHz swept-lobe interferometer at Potts Hill. Clearly, because of its engineering and scientific challenges, she felt more attracted to this promising new project. It is possible she had begun to feel as though she were placed away from the action where the "big boys" were at work.

Probably at McCready's instigation, on 18 May 1948, Payne-Scott wrote a four page handwritten summary of her work at Hornsby to Pawsey (at the ASRLO, Australian Scientific Research Liaison Office, Africa House, London).[21]

The letter began:

> Lindsay McCready is on holiday and has forwarded to us your last letter concerning time delays. I thought I should write outlining the present situation and asking your opinion, particularly as I do not think the problem will be settled by the spectroscopic analysis equipment [the Penrith swept-frequency spectrograph of McCready and Wild] you suggest.

Next she provided a thorough description of the results of the first 6 months of observations at Hornsby. She described the circular polarization of the storm bursts (now Type I) and the lack of polarization of the (Type III) bursts. She then pointed out that the storm bursts showed little correlation at adjacent frequencies. The unpolarized bursts did show correlation at adjacent frequencies; this result had been one of the major conclusions. She then discussed the delays between the three frequencies in the Hornsby data. The letter also contained a discussion of the time resolved unpolarized bursts recorded at high speed on an Easterline Angus recorder at the high speed of 12 in. per minute. The exponential form of the decay of the burst was fitted with a decay constant at the various frequencies; the constant at 85 MHz was "rather higher" than at 60 MHz. There was also a hand drawing of the burst form, showing a sudden rise and then exponential decay.

A year later she would describe this behaviour in her June 1949 publication in the *Australian Journal of Scientific Research*: "A characteristic unpolarized burst shows a finite rise time, rounded top, and shows decay, reminiscent of the transient response of a medium with a natural resonant frequency."

---

[20] NAA: C4659, 8.

[21] Ibid.

Finally it is clear that Payne-Scott's major purpose in writing this letter to Pawsey was a request for advice: "...I would like your opinion on whether to leave it there or to try anything more...I am convinced of the reality of the [delays?] ..."[22]

An important aspect of the letter was the expression of her frustration at not being able to begin her expected participation in the planning of the 100 MHz instrument (as described above, based on McCready's letter to Pawsey). She wrote[23]:

> You have probably heard that I am taking on 100 Mc/s interferometry and hence would like to get clear of all the [responsibilities at Hornsby?] ...

Eleven days later, on 29 May 1948, Pawsey wrote a remarkable response.[24] This letter was handwritten on the train (Pawsey's handwriting was fairly illegible even when written at a stationary desk!) and contained even-handed advice; we can suppose that this communication represented some comfort to Payne-Scott after all the controversy since Pawsey had left Sydney, exactly 8 months previously. He began:

> I think the proposed subject [Type III bursts] would make an excellent paper subject to your being able to give sound evidence for the points you mention. You have to chose [sic] between doing work on bursts or starting interferometry. I think you should realise on what you have done—Do only what is necessary to make it a good plan. In the interferometry field you start behind Ryle and will probably take some time to catch up. Therefore go ahead with it in second place—complete the burst story first.

Eleven days later Pawsey wrote to Bowen with a detailed summary of his letter to Payne-Scott. He emphasised that Payne-Scott should publish her results "only... [if] the 'facts' she mentioned could be considered to be reasonably established".[25] It must have been trying for Payne-Scott to face the scepticism of so many, even Pawsey, on the reality of the data which had been carefully obtained over a long period.

On 22 July 1948 the Propagation Committee met with Piddington instead of Bowen in the Chair. Payne-Scott gave a report. In contrast to the plan to cease solar observations at Hornsby as previously discussed, the research was to continue for some months. As before, the main goal was the determination of burst delays using

---

[22] There were some other interesting pieces of news in the letter. Ruby thanked Pawsey for some contributions to her postcard collection. She told Pawsey that Owen Emery (from Mt. Stromlo and the Radio Research Board—RRB) had tuberculosis. She said she had seen Bernard Mills who was making good progress, having also had tuberculosis. At this time, Pawsey had been in the UK for about 5 weeks. It is striking that neither Payne-Scott nor Bolton mentioned their mutual conflicts of 1947–48 in any of their letters to Pawsey while he was overseas.

[23] NAA: C4659, 8.

[24] Ibid.

[25] NAA: C4659, 8. On 9 June 1948 McCready wrote Pawsey in London an enthusiastic report about Payne-Scott's reaction on receipt of the 29 May 1948 handwritten letter. "Her reaction was as I anticipated viz she still wants to keep interferometry. Nevertheless, I sowed a few seeds but I doubt whether they will be fertile unless you add [some fertiliser?] in a subsequent letter."

a fast recorder, in order to compare with Jaeger and Westfold's[26] theoretical predictions. A description of the continuous monitoring of the short-wave, broadcast radio station VLQ3, a 10 kW station at 14.4 MHz, was given; ionospheric fade-outs of this station could be used as a proxy for the onsets of major solar flares.

At the next meeting of the Propagation Committee on 6 September 1948, Bowen was in the Chair. Payne-Scott stated that she was busy analysing the Hornsby data. These series of observations were coming to an end; a number of the existing Yagis at other field stations were being moved to Potts Hill, including the 97 MHz system from Dover Heights and the 62 MHz (probably the 60 MHz system) from Hornsby. These aerials were used for continuous monitoring of the sun for a "total power" determination of the solar intensity during the day. These total intensity records were then compared with the high resolution new Michelson swept lobe interferometer at Potts Hill in the years 1949–1951.

Publication of the Hornsby Observations

Based on the Hornsby data from 1948, Payne-Scott found that the unpolarized bursts tended to occur nearly simultaneously over a range of frequencies. In many cases the higher frequencies arrived first, followed by a delay on the order of seconds for the lower frequencies. No clear-cut cases of minute-scale delays— such as those of the famous 8 March 1947 Type II giant outburst— were observed, and Payne-Scott even began to doubt the reality of their existence.

Payne-Scott's uncertainty about the ordered delays on the minute-time scales was also expressed in additional correspondence with Pawsey in an official letter (which was *typed*), written a month after the handwritten letter of 18 May 1948, discussed above. Payne-Scott wrote to him again at the Australian Scientific Research Liaison Office in London on 11 June 1948.[27] She was clearly worried that Pawsey would emphasise the "well behaved" (predictable) frequency delays for outbursts (e.g., the March 1947 Type II event) in the solar survey paper that Pawsey was preparing at this time (Pawsey 1950a). Payne-Scott wrote:

> I am still rather worried by the use of this figure [the 8 March 1947 giant outburst] in so far as it shows delays. The Dover records were very messy and Martyn at the time criticised our interpretation of them. Of dozens of outbursts during the last six months, I have only once seen an analogous case, in which outbursts appeared to have a markedly different appearance on different frequencies. This occurred on May 14 last, when a fade-out was

---

[26] At this meeting, Westfold gave a report on "Solar Noise (Theory)"; he described "...what happens to the radiation... after a burst has been produced... The magneto-ionic theory has been worked out for the case of a non-uniform distribution of ion density with height, and the results applied to solar bursts. By measuring the rate of decay of bursts and knowing the [radiofrequency] of the observations, it is possible to calculate the collision frequency at a given electron density".

[27] NAA: C3830, F1/4/PAW/1, Part 2. In addition her doubt was also expressed in the unofficial (handwritten) letter of 19 May 1949: "[I have] never seen anything like this [the 4 min delays for various frequencies during the outbursts] since. There is the possibility that it has a bad interpretation."

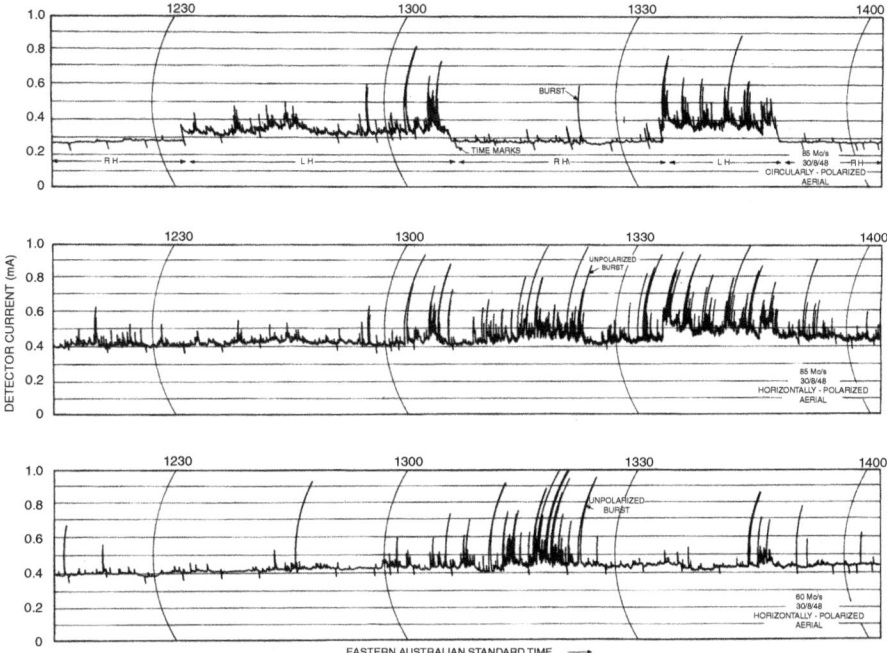

Fig. 9.3 An example of a Type I event (called enhanced radiation by Payne-Scott) observed at Hornsby at 60 and 85 MHz. The 85 MHz data are shown in the *top two panels*; the polarized nature of the 85 MHz radiation is evident in the *top panel* as the observations are switched between right hand and left hand polarization. Data from 30 August 1948. Figure 3 of Payne-Scott 1949 (CSIRO Radio Astronomy Image Archive B1640-2)

accompanied by outbursts on 200, 85 and 60 Mc/s ... when an outburst began on 200 and 60 there was nothing at all on 85 [her underlining] for over a minute ...

Later in the letter she repeats her objections in a section on relations between solar intensity variations at different frequencies:

Delays—same for all types of bursts including outbursts—usually in order of decreasing frequency, occasionally reversed—of order of seconds.

However, Pawsey did use the figure (Fig. 8.4, Chap. 8) showing the observations of the giant outburst of March 1947 at 200, 100 and 60 MHz. Following the Penrith observations of Wild (1950a), the evidence for the characteristics of Type II bursts was well established; the major clincher for the role of plasma emission was the detection of the second harmonic emission for the Type II bursts by Wild et al. (1953, 1954). The major contribution of the 1949 Payne-Scott paper was a detailed description of the unpolarized bursts; few details are given of the enhanced radiation (Type I) bursts (Fig. 9.3).[28] She did remark that the Type I bursts were seldom

---

[28] Martyn (NAA: C3830, D5/4/62 from Sullivan archive) was most impressed by the circular polarization data at 85 MHz as shown in a letter of 4 January 1949 to Bowen in Sydney. This new data "cuts across the line that Bolton and I [Martyn] had planned to do, notably on **the building up**

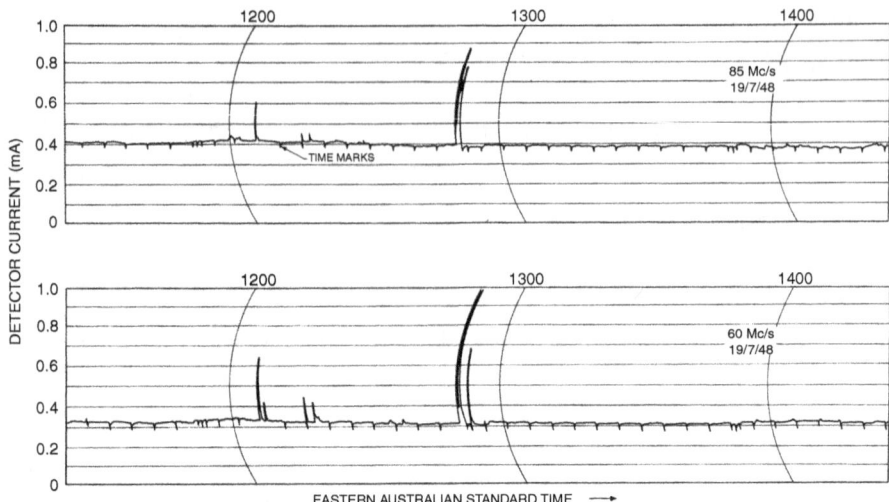

Fig. 9.4 An example of a Type III burst, called an unpolarized burst by Payne-Scott. Observations at 85 and 60 MHz on 19 July 1948. The isolated nature of this burst structure is obvious as compared to the bursts shown in Fig. 9.3. Figure 1 of Payne-Scott 1949 (CSIRO Radio Astronomy Image Archive B1640-4)

observed at the low frequency of 19 MHz, in contrast to the Type III bursts. She then summarised the spectral index of the Type III bursts by comparing the relative intensity of the 60 and 85 MHz data and found that the lower frequency intensity was usually twice that at the higher frequency. Figure 9.4 illustrates a Type III burst (unpolarized) from 19 July 1948.

The occurrence of time delays between the arrivals of the "corresponding" unpolarized bursts on different frequencies was confirmed, the higher frequency commonly arriving earlier, with delays of about 0.7 s between 85 and 60 MHz and 9 s between 60 and 19 MHz. There was good correlation between major radio fade-outs and large bursts on these frequencies. Payne-Scott interpreted this extensive data set (certainly the most thorough collection of burst data in late 1948) in terms of the theory proposed by Jaeger and Westfold (1949). These colleagues had suggested that the unpolarized bursts arose well out in the solar corona, with a common level of origin. They also proposed that the decay constant could be identified with the local collision frequency, with bursts at 85 MHz arising at heights in the corona of 5×10^5 km (Westfold 1949). There were serious disagreements between the observations and the theory.

---

of polarization on the various frequencies as a spot develops with a view to the special study of the **origin of bursts and general level**". (Martyn's emphasis). Martyn was convinced that polarization was the key to solving this problem. He wrote: "Miss Payne-Scott has made a very good job of doing it ... the main thing is that someone has done it, and we [Bolton and Martyn] can get on now to something else. I presume that Bolton too is quite happy. We shall certainly have to revise our ideas about what we should do now."

Sullivan (2009) has provided a description of the resolution of this discrepancy a year later by Paul Wild in his classic paper on the Type III bursts (Wild 1950b). Wild had continuous data from 70 to 130 Mc/s. Although the predicted intensity frequency spectra were in rough agreement with the Jaeger-Westfold theory, the predicted time-frequency spectrum was in disagreement with the slope of the dynamic spectra. Wild was thus forced to make the audacious suggestion that there were outward moving disturbances moving at relativistic velocities of 20,00–100,000 km/s (1/3 the speed of light), much faster than the fastest moving disturbances of about 1000 km/s known at this time. Sullivan (2009) has provided an anecdote of the interaction between Wild and Jaeger regarding the derivation of the equations governing the arrival times as a function of frequency from an ionised medium. Wild had derived a multiple page mathematical treatment of the major result. A few days later Jaeger produced a few lines of equations, that were incorporated (with credit) in the Wild paper. Ironically, Payne-Scott had given her colloquium on 9 February 1948 (Chap. 8) with a discussion of unpolarized burst velocities of about 25,000 km/s, a value so large as to be rejected!

Pawsey was anxious to ensure that the Hornsby experiment would provide a decisive result about the reality of the time frequency behaviour of both the Type II and III events. Two weeks after a meeting of the Royal Astronomical Society in London meeting he wrote by hand to McCready: "People [here in the UK] do not believe the high frequency precedes low." A month later, he made a proposal to establish a collaboration with the Cambridge group and also to ensure the priority of the RPL claim to time delays. This complex plan was only partially successful as can be seen from his letter to Bowen of 24 June 1948[29]:

> In my discussions at the Cavendish I found Smith [F. Graham-Smith, later Astronomer Royal, 1982–1990], one of Ratcliffe's men who works on solar noise, contemplating carrying out a series of observations on the form of bursts which is very similar to that done by Ruby. It is a clear case where the discouraging sort of duplication of work could occur if we do not get together before hand. I indicated the general lines of Ruby's program without giving much of the results and we agreed that I should write you with the following suggestions ... We should ask Ruby to prepare an outline of the paper she intends to write and send it to Ratcliffe for Smith's information in the near future, and when he has results to publish would of course expect to acknowledge any relevant work of Ruby's ... By this arrangement we gain freedom from fear of immediate prior publication by Smith and consequent undue haste in publication. His gain is obvious ... At the same time it emphasises that competition is likely and I should think strongly recommend Ruby to get on with the job promptly. I think it should take full precedence over any interference work [the 100 MHz interferometer project at Potts Hill] she may have on hand.

There is no record of the detailed response from RPL. Graham-Smith reported to Goss[30] in 2008 that he had had no idea of this surprising 1948 negotiation and had never heard of any of these discussions. He was

---

[29] NAA: C3830, F1/4/PAW/1, Part 2.

[30] Interview with Graham-Smith by Goss at the Jodrell Bank Observatory, 19 May 2008. Professor Sir Francis Graham-Smith was Director of the Royal Greenwich Observatory (1976–1981) and later Director of Jodrell Bank Observatory of the University of Manchester (1981–1988).

...happily immersed in the early days of interferometry and position finding [his participation in solar research decreased in the course of 1948–1949]. I am surprised that Pawsey was more concerned with overlap in the research agenda.... This was totally unnecessary. If two groups were involved in solar burst research, this was jolly good! Pawsey seems to have thought that there was limited science in this field and that one group would mop it up...with cut throat competition with the other group. This is ridiculous looking back from 60 years later on. Joe Pawsey was quite concerned...about the Australians being upstaged.... I had always thought of Pawsey as an "internationalist" in world-wide astronomy. [Later on Pawsey was a prominent leader in both in the International Astronomical Union (IAU) and in the International Union of Radio Science (URSI). Already in 1948, there were growing signs of distrust between the two main groups working on solar radio noise.]

A partial conclusion of this complex process was initiated later in 1948 following a letter from Ryle to Pawsey on 23 November 1948.[31] Apparently the contacts between RPL and the Cavendish had been minimal since Pawsey had left the Cavendish Laboratory about four months earlier. Ryle wrote Pawsey:

> During your last visit to Cambridge we discussed our programme and among other things the question of our continuing work on solar bursts. You were writing to find out how your experiments on the correlation of bursts on different frequencies were going. We have done nothing further on the subject, but I think there is much more important work to be done and if it does not overlap with your programme. I should like to start it up again with a new member of the team [Smith was then concentrating on the radio star position work. John Baldwin pointed out to Goss in 2010 that the new student was, in fact, Anthony Hewish, who joined the group in Cambridge in the summer of 948; Hewish had no intention of working on solar bursts, either.[32]] There is no immediate hurry, as he will be fully occupied for the next two or three months, but I should like to get him thinking about his own line of experiment soon, and the analysis of solar bursts is the first choice at the moment.[33] Please remember us to your wife. We have often wondered if your family recognised you again![34]

A few weeks later, on 9 December 1948, Payne-Scott replied to Ryle, sending a copy of her draft paper on the Hornsby campaign (the paper was submitted a month later to the journal). She wrote:

> You will see that one of the main points is the distinction between variations (bursts) in the circularly polarized "enhanced radiation" and unpolarized bursts. It is the latter that show good correspondence on different frequencies and time-delays. Much of the past confusion originated because no distinction was drawn between the two kinds of short-period variation in solar noise.[35]

---

[31] NAA: C3830, A1/1/1, Part 3.

[32] Email exchange with John Baldwin to Goss, July 2010.

[33] In pencil at the bottom is a note by Pawsey: "Reply when copy of Ruby's paper available."

[34] Pawsey and his wife Lenore had been away from the three young children (ages 11, 9 and 3 in 1948) for over a year. The two grandmothers (from Canada and Melbourne) had looked after Stuart, Margaret and Hastings in the Vaucluse (Sydney) home.

[35] NAA: C3830, A1/1/1, Part 3. Eight days later, on 17 December 1948, Pawsey wrote to Ratcliffe. (See above.) There was some confusion concerning who would answer whom! Pawsey wrote to Ratcliffe concerning the "investigation of bursts"; he had delayed writing until Payne-Scott's paper was available. The paper was to be ready in a few days and then he would send a copy to

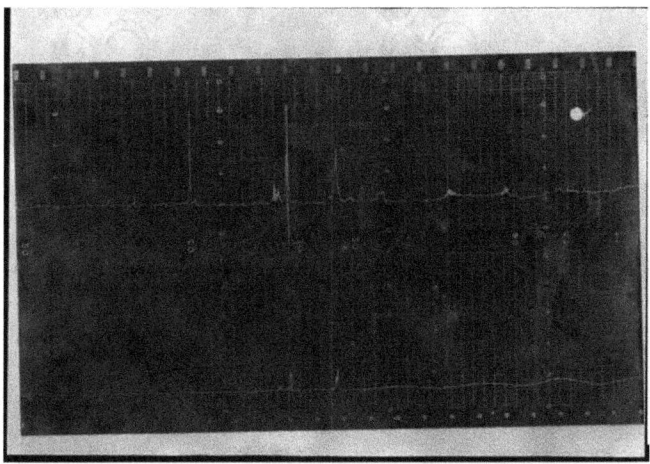

Fig. 9.5 Cambridge (UK) solar noise observations at 175 MHz from a letter from Ryle to Pawsey 21 March 1950. The figure was purported to indicate that aircraft were the major cause of solar bursts, leading Ryle to believe that at least the unpolarized bursts (Type III) did not arise from the sun. The original in the NAA has a poor quality (National Archives of Australia. NAA: C3830, A1/1/1 Part 5)

No record of a reply to her letter has been found in Australia or in Ryle's archives in Cambridge.

Payne-Scott's letter provided the solution to the confusion about the frequency time behaviour that had plagued the discussions in the previous year. The seconds of time delays were **only** relevant for the Type III (unpolarized bursts), not the Type I bursts (storm bursts).

The controversy over the reality of the unpolarized Type III bursts continued during 1949 and 1950. Some of the correspondence in this period has been described in *Under the Radar*.[36] Ron Bracewell (who returned in late 1949 from a 3 year visit to Cambridge where he did a Ph.D. with Ratcliffe) acted as a middle man, having traced some of the Cambridge 175 MHz records and mailed these to Pawsey, Fig. 9.5. Ryle claimed that the records were not solar bursts but aircraft flying near Cambridge. Ryle wrote Pawsey[37] on 21 March 1950 that "in all probability all the 'bursts' on this afternoon were caused by a light aircraft..." Bracewell told Goss a similar version.[38] Ryle told Bracewell in late 1948 or early 1949 that "These bursts do not occur in Cambridge, but any that did were due to aeroplanes, as he knew from having a loudspeaker on line."

Ratcliffe. Probably Pawsey was not aware that Payne-Scott had previously sent a draft to Ryle. Pawsey also wrote that Payne-Scott had included some theoretical speculations concerning the origin of bursts in the new paper (the comparison with the Westfold theory).

[36] Chapter 9 and Appendix D in *Under the Radar*.

[37] NAA: C3830, A1/1/1, Part 5.

[38] Interview 10 November 1997.

In the end, the results of Payne-Scott stood the test of time; the delays determined at Hornsby in 1948 agreed with the new data of Paul Wild and colleagues obtained in the next years at Penrith near Sydney. The confusion would continue for a year or two; but the decisive observations of Wild and McCready (1950) put an end to the controversy.

With a manuscript submitted on 5 January 1949, only 3 months after the Hornsby observations drew to a close, and very soon after this flurry of international correspondence, Payne-Scott's paper (1949), "Bursts of Solar Radiation at Metre Wavelengths" was published in June 1949 in the *Australian Journal of Scientific Research*. Pawsey wrote to J. A. Ratcliffe, at the Cavendish Laboratory of the University of Cambridge on 17 December 1948, "I have been delaying writing to you on this subject because I thought that Miss Payne-Scott's paper would have been available by now."[39] Thus it is clear that Pawsey was in a hurry to have Payne-Scott complete this publication.

The manuscript provided a discussion of the variable component of solar radiation at these four low frequencies over a period of 9 months. Two types of variable high intensity radiation were distinguished. One was the "enhanced level" and "storm bursts", which were circularly polarized; these were the Type I events also called "noise storms" in the early literature (i.e., Type I continuum plus Type I bursts). The other, a particular type of short duration, was the unpolarized burst, Type III.

At about this time, a major event occurred in the birth of the field of radio astronomy with the creation of a new name for the Propagation Committee. At the meeting of 7 March 1949, Pawsey was again the Chair, having arrived back in Australia in October 1948. A discussion was held about the future purpose of the committee, including its name, content and frequency of meetings.

On 5 April 1949, the first meeting of the "Radio Astronomy Group" took place with a 1 hour talk by Piddington on the derivation of the properties of the solar chromosphere. Thus "Cosmic Noise" and "Solar Noise" became the new "Radio Astronomy". The following week, on 11 April 1949, a general discussion of all projects for the RPL radio astronomy group took place. At this inaugural meeting, the Radio Astronomy Committee (Christiansen took over the secretary position from Kerr) presented an overview of all ongoing projects about 6 months after Pawsey's return from the long visit overseas.

In summary, in spite of the controversy with Bolton, and great doubt over the validity of her data, the 9 months in 1948 were very productive for Payne-Scott. The results of this period provided a solid basis for the future of solar noise research at RPL. She was then poised to begin the final stage of her scientific career at Potts Hill with the 100 MHz interferometer. For the first time RPL scientists could

---

[39] NAA: C3830, A1/1/1, Part 3. A personal note was added by Pawsey: "I sent you a food parcel a couple of weeks ago and I hope you receive it by Christmas. Wishing you and your family the compliments of the season." All correspondence between them was on a familiar basis, beginning either "Dear Jack" or "Dear Joe", in contrast to most other correspondence, such as that between Ryle and Pawsey, which began "Dear Pawsey" or "Dear Ryle".

observe the solar bursts and outbursts at any time of the day, not just at sun-rise. Payne-Scott had clearly shown that Type III bursts (her unpolarized bursts) showed rapid changes in frequency drift, from high to low frequencies. Thus a fast exciter speed for Type III bursts was clearly required; within a few years this was the accepted explanation.

Additional Notes

1. McCready wrote to Pawsey about 6 months later on 9 June 1948 (NAA: C4659, 8): "... the 'feud' between Bolton and Ruby is still on, and leading to stupid and undesirable secrecy". This open infighting probably continued until Pawsey returned in October 1948. McCready showed weak leadership in handling this major disruption: "If you are wondering why we can't solve our own problems I would point out that everyone rightly regards you as leader and my impression is that Taffy has too many other worries at the moment to be bothered with things I should attend to ... far better to handle it the way we are attempting than getting Taffy to give orders ... Please don't imagine that people have gone completely haywire in your absence. As I said before I think I could patch it up and would have, but for many reasons decided to take no direct action as the matter is by no means urgent now that John [Bolton] is occupied in N.Z. [Cosmic Noise Expedition to New Zealand from June to mid-August 1948] ... and it has one advantage viz. Ruby is greatly spurred on!" In this letter McCready was also quite critical of Bolton for the first time: "There is evidence that he is attempting to go beyond his 'sphere of influence' and terms of reference as laid down before your departure ... it must be done without the leader's approval and the knowledge of other members." Clearly McCready was out of his depth in dealing with these two strong personalities!
2. The origin of the Hornsby Valley field station was described in a PC committee meeting on 11 June 1946 (rubric—High Power Vertical Radar) where Kerr outlined the requirements of a new site for 10–43 MHz systems for the study of cosmic and solar noise. A striking fact was that he required a non-cliff site (in contrast to Dover Heights), a deep valley or "hollow" to avoid ground echoes due to the Lloyd's mirror effect from sea reflections. The first discussions of radar observations concerned radar reflections from the sun! At meetings in July and October 1946, sites at Hornsby were discussed with the "Old Man's Valley" site chosen after another site was found to be unavailable. By mid-November, work had started on this site. John Murray has told us that most of the equipment had been transferred from Hornsby to Potts Hill, probably towards the end of 1948, while the remainder was destroyed in a bush fire near Hornsby later in the 1950s.

Chapter 10
1949–1951: Radio Astronomy Blossoms as a Field, but Ruby Must Resign from the Radiophysics Laboratory

Payne-Scott's Career in 1948, Choice of the Swept-Lobe Interferometer Site

As we know, much of Payne-Scott's professional life in 1948 was taken up with diligently observing solar bursts at the Hornsby Valley site to determine the nature and even validity of the time delays between detection of different frequencies from the bursts. Concurrent with those 9 months of careful observation, plans were being put in motion for the creation of the swept-lobe interferometer at the new Potts Hill location, a level and more spacious site.

Plans for the creation of the Potts Hill observatory had been laid before Joe Pawsey left for his long, overseas trip, with many of the details and justifications for the project worked out at that 23 September 1947 meeting of "Dr. Pawsey's Solar Noise Group".[1] Ruby naturally had been counting on a leadership position in the creation of the site, and as has been previously discussed was dismayed that the seemingly endless, and increasingly disheartening Hornsby work might keep her from overseeing this exciting new engineering project.[2]

These Potts Hill observations had a lasting impact on the progress of solar physics in the latter half of the twentieth century. This new instrument was the first Michelson interferometer constructed at RPL; it was also the first swept-lobe interferometer in radio astronomy. The term "swept-lobe" refers to the way in which the narrow beam that characterised the detection of radio frequencies was swept across an area—in this case, the surface of the sun—at a fast rate, making

---

[1] NAA: C3830, A1/1/7.

[2] NAA; C3830, B2/2, Part 1. The PC minutes of 23 September 1947 contain a thorough discussion of a new solar interferometer for a new field station at Bankstown-Sydney. As had been the experience at Dover Heights, this site initially had problems with robbery and vandalism. By April 1948, the location was moved to Potts Hill Reservoir. At that time Payne-Scott was in charge of the project. The new instrument was to be "capable of yielding interference patterns in a fraction of a second with a view to extending this technique to bursts".

radio astronomical positional determinations on the sun at the rate of 25 times a second. In fact, the speed of these sweeps was so high that the old method of using electrical pulses to move a pen on paper to display the data became obsolete due to the inability to keep up with the data transmission rate. A new recording system was devised that turned to the speed of film in a movie camera to more accurately record the incoming data. As the interferometer data came in, a cathode ray tube would light up in direct reaction, in much the same way the war-time PPI tubes would have lit up upon detection of incoming aircraft. During periods of intense solar radio emission, the camera would be pointed directly at the cathode ray tube, and the complex and rapid changes in luminosity would be recorded on film, which could then be analysed at slower speeds later on. The technique was crude by today's standards, yet a brilliantly elegant solution to the inability of the human eye to detect and analyse the incoming data in real time. In addition, the polarization state of the incoming radiation could be determined once per second; this feature of the swept-lobe interferometer was likely also a first in radio astronomy.

For slower time scale observations (e.g., the enhanced solar radio radiation or cosmic radio sources), a drift scan mode was constructed using the slowly varying lobe motions due to the rotation of the earth. The initial system consisted of two antennas on an east-west line; thus only the right ascension (the east-west displacement) of the sources could be obtained, since observations were carried out near transit. Thus the imaging was only a one-dimensional scan. Possible extensions to addition antennas were discussed in order to obtain two-dimensional positions or even distributions for extended sources (i.e., aperture synthesis), again a facility that was never implemented. A third antenna on the east-west line was added later on.

Pawsey was quite impressed with the status of the project in late September 1947, 2 days before his departure. He wrote in the solar group minutes: "This plan looks further ahead than the others. Many aspects will not be touched for some time [this was indeed true], perhaps never."[3] The other great advantage of the Michelson interferometer was the ability to use it at almost any time of day since the sun could be tracked for a period of about 4 hours a day, centred on local noon.

The solar radio astronomers in Sydney had recognised the awkward nature of the sea-cliff interferometer with observations only possible for about an hour at dawn. Also the use of eclipses of the sun by the moon (Hey 1955) to study small scale solar radio structures had proven to be an awkward technique[4]; the times of observation were infrequent and the results were often ambiguous, leading to uncertain interpretations. Thus the ability to make high resolution observations at most times during daytime represented a major step forward. As Wild (1968) has pointed out: "...another Pawsey-inspired experiment was put into operation and brilliantly performed by Payne-Scott and [Alec] Little [1925–1985]. The idea was to locate...the instantaneous position of the dominant source on the sun at any one time".

---

[3] NAA: C3830, A1/1/7

[4] An example was the eclipse of 1 November 1948 observed in eastern Australia.

The detailed scientific motivation for the 100 MHz swept-lobe interferometer was presented at the 23 September 1947 meeting of "Dr. Pawsey's Solar Noise Group",[5] just prior to Pawsey's departure for the trip to the US and Europe. Under the rubric, "Solar Interferometer Research Programme—Bankstown—Treharne and Little" a detailed rationale was presented: (1) a determination of the sizes of the "bursts" (Type I storm bursts) as compared to the "enhanced radiation" (Type I background continuum). At that time, the expectation was that the bursts were smaller than the background during "noise storms", (2) the determination of the positions of the "bursts" and "enhanced radiation" sources in comparison with the positions of sunspots, (3) the determination of the correlation of optical flares with "outbursts" (likely Type II and Type IV), (4) a study of the thermal emission of the quiet sun[6] in both total intensity and polarized intensity, (5) the determination of the properties of the radio source in Cygnus and other "circumpolar" cosmic sources, and (6) the extension of this technique with additional frequencies 65, 200, 1,200 and 3,000 MHz—a project only partially carried out later in the 1950s by Paul Wild and colleagues.

In June 1948, there was a flurry of correspondence between the Cambridge and Sydney groups while Pawsey was in the UK. The relative merits of the sea-cliff interferometer and the Michelson (by now called "spaced" interferometer) were discussed by Pawsey, Ratcliffe, Ryle and F. Graham-Smith. Smith (1948) had prepared a report: "The Determination of the Positions of Discrete Galactic Sources".[7] The discussions applied, of course, to both cosmic noise sources ("radio stars", at this time) as well as solar noise, and went into the details of the work concurrently being done in Australia. Pawsey promptly sent Smith's report to his group.

The exchange of letters and the report from Smith of June 1948 set the stage for a follow-up discussion that ensued a year later. Clearly, the Cavendish report had had an impact on the RPL group, who were learning the details of spaced interferometry for the first time. By mid-1949, the Potts Hill interferometer was coming on-line. Payne-Scott and Mills worked on the theory of position determinations and had carried out a detailed analysis of the possible errors. On 28 June 1949, Pawsey wrote Ratcliffe[8]:

---

[5] NAA: C3830, A1/1/7.

[6] With the completed instrument, this observation was difficult due to limited sensitivity. The quiet sun has a total flux density of about 2×10^4 Jy at 100 MHz; the flux density sensitivity of the interferometer was a few thousand Jy for observations of a few hours and only a few hundred thousand Jy for the sub second observations. In addition the fringe size of about 40 arcmin was comparable to the size of the radio quiet sun of 40–50 arcmin; thus the interferometer response to the quiet sun was quite attenuated since the sun was "over-resolved".

[7] The report by F. Graham-Smith was not published until 4 years later under the title, "The Determination of the Position of a Radio Star", Smith 1952.

[8] NAA: C3830, A1/1/1 Part 4. The report "Notes on Interferometer Errors" (Payne-Scott and Mills 1949) was enclosed in the letter to Ratcliffe.

Some of our people have been working over the past few months using Ryle's spaced aerial interferometer technique to observe both sun and discrete cosmic sources. They have been impressed with the rather tricky nature of the possible sources of error in the case of attempts at precise position finding....might be a good thing to maintain rather close contact with Ryle and his group with the idea that each might be able to help the other. Ruby Payne-Scott and B.Y. Mills have prepared the enclosed notes which you might pass on to Ryle with a request from us for his collaboration. Incidentally the "dawn" [sea-cliff] technique is just as bad or worse but the errors are different.

The Payne-Scott and Mills report, "Notes on Interferometer Errors" contained an extensive discussion of errors due to both atmospheric and ionospheric refraction, as well as alignment and timing errors and the effects of changes in cable lengths due to temperature fluctuations. In this report, Mills provided a preliminary position for Cygnus A obtained at Potts Hill; observations of this northern source were quite difficult from Australia since the source only reached an elevation of 15° above the northern horizon. Both Ratcliffe and Ryle answered—the latter in detail.[9]

What is important to note is that the Australian group was successfully working out the engineering and physics questions for the creation of a Michelson interferometer, while the British group struggled with creating their own. While Pawsey had gone on his 13 month odyssey to the U.S. and Europe in order to bring credence to the work his people were doing in Australia, it is clear that he succeeded by the professional equality with which the groups were conferring by the time these letters were being exchanged.

Based on seven letters from Lindsay McCready and Don Yabsley to Pawsey[10] during the latter's trip in 1947–1948, additional facts about the planning of the swept-lobe interferometer, the choice of sites and the conflicts associated with Payne-Scott's participation can be established. During Pawsey's absence, numerous frustrations with equipment and personnel were experienced.

In mid-1948, Payne-Scott was highly motivated to start the 100 MHz interferometer project *before* the Hornsby research was finished; due to the doubts over the validity of this project, she thought there was a possibility that the new exciting Potts Hill work would begin without her being able to work full-time on the new interferometer. On 6 June 1948, McCready wrote Pawsey a troubled letter about the Potts Hill project.[11] On the one hand he was pleased with the negotiations with the Sydney Water Board: "[The] equipment losses thru [sic] vandalism not yet overcome but pretty close to it." On the other hand, McCready was worried whether Payne-Scott could handle all the engineering planning due to lack of time; she would possibly need "an army of assistants".

McCready thought the logical choice would be Don Yabsley, since Chris Christiansen would have the automatic multi-frequency recording apparatus working at 200, 600 and 1,200 MHz at Georges Heights, thus freeing up Yabsley. And furthermore, Yabsley and Payne-Scott worked well together. But Yabsley was quite worried about the "terrific amount of work" already on hand. McCready wrote that

---

[9] NAA: C3830, A1/1/1 Part 4.
[10] NAA: C4659, 8.
[11] Ibid.

Payne-Scott would not react well to this suggestion of joining forces with Yabsley. The situation was becoming chaotic for McCready:

> I have not discussed it with Ruby yet but I know her reaction will be violent hostility and complete assurances that Alec [Little] can do the work. I am not expecting you, 12,000 miles away, to make decisions or do anything. I will be my best and of course watch relations and the group's welfare.... you suggest in a future letter that there is a better chance of Ruby getting one or two important papers out quickly by concentrating entirely on the loose ends of her present work [Hornsby] and she should seriously consider letting someone else take on the interferometry at 100 Mc/s. I would prefer of course that you did not mention that I have written this.[12]

The group in Sydney was clearly floundering due to the missing leadership of Pawsey. Within a week, a welcome solution to the crisis arrived when Pawsey's hand-written letter of 29 May 1948 to Payne-Scott arrived (Chap. 9), urging her to tie up the unfinished research from the Hornsby site completely before beginning at Potts Hill.[13] Both McCready and Pawsey saw that the Hornsby project could be completed on time, allowing for an active participation of Payne-Scott in the planning of the new Potts Hill interferometer project.

Thus in the end Payne-Scott was able to finish the Hornsby research and could write the paper describing the properties of solar bursts before the end of 1948. More importantly, she did manage to maintain her control of the scientific leadership of the 100 MHz interferometer project, especially after Treharne left RPL sometime between April and July of 1948.

By 23 April 1948, McCready had described the adopted site to Pawsey. This location was the Potts Hill Reservoir site, controlled by the Sydney Water Board.[14] Given the problems with security at some of the field stations near Sydney, the fact that this location was "fenced and guarded" was a major advantage. Also the contact person for RPL was a Water Board engineer, H.A. Stowe, who was "most anxious to help us". He had taken a tour of the RPL; he knew "all about solar noise" and was even a ham radio operator. McCready pointed out that the location of the new site— between Lidcombe and Bankstown—only required 5 min longer to reach by car from the RPL at the Sydney University grounds, as compared to the travel time to reach the

---

[12] In the 6 June 1948 letter, McCready also seemed to blame himself for the conflicts which he had been unable to resolve. The excuses for the imbroglio were (1) Payne-Scott was very keen to start the new Potts Hill project and (2) the work at Hornsby was going nowhere. Even Bowen had suggested that she "fold it up and get into something more profitable". As explained in Chap. 9, the pessimism about the Hornsby endeavours eventually was shown to be unjustified. NAA: C4659, 8.

[13] NAA: C4659.

[14] Ibid. The history of the site selection and some aspects of the scientific research programs at the Potts Hill site were described by Frank L. Kerr (1918–2000) in the January 1953 issue of the *Sydney Water Board Journal*(Sullivan archive) Although by this date, the swept-lobe interferometer was no longer in use, the determination of solar noise positions from active regions was described in detail. The opening sentence of the article (Kerr 1953b) was: "In 1948, the Chief of the Division of Radiophysics at C.S.I.R.O., Dr. E.G. Bowen, sought and the Board was pleased to grant, permission for the Division to install certain equipment on the Board's land at Potts Hill for the investigation of solar radio activity... The Board has been very happy to have been able to provide the Division with the space and other activities needed."

field station (used earlier) at Georges Heights, north of the main Sydney harbour. In a later letter, McCready was quite pleased with the continued level of cooperation from the Sydney Water Board, as the negotiations continued for the use of the new site.[15] He did suggest that a prototype of the interferometer would be constructed at the site; later numerous engineering details still would need to be completed.

By mid-April 1948, Ruby Payne-Scott was associated with the project. McCready wrote Pawsey[16] on 23 April 1948 that Payne-Scott was "tapering off at Hornsby and orienting to [the new interferometer at Potts Hill]". As she wound down her activities at the Hornsby Valley field station, the PC minutes indicated that she was taking over the scientific leadership of the project as well as the development of an ingenious, simple calibration scheme. By the meeting of 30 April, only Payne-Scott and Little were mentioned; at the next meeting on 22 July 1948, the announcement was made that Treharne, the first engineer who had started work on this project in 1947, had left the project. We know that Payne-Scott's salary was raised on 1 July 1948 to £A700 per annum, a solid sign of her growing professional stature. Subsequently throughout the period from mid-1948 to the time of her resignation in July 1951, the reports on this project were given by Payne-Scott. Active observations were carried out in the period May 1949 to August 1950.

As Payne-Scott finished her work at Hornsby in September 1948, the move to the Potts Hill site was certainly more convenient for Payne-Scott. She and her husband Bill Hall were thinking of building a house in the nearby suburb of Oatley, which is about 14 km from Potts Hill.[17]

The layout of the Potts Hill field station (about 16 km from the centre of Sydney, near the Sydney suburbs of Regents Park, Chullora and Birrong) is shown in Fig. 10.1, taken from Wendt (2009), based on the site layout in the early 1950s. In Fig. 10.2, a RPL map from the planning stages is shown. The sketch, dated June 1948, gives the placement of aerials number 3 and 4, which were never constructed. An aerial view from the 1950s is shown in Fig. 10.3. The Potts Hill site was closed in 1962, when other RPL sites and use of the Parkes radio telescope became prominent—the 64 m telescope opened on 31 October 1961; Wendt (2008) has provided a thorough history of the site in the period 1948–1962.

By the time of the 22 July 1948 PC meeting,[18] Payne-Scott was clearly in charge of the new instrument. She presented again a summary of the purpose of the project and began to provide regular progress reports at the PC meetings. She urged other groups to consider the use of the Potts Hill field station for other projects; in the end a number of additional groups did move to Potts Hill (e.g., the Christiansen grating array, the prototype Mills Cross, the previously used Georges Heights antenna,

---

[15] NAA: C4659, 8; 6 June 1946.

[16] Ibid.

[17] This house was actually started in 1950–1951, with major construction done by Payne-Scott and her husband. Bill and Ruby Hall (she started to use her married name when she retired in 1951) moved to the new house in August 1951.

[18] NAA: C3830, B2/2, Part 1.

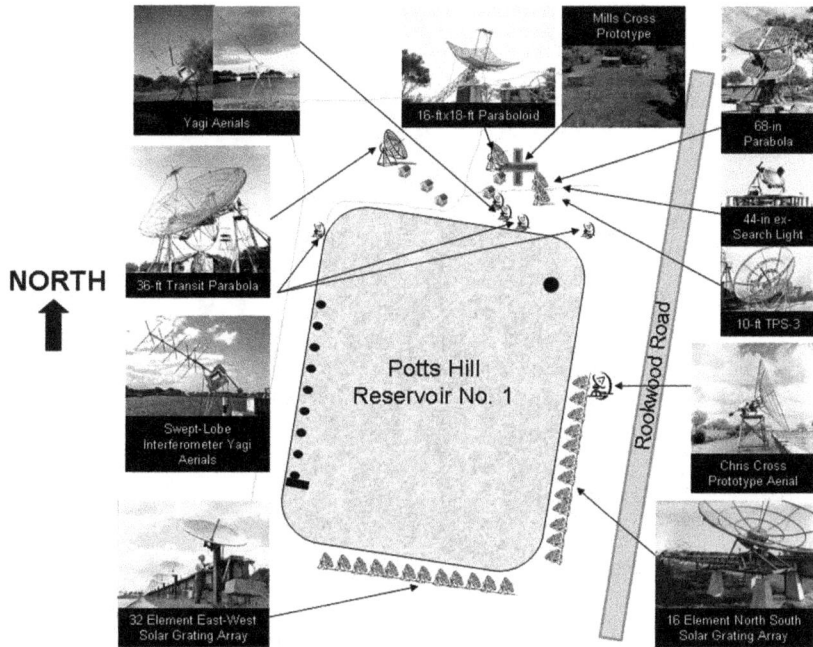

Fig. 10.1 Detailed site layout of the Potts Hill Reservoir RPL field station in the early 1950s. The instruments used by Payne-Scott and Little were the swept-lobe interferometer and the solar radio noise Yagis. The east-west solar grating array used by Christiansen et al. was completed in about November 1951 and the north-south array in mid-1953; the 36 ft. (11 m) transit telescope was completed in January 1953. The east-west extent of Reservoir Number 1 is about 280 m (The insert photos are from the CSIRO Radio Astronomy Image Archive. Figure provided with permission by Harry Wendt)

numerous small high frequency antennas and the new 36 ft. (11 m) transit antenna that was constructed in 1952–53[19]).

On 16 August 1948, McCready wrote Pawsey extolling the qualities of the Potts Hill location: "Potts Hill field station is [a] beautiful sight with all the wattle trees in full bloom." (It was early spring in Australia.) On this same date, Don Yabsley wrote Pawsey with an optimistic assessment of the status of the new instrument; first lobes (or first fringes) had been obtained by Payne-Scott and Little; the group was beginning to worry about methods to improve the calibration of the instrument.[20] At the PC meeting on 6 September 1948, Payne-Scott gave a progress report on the initial tests of the new interferometer (see Additional Notes, No. 1, end of this chapter).

After Pawsey returned at the end of October from his overseas trip, three meetings of the PC were held in November. Pawsey was catching up on the activities of the various groups at RPL; in addition he gave several talks about his

---

[19] Wendt (2008) has provided a detailed description of the various projects carried out at Potts Hill in the years 1948–1962.
[20] Ibid.

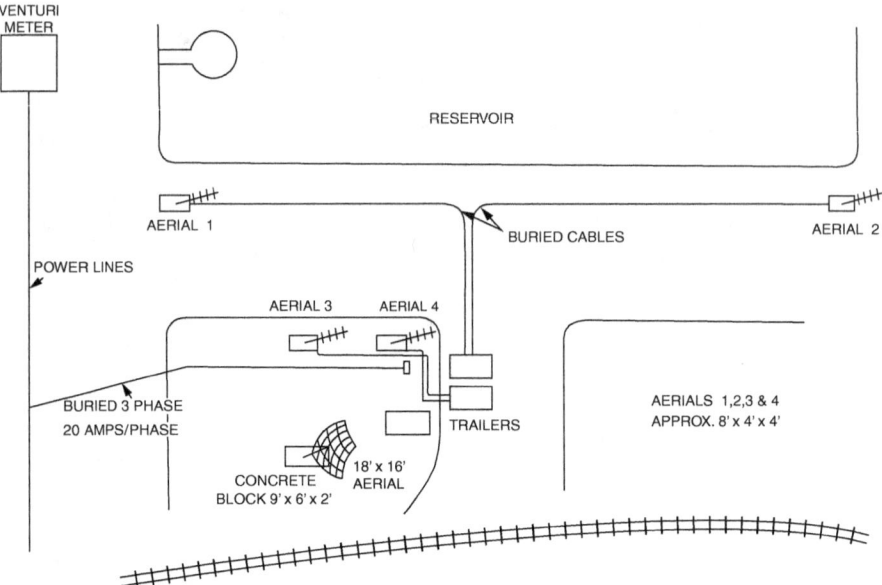

SKETCH PLAN OF PROPOSED SOLAR NOISE STATION AT POTT'S HILL

Fig. 10.2 A sketch map of the planned Potts Hill RPL field station. South is to the *top* of the map, East to the *left*. Note that in this plan there were four antennas for the planned swept-lobe interferometer at 97 MHz. In the end only three were built with an East-West orientation (Fig. 10.8). From June 1948 (-National Archives of Australia-NAA: A/1/1/1, Part 3)

impressions of North American and European radio astronomy groups. He brought a new sense of urgency to the activities of the radio astronomy group at RPL.

Early Testing of the 97 MHz Interferometer

After Pawsey's return in the last days of October 1948, an increasing energy could be detected in the activities of the newly named Radio Astronomy Group.[21] On 8 February 1949 at a PC meeting,[22] Payne-Scott reported that 97 MHz—changed

---

[21] The change of name has been described in Chap. 9. As discussed by Sullivan (2009), the first use of the term **radio astronomy** was by Pawsey in January 1948 in a letter and by Ryle in April 1948 at a meeting of the Royal Astronomical Society (he did use the term inside inverted commas, however). Ryle wrote: "More refined observations in this new 'radio astronomy' should provide us with much new information on such processes." The new term for this expanding research field caught on quickly.

[22] NAA: C3830, B2/2, Part 1.

Early Testing of the 97 MHz Interferometer   175

Fig. 10.3 Aerial photo from 19 March 1954 taken from the north looking south. The main portion of the RPL field station is shown in the foreground of Reservoir No. 1. The 36 ft. (11 m) transit telescope is visible as well as the rail line in the foreground. The rail line was used to bring coal to the nearby power plant; the smoke stacks are visible in Fig. 10.5 (CSIRO Radio Astronomy Image Archive B3253-1)

from 100 MHz to avoid external interference—interferometry was ready to start at Potts Hill and the Yagis for continuous monitoring had been moved from Hornsby (62 MHz) and Dover Heights (97 MHz). By the time of the next meeting of the PC committee on 7 March 1949, Payne-Scott had exciting news; continuous recording with the two Yagis was in operation and some successful observations had been made with the interferometer. "On 1 day, while unpolarized radiation [Type III] was being received all day, an isolated burst [maybe a Type I] appeared to come from a quite different place on the sun." In Figs. 10.4 and 10.5, two photos of the swept-lobe interferometer are shown. The former shows the details of the western-most Yagi with vertical and horizontal polarization; the latter shows the layout of the three elements next to the northern boundary of the reservoir. In addition there was a total power Yagi operating at 98 MHz, used to calibrate the data from the swept-lobe interferometer at 97 MHz. This polarization state of this single Yagi was also switched at a rate of two times per minutes to check the polarization state of the solar radiation at radio frequencies.

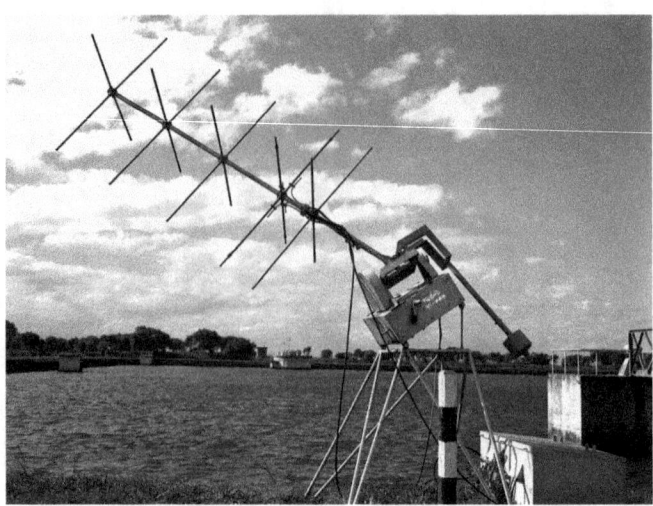

Fig. 10.4 The western most antenna of the three elements of the swept-lobe interferometer; this 97 MHz radio telescope determined the motions of solar bursts in a one-dimensional scan many times per second. Used by Payne-Scott and Little in 1949–1950. With the crossed dipoles circular and linear polarization could be determined. From 28 July 1950 (CSIRO Radio Astronomy Image Archive 2217)

POTT'S HILL INTERFEROMETER SITE LOOKING NORTH-EAST
PLATE I

Fig. 10.5 Photo of the swept-lobe interferometer site on 12 November 1950. The Western antenna is clearly visible; the signals from the three antennas were combined in the "Hut" (CSIRO Radio Astronomy Image Archive 2312)

In Fig. 2.14, one of the single total power Yagis at 65 MHz is shown, located close to the central hut; this antenna was used to determine the total emission from the sun, presumably while the interferometer was in use. A record from this

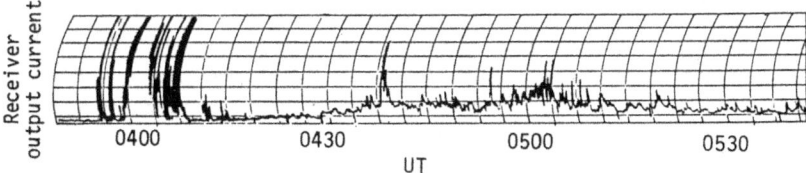

Fig. 10.6 Example of the type of data obtained at 62 MHz, taken from Wild (1985). Wild wrote: "Single frequency (62 MHz) record of a flare-associated outburst followed by a prolonged period of storm [Type I] on 4 March 1949" (CSIRO Radio Astronomy Image Archive B3045)

instrument is shown in Fig. 10.6, an example used by Wild in 1985 to illustrate the nature of fixed frequency solar noise observations in circa 1950.

By the time of the 11 April 1949 meeting of the Radio Astronomy Committee,[23] Payne-Scott was pleased with preliminary observations; two construction projects were underway, adding both circular polarizations as well as a third aerial on the east-west axis—the latter was required to sort out lobe ambiguities (i.e., to identify the correct lobe; without this feature the location of the solar radio emission source could arise from at least two possible positions). Mills was also busy with initial observations on three of the cosmic radio sources looking for polarization with the 97 MHz interferometer; none showed any detectable polarization signals. He already realised that the ability of the Potts Hill interferometer to detect weaker radio objects was not suitable for systematic observations of cosmic radio sources, due to limited sensitivity.[24] A few weeks later[25] on 28 April 1949, Payne-Scott was complaining about the lack of solar activity even though the continuous recording (two single Yagis) continued. She was then working on the scheme to achieve an accurate phase and amplitude calibration for the system; by 16 June 1949, a calibration scheme was almost in place. The first main results were then available; a Type I noise storm (continuum plus noise bursts) was located 300,000 km above the visible photosphere of the sun.

Not surprisingly, Payne-Scott's salary was raised by the standard £A25 up to £A725 per annum on 1 July 1949.

Ruby was by now a senior member of the scientific staff. She also was still prone to nearly rude behaviour when she found her colleagues' work to be less than

---

[23] NAA: C3830, B2/2, Part 2.

[24] At this meeting, Mills and Thomas announced the results of their first Cygnus A position determinations. Discrepancies were found with both Bolton's and Ryle's previously determined positions. At the next meeting on 7 July 1949, Mills announced that he was giving up on the Potts Hill interferometer for cosmic source positional determinations due to the limited sensitivity. The flux density of Cygnus A at 100 MHz was observed to be 15,000 Jy with a measured uncertainty of 3,000 Jy. In 1951, he and Thomas published the Cygnus A position, followed by a refinement by Mills (1952) with an error 0.5 by 1.5 arcmin; even with this small error rectangle the optical object eluded a definitive detection. The optical identification was only made by Baade and Minkowski (1954) based on the more accurate position determined by Smith (1951). It is striking that the position determined earlier by Mills and Thomas was quite close to the more accurate determination made by Smith.

[25] NAA: C3830, B2/2, Part 2.

excellent. Chris Christiansen, who worked with her at Potts Hill in 1948–1951, recalls in a 1997 letter to Goss a memorably ornery exchange at the end of a meeting of the Propagation Committee on 28 July 1949. At the meeting, John Bolton and Kevin Westfold presented a progress report on their observations of the entire southern sky at 100 MHz with a resolution of 17°.[26] Bolton summarised the observations while Westfold continued with a lengthy discussion of the observed radio background emission of the Milky Way galaxy. At that time, before the emission mechanism was understood to be synchrotron emission, the intensity of the background radiation was unexplained. Westfold demonstrated that free-free emission could not provide an interpretation. He then tried to provide an explanation using radiation from *visible* stars, which was rejected because of the implausibly high effective temperatures that would be required due to the small filling factors. The final model, which had been favoured by Westfold and Bolton, explained the emission as originating from galactic radio stars with large angular sizes. This model could apparently fit the data but strong objections were raised at the meeting. As secretary of the committee, Chris Christiansen wrote in the minutes, "The detectability of [radio] stars was to be treated but the validity of the initial assumptions of the analysis were challenged by the members of the Committee and the section was withdrawn for modification."

Mills and Christiansen have both provided us with parallel accounts of what happened. According to Christiansen's letter,

Ruby said "Kevin, where did you get that?" Kevin said, "From John Bolton." Ruby said, "well, it is utter nonsense" upon which Kevin said, "then so is my paper" and sat down. The meeting ended.

Mills has confirmed this exchange in detail in a letter to Goss (see Additional Notes, No. 2, end of this chapter).

At the 15 August 1949 Radio Astronomy (the Propagation Committee had a new name) meeting, Payne-Scott announced that the third antenna was almost ready for use. The recording mechanism using a film camera for strong bursts was then in operation. The next report on the Potts Hill interferometer occurred at the 17 November 1949 meeting; the complete system was working with three antennas as well as crossed polarized feeds to determine the complete state of polarization. Payne-Scott reported that both noise storms (later Type I) and unpolarized (later Type III) bursts were being observed. After reporting a large outburst at the 7 February 1950 meeting (surprisingly with no visually prominent sunspots, perhaps the relevant active region was over the limb of the sun), a detailed report of a comparison of 97 MHz and 600 MHz (from Don Yabsley) data was carried out for solar bursts from dates shortly preceding the meeting of 14 March 1950. At some periods there was good agreement, at other times poor. A thorough summary (Payne-Scott 1950) of the status of the results from the swept-lobe interferometer (as well as a summary of the Hornsby results) was given in late January 1950 by

---

[26] Described in three papers in the *Australian Journal of Scientific Research* (Bolton and Westfold 1950a, b, 1951).

Payne-Scott at ANCORS (Australian National Committee for Radio Science), an Australian meeting of URSI (International Union of Radio Science).

A short report was given by Payne-Scott on the 97 MHz interferometer at a heated meeting of the Radio Astronomy Committee on 27 July 1950.[27] Wild gave a long summary of the first campaign from the Penrith instrument, the swept-frequency spectrograph (Wild and McCready 1950). During a detailed discussion of the revolutionary results of the short campaign from February to June 1949 (249 hours) reported later on by Wild and McCready (1950) and Wild (1950a, b, c, 1951), Christiansen (as secretary of the committee) reported:

> During the discussion on this report [of Wild] the use of the term "isolated burst" [meaning Type III] was responsible for an outburst from Miss Payne-Scott.[28]

Wild reported on his results concerning Type I, II and III bursts, making the strong claim for the latter that group retardation did not explain their spectral characteristics. He suggested instead (as Payne-Scott had in the 9 February 1948 colloquium) that outward moving disturbances at 10 % of the speed of light were required. No record was given as to how this remarkable claim was received by Wild's colleagues. We can assume that the improved nature of Wild's determination of these speeds lead to an increased credibility.

Payne-Scott's career was progressing steadily, and on 1 July 1950 her salary was raised to £A750 per annum—the last raise she would ever receive prior to her forced resignation on 20 July 1951 due to the upcoming birth of her son.

At subsequent meetings of the Radio Astronomy Committee[29] in early 1951, Payne-Scott gave detailed summaries of the results from the 97 MHz interferometer. A cursory description of the planned publications was given on 23 January 1951, followed by a detailed description of all the results for noise storms (later Type I) and outbursts (interpreted at the time as the slow moving Type II outbursts, but in fact later shown to be moving Type IV outbursts, Type IVM[30]) on 6 February 1951. For the Type III bursts, Payne-Scott remarked that bursts appeared to originate around sunspots but with a large positional scatter; this result was never published. On 8 May 1951, the minutes reported that discussions had been held at Mt. Stromlo in view of working more closely with the group of astronomers there. The minutes stated that "several people here went to a colloquium there. This might be reciprocated. Ruby Payne-Scott to talk early in July". It is not clear if she ever gave the talk, as she resigned on 20 July 1951. Ron Bracewell was the secretary of the Radio Astronomy Committee on 23 July 1951, 3 days after Payne-Scott's resignation from CSIRO. No mention was made of her departure in this or later minutes of the Radio Astronomy Committee.

---

[27] NAA: C3830, B2/2, Part 2.

[28] Ibid. Payne-Scott's preferred term was "unpolarized".

[29] Ibid.

[30] The "M" provides the distinction for Type IVM outbursts in contrast to the stationary component of Type IV continuum. The latter term includes all continuum emission (in the metre, decimetre and centimetre regions) produced during a flare event.

Publication of the Results at 97 MHz: Paper I- The Instrument

The first in a set of three publications, "The Position and Movement on the Solar Disk of Sources of Radiation at a Frequency of 97 Mc/s. I. Equipment" provided a description of the instrument (Little and Payne-Scott: December 1951).[31] The paper was submitted about a month before Payne-Scott's resignation on 20 July 1951. The second and third papers were submitted on the same date, 1 July 1951.

This swept-lobe instrument remained a unique instrument until the late 1950s when Wild et al. (1959a, b) extended the Dapto swept-frequency interferometer to include the facility to determine positions of solar bursts in the east-west direction with a precision of about 1 arcmin over a frequency range of 40–70 MHz.

A restatement of the motivation for the construction of the new interferometer was described by Little and Payne-Scott (1951):

> In these existing interferometers [previous instruments] the position of the source is calculated from the interference pattern, which is produced by motion of the source through the lobes patterns of the aerials. But if, as in the case of bursts, the source has a life of only a few seconds, it will not have moved sufficiently during this time to produce the required interference pattern. Consequently, these methods are not directly applicable where the disturbance is of short duration or is rapidly changing its position, and this paper will describe a modified spaced-aerial interferometer which has been built to measure the position and polarization of such a variable component.

In this system the lobes were swept over the sun 25 times a second with two long baselines (240–280.5 m); the use of these two baselines with fringe spacing of about 44–38 arcmin enabled lobe ambiguities to be resolved (the unique position was required to be within a few solar radii of the solar centre). The simple theory of the two element interferometer was sketched in Fig. 10.7. (See Fig. 10.8 for the lay-out and a simple sketch of the layout of the interferometer.) The quiet sun, with a size of 35–40 arcmin, was not detected with this fringe spacing; in addition, the sensitivity was inadequate. In addition the polarization state of the radiation was determined once per second. The E-W position of the bursts and outbursts could be determined with a precision of about 2 arcmin.

The observing sequence was to observe on the 240 m baseline with parallel hand correlation (vertical versus vertical, etc.) followed by cross hand correlation (vertical versus horizontal, etc.) in order to determine the polarization state of the solar bursts. Finally the parallel hand correlations were determined on the 280.5 m baseline.

The total time to complete the sequence was about 1 s. A film recorder was used (a modified 16 mm ciné-camera) which was triggered automatically when the

---

[31] RPP 135. The first draft was prepared on 10 August 1950 (close to the cessation of data taking) and the complete manuscript was sent to the *Australian Journal of Scientific Research Series A-Physical Sciences* on 28 June 1951. *RPP 136* was paper II in the series and prepared and submitted a week later. The papers appeared back to back in the December 1951 journal on pages 489–525.

Publication of the Results at 97 MHz: Paper I- The Instrument

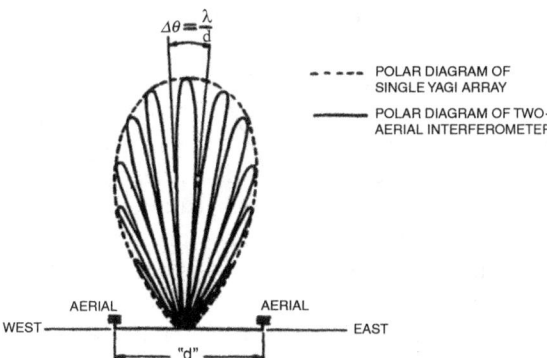

Fig. 10.7 Little and Payne-Scott provided a tutorial on interferometry in their publication on the equipment used for the swept-lobe interferometer. (Little and Payne-Scott 1951, *Australian Journal of Scientific Research, Series A*, vol.4, page 489, 1951, "The Position and Movement on the Solar Disk of Sources of Radiation at a Frequency of 97 Mc/s. I. Equipment", Fig. 2) (CSIRO Publishing, Copyright © CSIRO http://www.publish.csiro.au/nid.17.htm)

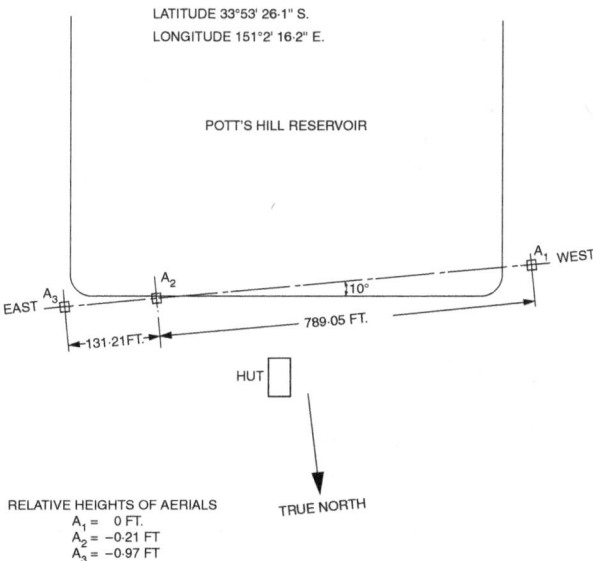

Fig. 10.8 Layout of the interferometer used for the swept-lobe 97 MHz instrument at Potts Hill. The lobe spacing for the long baseline of 280.5 m was 38 arcmin; the lobes were swept across the sun 25 times per second; the circular polarization state was determined once per second. The novel calibration method was developed by Payne-Scott. (Little and Payne-Scott 1951, Fig. 3) (CSIRO Publishing, Copyright © CSIRO http://www.publish.csiro.au/nid.17.htm)

intensity of the radiation reached a suitably high level. For circumstances when there was a steady level of solar bursts ("enhanced radiation"), the system could be used as a normal fixed lobe interferometer (the mode of observing used by Mills and Thomas (1951) for the Cygnus A observations). The flux density uncertainty in this case was reduced due to the longer integration times, a few thousand Jy. Any state of the radiation polarization could be determined, i.e. random, linear, elliptical or circular. In fact Payne-Scott had worked out the equations for the determination of the polarization state (modern usage would be "Stokes" parameters to describe the polarization state of the radiation) in 1948.

The calibration of this instrument was quite straightforward; due to the low sensitivity, the use of known point sources for calibration was not possible as is the case for modern interferometers. A large part of the Little and Payne-Scott paper consisted of a detailed description of the determinations of the positions, flux densities and sizes of the solar sources.[32] Payne-Scott designed the simple calibration system, which relied on a set of precise calibration cables and the injection of test signals from portable Yagi systems. The observations occurred every week-day for about 2 hours on each side of meridian transit of the sun; during periods of high solar activity, observations were continued during the weekend. Data were taken for 30 noise storms (Type I storm, Paper II), 6 outbursts (Possibly Type IV events, Paper III) and 25 unpolarized bursts ("isolated" bursts, Type III). Later Little tried to analyse the Type III bursts; however the positions did not yield consistent results and these results were never published (Sullivan 2009). The detailed description of the direct determination of the motions of the Type III bursts had to wait for the enhanced Dapto interferometer data presented by Wild et al. (1959a, b).

Paper II- Noise Storms (Type I)

In paper II (Payne-Scott and Little 1951, "II. Noise Storms"),[33] the authors described their results of observations of 30 noise storms (Type I continuum), with detailed descriptions of the progress of six of the storms. Payne-Scott and Little determined in an elegant fashion (the reduction of these records from the movie must have taken Payne-Scott many hours of tedious work) that the Type I continuum originated high in the corona over large sunspots. By locating the position of the source of radio emission for these six storms of long duration over

---

[32] It is likely that the publication by Little and Payne-Scott contained the first published recognition that atmospheric refraction can be neglected (to first order) for a conventional Michelson interferometer. For a plane parallel atmosphere, the equality of the phase change in both arms of the interferometer leads to a cancelation of the effect : "A uniform plane sheet of refractive material does not introduce any differential path difference between parallel rays, and hence the refraction due to an uniformly stratified regions above a plane Earth can be ignored..."

[33] NAA: C3830, B2/2, Part 2, Radio Astronomy Committee of 18 July 1951: All three papers in the series had been submitted.

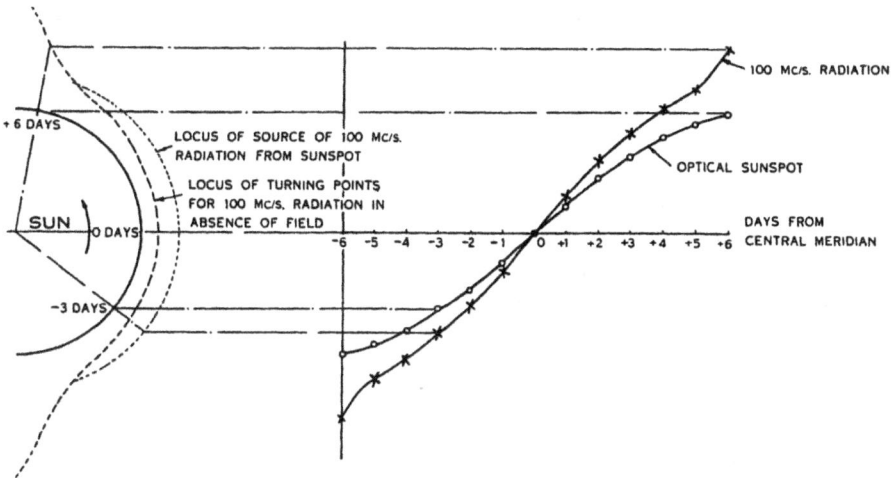

Fig. 10.9 The relation between the motions of the 97 MHz Type I solar storms and the associated sunspot group over a solar rotation proved that the radio emission arose high in the corona; the observed range was 0.3–1 times the photospheric radius above the sun's visible surface. This deduction is one of the more important conclusions of the collaboration of Payne-Scott and Little. (Payne-Scott and Little 1951, *Australian Journal of Scientific Research, Series A*, vol.4, page 508, 1951, "The Position and Movement on the Solar Disk of Sources of Radiation at a Frequency of 97 Mc/s. II. Noise Storms", Fig. 8) (CSIRO Publishing, Copyright © CSIRO http://www.publish.csiro.au/nid.17.htm)

periods of 10–16 days, Payne-Scott and Little could show that the angular rate of change of position for the radio bursts was faster than that of the optical sunspots (Fig. 10.9). If the assumption was made that the source of the noise-storm lay at some height above the visible surface of the sun (near the position where the plasma frequency was 97 MHz) and that the noise storm was radially displaced above the relevant spot group, it was then possible to calculate the radial displacement of the two. For most of the observations, this displacement was found to be 0.3–0.4 radii above the sun's visible surface. For the storms starting 10 Feb 1950 and 10 June 1950, the displacement was found to be 0.8–1 solar radii. The disagreements with theory for the location of the plasma frequency (the "turning points") were attributed to the suggestion that the expected values for the coronal electron density might well be roughly a factor of two larger in these periods, at least near solar maximum.

Next Payne-Scott and Little described the detailed relation between sunspots and noise storms. They carried out a comparison of both the number of sunspots and the maximum area of the largest sunspot with the associated Type I bursts. The Mt. Wilson optical data were used for the comparison. The major result was: "…the size of the largest spot in the group is a more certain indication than the size of the whole group of the chance that a noise storm will be associated with the group". Thus a large spot group would not likely produce a noise storm unless it contained at least a single large spot with area of at least 0.04 %. Since the close relationship between the size of a sunspot and the associated magnetic field was well known at this time, the direct association of the Type I bursts and the presence of a strong magnetic field was suggested.

The final result described in detail was the behaviour of the circular polarization of Type I bursts. The radiation tended to be circularly polarized with linear and elliptical states occurring only on rare occasions. The radiation was often found to be nearly 100 % polarized (i.e. RH, right-handed or left-handed, LH). The direction of the sense of rotation of the circular polarization was found to depend on the magnetic polarity of the associated spot, with LH being present when the largest spot was a north seeking pole (positive polarity) and RH for a south seeking pole. (Positive indicated that the magnetic field was directed towards the observer.) The conclusion of the Payne-Scott and Little (1951) paper was: "The observations quoted here point to magnetic fields as the important factor in the production of solar noise storms at a frequency of 97 Mc/s." Then the paper ended with a comparison of the data with theories of the emission mechanisms, prevalent in 1951. The continuation of numerous controversies with the Cavendish astronomers can be anticipated in the final sentences of the paper:

> Unfortunately most of the theories put forward to explain the origin of storm radiation do not lead to any definite predictions of region of origin, relation to sunspots, or polarization, and hence are not capable of verification or refutation by the observations recorded here...The exciting mechanism must then be non-thermal. ... Ryle (1948) has suggested that storm radiation is due to local temperature increases, but the heights of origin observed seem much greater than his theory would indicate and the polarization appears to be in the wrong direction.[34]

The importance of this publication concerning the detailed description of the angular and polarization properties of the Type I bursts observed from May 1949 to August 1950 can be judged by two major reviews published two and then 34 years later.

In 1953, the influential review article, "Solar Radio Emission", (Chap. 8, by Pawsey and Smerd) was published in the book *The Sun* (*The Solar System* series) edited by G.P. Kuiper and B.M. Middlehurst. This article of 66 pages summarised the pioneering work of the RPL group from 1945 to 1952; the article was widely quoted and had a major impact in the astronomy world. This was one of the most thorough of the early review articles concerning solar radio astronomy. In this article Pawsey and Smerd summarised in detail the Payne-Scott and Little interferometer results. They pointed out that the 97 MHz data indicated that the sense of rotation of the circular polarization corresponded to the ordinary mode of propagation,[35] if the magnetic field of the sunspot extended into the corona. Pawsey and

---

[34] Sullivan (2009) in Chap. 14, "The Radio Sun", discusses the *thermal* theory proposed by Ryle to explain solar bursts. The problem was that a fatal flaw was incorporated into the theory; he had assumed that the collision cross sections and rates of emission for the high temperatures of the solar corona were the same as the earth's ionosphere at a modest temperature of only 300 K.

[35] The magnetoionic theory deals with the propagation of waves in a cold magnetised plasma; two solutions of the equations are possible, the ordinary and extraordinary. For the ordinary mode, the electric field accelerates parallel to the magnetic field, while for the extraordinary mode the

Smerd then discussed the significance of the high intensities observed by the interferometer. The true brightness temperatures were uncertain since the angular sizes were generally unknown; upper limits to the angular sizes of several arcmin could be determined. Thus typical brightness temperatures of 10^8–10^{10} K were common. Based on the burst duration of only seconds, they also concluded that a simple thermal origin was not likely and that "enhanced radiation probably originates as coherent radiation from groups of charged particles".

In fact after observations over several decades by numerous solar radio astronomers, Kai et al. (1985) in the monograph "Solar Radiophysics" wrote:

> Although Type I storms have been extensively studied for more than 30 years [written in 1985], it is only in recent times that any convincing theories have been advanced to explain their occurrence. It is now generally accepted that Type I bursts must be some form of fundamental frequency plasma emission.

Pawsey and Smerd then reiterated the importance of the threshold effect (sunspot size related to the presence of Type I noise storms) discovered by Payne-Scott and Little. Pawsey and Smerd emphasised that this was further evidence that if the threshold were surpassed, Type I emission would be emitted due to the presence of a high magnetic field.

Kai et al. described the three major conclusions of the Payne-Scott and Little 1951 publication: (1) the determination of the source height above the visible sun, (2) the association of polarization with sunspots indicating that the storms are polarized in the ordinary wave mode and (3) the association of the Type I bursts with the prominent sunspot in a group. Kai et al. wrote: "It is now generally accepted that large well-developed spot groups tend on the average to give rise to Type I storms more frequently than smaller spot groups."

Paper III- Outbursts. Type IV

The final paper (RPP 137) in the series of Payne-Scott and Little was "III. Outbursts" (1952). The date of reception at the journal (20 July 1951) was noteworthy—the final day of Payne-Scott's employment at CSIRO, Division of Radiophysics. At the time of writing the paper, the authors made the assumption that these six outbursts were "outbursts" (later Type II bursts). However, as Wild (1968) has pointed out, the observations reported were in fact Type IVM (moving Type IV events) bursts, the rarest of the burst and outburst types.[36] The evidence

---

acceleration is perpendicular to the magnetic field. For the ordinary wave the magnetic field has no influence on the motions of the electrons, while for the extraordinary wave the motions of the electrons are modified.

[36] Stewart (1985) has pointed out that in the years 1966–1980 the Culgoora Radioheliograph only recorded 56 Type IVM bursts, in comparison to 560 Type II events. The stationary components of Type IV events are associated with non-storm, flare-related continuum emission (Robinson 1985). Indeed the Type IV class is confusing!

from the Potts Hill 1949–1950 data was not consistent with Type II bursts: (1) the movement took place later in the lifetime of the disturbance than was expected for a Type II burst and (2) the motions observed at a fixed frequency were not expected for Type II bursts, since only one level of the corona would radiate at this frequency, according to the plasma emission hypothesis. The Type IV event is now known to occur after a major solar flare (about 20 % of Type II events are followed by an associated Type IV) and lasts for some tens of minutes, with some durations of up to 2 hours. The radio spectrum of Type IV bursts is characterised by a smooth continuum possibly due to the synchrotron process. The typical angular size has been determined to be 8–12 arcmin (at 169 MHz, Pick and Vilmer 2008). The solar source associated with the moving Type IV bursts travels outwards through the corona after a solar flare or an eruptive prominence mass ejection.

In this paper by Payne-Scott and Little, the velocities deduced were comparable to those of Type II outbursts, as is now known to be the case for Type IV moving bursts (in excess of some hundreds of km/s). The moving Type IV bursts (IVM) were only recognised a few years later based on data obtained with the 32 element grating interferometer at Nançay in France. Boischot and Denisse (1957) suggested that the emission mechanism was synchrotron radiation from electrons with energies in the range of 0.1–1,000,000 electron volts.[37] Subsequent Sydney observations by Wild, Sheridan and Trent (1959b) showed that the Type II bursts as observed at *different* frequencies did arise from different positions in the corona, as expected from the plasma hypothesis. For the moving Type IV events all frequencies arose from about the same position; the position did move outwards throughout the period of the event. Both the height and the motions of the Type IV events were thus inconsistent with the origin via plasma waves. It was soon found by later investigations in the 1950s that the radiation of the Type IV events was partially circularly polarized; this was also observed in the Potts Hill data from 1949 to 1950. Based on the relation to the magnetic field of the leading sunspot of the active region, the sense of polarization was found to correspond to the extraordinary mode of propagation.

Nevertheless, the Potts-Hill determination of the velocities of the Type IVM events in 1949 and 1950 provided an important characterisation of the velocities, positions and polarization of these events of Type II followed by Type IVM. Often these events were associated with terrestrial aurora and magnetic storms about a day later. Six large events were observed with five being associated with large optical flares and short-wave fade outs (likely based on observations of short wave broadcast stations near 10 MHz) (5 September 1949, 17 and 22 February 1950, and 5 and 11 August 1950; the latter dates are close to the dates of the cessation of

---

[37] The radiation had the characteristics of a smooth continuum, consistent with the gyro-synchrotron theory. However, a number of problems existed with the model. Stewart (1985) has summarised the pros and cons of the gyro-synchrotron model; a major problem was the presence of substantial circular polarization. He suggested a hybrid model consisting of second-harmonic plasma emission followed by gyro-synchrotron emission. On the other hand, Smerd and Dulk (1971) made a strong case for gyro-synchrotron emission.

observations at Potts Hill). The additional event of 29 June 1950 had no optical data; short-wave fade outs were observed; all of these outbursts were associated with sunspots that were about a factor of 4–5 smaller in area compared to the major solar events of February 1946 and March 1947. For two of the events (5 August and 22 February), sudden commencement magnetic storms were observed by the Watheroo Observatory in Western Australia.

By a stroke of good fortune, Ron Bracewell had recently returned from the UK where he had obtained a PhD from Cambridge in late 1949, where he also had worked with Ratcliffe carrying out ionospheric research. He was frustrated with the continued isolation of his RPL colleagues in Australia; he began to write a series of five short articles in the UK journal *Observatory*. The purpose was to provide publicity for the Australian results. Bracewell was convinced that the RPL group was suffering from isolation within the world astronomical community. The *Observatory* articles started in October 1950 and ran through August 1954. The first *Observatory* article provided advance publicity for the swept-lobe interferometer at Potts Hill. Bracewell's article (Bracewell 1950) was "An Instrumental Development in Radio Astronomy", a title that gave few clues to the subject discussed. The new 97 MHz swept-lobe interferometer was discussed in the one page article, written in a succinct but clear style. Bracewell wrote:

> A new instrument has been developed and put into use by Ruby Payne-Scott and A.G. Little ... which can determine the place of origin and polarization of bursts of radio energy from the Sun, virtually instantaneously. It is thus capable of locating the source of the radio energy emitted at the time of solar flares, and following its movement through the solar atmosphere. That this instrument is a triumph for radio astronomy will be readily agreed on contrasting the resolving powers obtainable at optical and radio wave-lengths.

The figure shown in the publication by Bracewell shows the dramatic motions of a Type IVM event on 17 February 1950. As the event was initially observed, a Type II radio outburst may have accompanied an optical flare. It shows that the radio emission began near the flare at a displacement of about 0.5 solar radii; the right ascension of the object was followed for the next 30 min as the radio emission moved to the west ending up high in the corona after crossing the solar limb. The final position (when the radio emission was no longer detected) occurred at a position more than a half a solar radius beyond the solar limb. The polarization state began as random polarization (possibly the Type II event) and evolved into left hand circular polarization (the Type IV event) at a time of 0150 UT. The optical flare reached a maximum at about 0210 UT and ended 50 min later. This data were discussed in detail by Payne-Scott and Little in paper III; the Bracewell publication was likely effective in providing publicity for a wider audience, since the final publication of the Payne-Scott and Little paper was to appear more than a year and a half later in the *Australian Journal of Scientific Research*, a publication with limited circulation in the scientific world. The figure for this event of 17 February 1950 as presented in the publication of Payne-Scott and Little (1952) is shown in Fig. 10.10.

Additional observations were reported by Payne-Scott and Little for five additional outbursts. As Pawsey and Smerd (1953, Kuiper and Middlehurst, eds.) pointed out in their review article: "This expectation [the movement of the

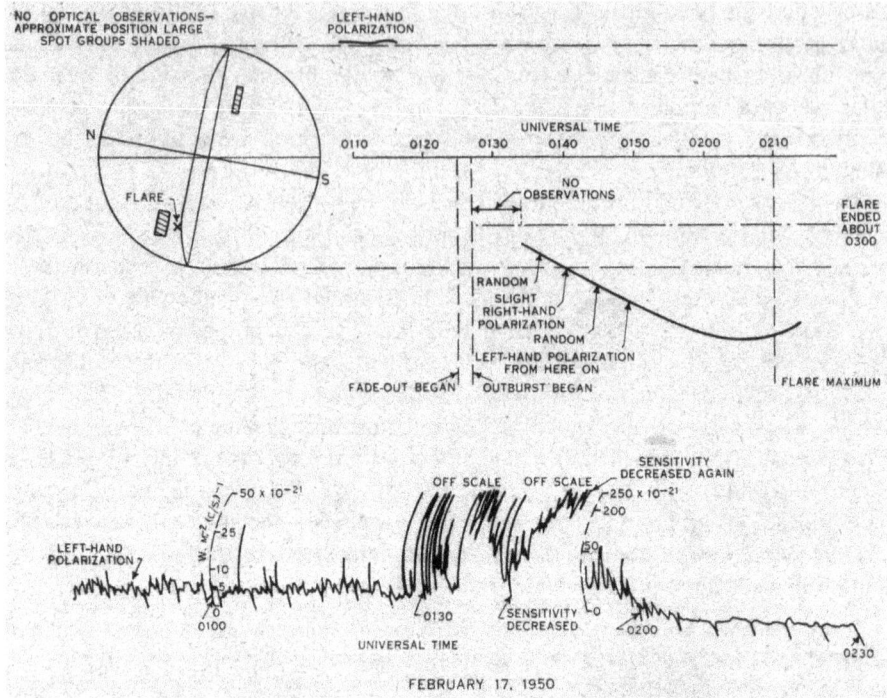

Fig. 10.10 The dramatic outburst (likely Type IVM) of 17 February 1950 showed that the solar emission position moved a projected distance of a solar radius in a time span of only 30 min. A prominent flare began a few minutes before the radio observations began. The total power data at the *bottom* of the figure were observed with a single 98 MHz Yagi (similar to the 62 MHz Yagi in Fig. 9.6a). (Payne-Scott and Little 1952, *Australian Journal of Scientific Research, Series A*, vol.5, page 32, 1952, "The Position and Movement on the Solar Disk of Sources of Radiation at a Frequency of 97 Mc/s. III. Outbursts", Fig. 1b) (CSIRO Publishing, Copyright © CSIRO http://www.publish.csiro.au/nid.17.htm)

outbursts towards the edge of the sun for non-central flare events] has been beautifully verified in the 'swept-lobe' interferometer observations of Payne-Scott and Little (1952)." The main conclusion of the Payne-Scott and Little outbursts publication was:

> The evidence…establishes that outbursts of radio noise are another manifestation of the solar disturbances already known to produce optical flares, radio fade-outs, and terrestrial magnetic storms. The apparent changes in position…can most readily be interpreted as due to the passage through the solar corona of an exciting agent moving with a speed of from 500 to 3,000 km/sec. It seems likely that normally the exciting agent is the corpuscular stream which is assumed to cause terrestrial magnetic storms.

In summary, the impact of these Potts Hill observations carried out in this 15 month period in 1949–1950 was evaluated almost two decades later by Paul Wild (1968):

Fig. 10.11 A photo of a number of the RPL staff at Potts Hill circa 1950. From left to right, Ruby Payne-Scott (*drinking a cup of tea*), Alec Little, George Fairweather (*in the doorway, a technician*), possibly Alan W. L. Carter and Joe Pawsey. The group are in front of one of the trailers used for instrumentation of the radio telescopes at Potts Hill. The photo is slightly damaged with numerous spots, likely ink. Some colleagues, who knew the group, disagree with the identification of Alan Carter (Bill Hall family collection, used by permission of Peter Hall)

> Following the first spectroscopic results another Pawsey-inspired experiment was put into operation and brilliantly performed by Payne-Scott and Little. The idea was to locate...the instantaneous position of the dominant source on the sun at any one time.[38]

There is no doubt that these 3 years at Potts Hill represented the high point for Payne-Scott's career as a radio astronomer. The work here laid the basis for the interferometer observations made in future years at Dapto and later at Culgoora from 1967 to 1984. A typical afternoon tea scene from this period is shown in Fig. 10.11, including Payne-Scott, Alec Little, George Fairweather, and Joe Pawsey.

In 1951, Payne-Scott's career as a solar radio astronomer came to an end. She had remarkable success, since she began her career as a radio astronomer, starting in March 1944. Her contributions to the discovery of Type I, II and IV bursts and the key role she played in elucidating the properties of Type III bursts, especially the seconds of time delays, remains an impressive achievement. Finally, she provided the first evidence for the direct determination of the motions of solar outbursts based on the swept-lobe observations of 1949–1951. She was also one of the initial pioneers in understanding radio interferometry. Her contribution to the recognition of the principles of radio aperture synthesis was decisive.

---

[38] 1968 Pawsey Lecture presented by J.P. Wild in Brisbane Australia, 30 April 1968 (Wild 1968).

Red Ruby: The Difficulty of Communist Ideology in Post-war Australia

Though it did not impact her career at RPL, Ruby Payne-Scott was ideologically aligned with communism in the 1940s–1950s, and suspected by the Australian government of being a member of the Communist Party of Australia (CPA). Her left-wing political orientation was well-known; as Rachel Makinson has pointed out, Payne-Scott was known as "Red Ruby" at RPL, though no-one was brave enough to use this term in her presence.[39] Since the CPA was a proscribed organisation under National Security Wartime Regulations between 15 June 1940 and mid-1943, Ruby was highly unlikely to divulge her CPA membership to anyone outside her party branch. Still, she never seemed to curtail her opinions around her co-workers, no matter how unpopular these might be.

This brings us to an interesting little side story in our biography. During the year that Joe Pawsey was abroad gathering allies in the US and UK for his nascent radio astronomy group, the scientists back in Australia decided to take a stand on an issue that had remained unsettled since the end of World War II. During the war, many research activities conducted by scientists at the CSIR, including that done at RPL was deemed a possible security hazard if leaked to the enemy. The research associated with the war effort was therefore marked as classified and could not be shared with the international scientific community.

Ruby Payne-Scott was one of 40 scientists[40] who signed an open letter to the editors, published in the 29 July 1948, *Sydney Morning Herald* and *Daily Telegraph* (Hindman et al. 1948) supporting the chairman of CSIR, Sir David Rivett in his effort to move all classified, military research out of the country's main scientific organisation (CSIR), and restrict it to non-CSIR military research institutes. Such a move would leave Australia's government-sponsored, civilian scientists at CSIR to work solely on non-secret research, and thus publish their findings, collaborate on projects with colleagues in other countries and generally increase the breadth, depth, and evolution of their work. The final paragraph of the letter to the editor summarised the issue:

> Those who advocate changes in the organisation of CSIR to enable work to be directed to defence projects would do well to consider the resulting detrimental effect on our largest research organisation and ultimately on applied science in Australia for both civil and defence purposes.

---

[39] Based on several interviews from 2 July 2003 and 21 February 2007 together with correspondence from 2003 to 2007 with Dr. Rachel Makinson and her son, Robert (Bob) Makinson.

[40] Roughly half of the signatories were from RPL with the other half from the Division of Physics. The RPL signatories included Payne-Scott, H. Minnett (a future Chief of RPL), J. P. Wild (a future Chief of RPL and future Chairman of CSIRO from 1978 to 1985), Cooper, Adderley, Humphries, McCready, Warner, Alec Little, Marie Clark, Hindman, Christiansen, Smerd, Yabsley, Coulson, and Labrum.

Ruby went even farther than just signing this article. She also submitted an article to the September 1948 CSIR OA (Officers' Association) *Bulletin* claiming, "If we try to carry over such rules into the peace-time lives of our scientists, it spells death to any form of international co-operation, hence death to the benefits we get from contact with scientists abroad, and it spells death to our own activity.... Frightened men do not produce great research."

The reader must keep in mind that although World War II had ended in 1945, by 1948 when these views were being published, the Cold War between the USSR and the major, non-communist states of the world was in full swing. The Commonwealth Investigation Service (CIS), which would be renamed the Australian Security Intelligence Organisation (ASIO) in March 1949, opened a file to investigate possible communist activity by Ruby Payne-Scott shortly after she and the other 39 scientists signed their names to that very public letter to the editor. It is likely that all the scientists who signed were checked, but Red Ruby's vociferous support of left-wing ideals lead to increased scrutiny.

The Payne-Scott file was started on 3 August 1948 with a letter by a CIS informer listing five RPL employees considered to be communist sympathisers, including Ruby, Rachel Makinson and John (Jack) Warner. Her file has periodic reports up until 25 August 1959, 8 years after she was no longer employed at CSIRO.

Her son, Peter Hall, has confirmed that his mother voted for communist party politicians, while Betty Hall recalls being asked by Ruby's husband, Bill not to deliver the CPA newspaper, *Tribune*, to their post box on the street, but rather straight to their house. Bill did not want the neighbours to be aware of the communist sympathies of the Hall family. The irony is that CIS/ASIO never confirmed that Ruby was in fact a member of the CPA, only that she was a left-leaning, free-thinking young woman. The truth came out many years later during research by Goss and Dick McGee for *Under the Radar*.

During interviews, both Rachel Makinson and B. Y. Mills confirmed that Ruby Payne-Scott indeed was a member of the CPA, and Len Hibbard wrote in a letter[41] that he "knew her as a fellow member of the 'Lab' communist party branch..." In addition, Mal Andrews, a former member of the CPA, contacted Peter Murphy of the SEARCH Foundation with a report of Ruby's membership in the CPA.[42]

Many Australian CPA members broke with the party after Russia invaded Hungary in November 1956, but it is uncertain just how long Ruby's idealism aligned her with the CPA. Certainly her atheism, feminism and belief in social justice remained as integral parts of her identity throughout her life.

---

[41] Letter to McGee, 14 October 1998. In addition, Bob and Christa Younger wrote Goss 15 March 1999 that they knew Bill and Ruby during WWII in the Sydney Bush Walkers. The teen agers were "reluctant to inquire into the lives of Bill and Ruby, being aware of the reasons for their secrecy in relation to their employment and association with the Communist Party of Australia".

[42] At the time of the dissolution of the CPA in 1991, the assets of the party were transferred to the Social Education and Research Concerning Humanity (SEARCH) Foundation.

Transition from CSIR to CSIRO; Discovery by CSIRO of Marriage of 1944

The proclamation of the law establishing the new Commonwealth Scientific Industrial Research Organisation was made on 19 May 1949, as the CSIR transitioned into the CSIRO. This transition necessitated a new oath or affirmation. Payne-Scott signed the oath shown in Fig. 10.12, witnessed by Arthur Higgs, the Technical Secretary of RPL. There were two choices for the oath, differing in a key phrase. "*So help me, God*" was one option, while the other option omitted this sentence. Ruby chose this latter option on 29 June 1949, bearing allegiance to the British King: "... do solemnly and sincerely promise and declare that I will be faithful and bear true allegiance to His Majesty King George the Sixth, His heirs and successors according to law". Thus, there is no mention of "God" in this alternative version of the oath signed by her.

The confused and unpleasant events of 1950 began with the official discovery that Payne-Scott had been married since 1944; the consequences were severe for her career at CSIRO. On 8 September 1944 William (Bill) Holman Hall and Ruby Violet Payne-Scott were married. His address was given on the marriage certificate as 6 Stanley Street, Arncliffe, with profession "telephone mechanic", while her address was listed as 5 Farleigh Street, Ashfield, with profession "physicist". The witnesses were Eva Neale, her aunt, and Mary Pilgrim, Hall's sister. Both of Payne-Scott's parents, as well as Hall's father, were deceased. B.Y. Mills has told us about this day in a letter from 14 September 1997:

> An interesting event occurred early in this work when, quite uncharacteristically, Ruby left early one afternoon without explanation. She later [in fact about a year later] told me that this departure was to marry Bill Hall, something which she found necessary to keep secret
> ...

From a conversation with a friend of Payne-Scott at the RPL in this period of mid to late 1940, there is the suggestion that she wore her wedding ring on a necklace instead of on her ring finger.

Six years later the CSIRO bureaucracy discovered the fact that Payne-Scott was married; this investigation started from the top. In early 1950, Ian Clunies Ross,[43] Chairman of CSIRO since its inception in May 1949, wrote to Payne-Scott about her marriage. No copy of this letter is extant. The handwritten reply (from her personnel file) from Payne-Scott of 20 February 1950 reads:

> Dear Dr. Ross
>   Your letter [someone else has written in the word "personal" at the top of the letter to indicate the nature of the letter to Clunies Ross] to me came while I was away on holidays, and in the rush of unprecedented solar activity of the last few weeks I seem to have mislaid

---

[43] Clunies Ross succeeded Sir David Rivett. Previously he had been Director of Scientific Personnel in the Commonwealth Directorate of Manpower after a career in the field of veterinary medicine at both the University of Sydney and the CSIR starting in 1926. See the biography of Ian Clunies Ross by Marjory Collard O'Dea (1997). Clunies Ross is described as one of the major architects of the post-war scientific revolution in Australia.

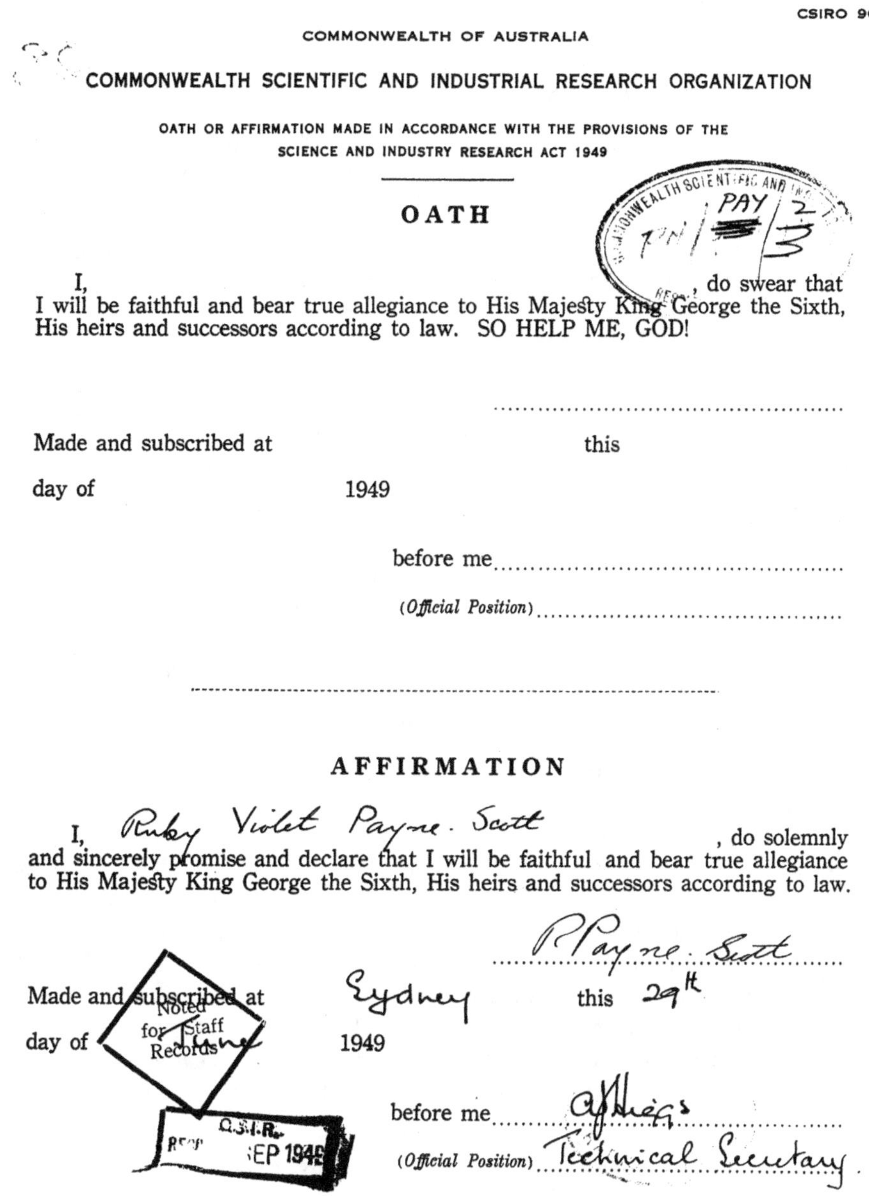

CSIRO 90

COMMONWEALTH OF AUSTRALIA

COMMONWEALTH SCIENTIFIC AND INDUSTRIAL RESEARCH ORGANIZATION

OATH OR AFFIRMATION MADE IN ACCORDANCE WITH THE PROVISIONS OF THE SCIENCE AND INDUSTRY RESEARCH ACT 1949

OATH

I, , do swear that I will be faithful and bear true allegiance to His Majesty King George the Sixth, His heirs and successors according to law. SO HELP ME, GOD!

....

Made and subscribed at this

day of 1949

before me..

(Official Position)..

AFFIRMATION

I, Ruby Violet Payne-Scott, do solemnly and sincerely promise and declare that I will be faithful and bear true allegiance to His Majesty King George the Sixth, His heirs and successors according to law.

R Payne-Scott

Made and subscribed at Sydney this 29th

day of June 1949

before me A J Higgs

(Official Position) Technical Secretary.

Noted for Staff Records

C.S.I.R. REC'D SEP 1949

COMMONWEALTH SCIENTIFIC AND IND... PAY 2/3

Fig. 10.12 Affirmation required by the newly formed CSIRO, signed by Ruby Payne-Scott on 29 June 1949. She did not sign the other option, an oath including the phrase "So help me God". Witness was Arthur Higgs, the Technical Secretary of RPL (From NAA: A8520, PH/PAY/002, page 33 [Personnel file])

it, so I will have to reply from memory. Thanks for your inquiries on my behalf, but when I spoke to you about my marriage I was in effect asking you whether the Executive realises that the customary demoting of women officers on their marriage to the status of "temporaries" does not appear to be required in the Act and whether the Executive agrees with this procedure or not. Whether or not there are material disadvantages to the women concerned in this procedure, all the married women research officers I have met feel that their classification as "temporary" puts them at a considerable psychological disadvantage in their work. Personally I feel no legal or moral obligation to have taken any other action than I have in making my marriage known. I have never denied to anyone who has asked me the fact that I am married, and it has gradually become common knowledge in the laboratory, particularly as many of the staff are my close neighbours at home. More recently I have stated the fact that I am married, among other information, in a form that I was asked to complete and return to Head Office. I have, of course, also stated it on my income-tax forms . . . I should still be very interested to know whether the Executive is in agreement with the present procedure with regard to married women, which seems to go far beyond the simple statement in the Act that employment may be continued when it seems desirable. I told you my story not in order to implicate you in any way but to demonstrate that the present procedure is ridiculous and can lead to ridiculous results.

The implication in this letter is that a personal meeting (". . . when I spoke to you") had occurred, given the report in the CSIRO Officers' Association (CSIROOA) *Bulletin* of December 1949 in which Payne-Scott wrote:

> The sub-committee on women's pay took the opportunity of the recent visit of Dr. Clunies Ross to Sydney to discuss the question of women's pay with him. He stated that the Executive favoured the principle of equal pay but might be in a difficult situation if it attempted to set a precedent in the Public Service. . .Dr. Ross promised to investigate the position. . . (see Additional Notes, No. 3, end of this chapter).

Thus in late 1949 she and Clunies Ross did discuss these issues in person at a meeting in Sydney.[44]

On 3 March 1950, Clunies Ross wrote a long, scathing response to Payne-Scott's 20 February 1950 letter, in which she was chastised for her behaviour. Clunies Ross did admit that in the act establishing the CSIRO, no mention of married women appeared in the legislation itself. But procedures were established under the Act which did imply that "a female officer shall be deemed to have retired from the service of the Council [still operating under the old terms of the CSIR], unless the Minister, upon recommendation by the Council, certifies that there are special circumstances which make her employment desirable". The negative result, as far as Payne-Scott was concerned, was the clause in the Superannuation Act which read:

> A female officer who marries after the commencement of this Section shall for the purposes of this Act be deemed to have resigned from the date of her marriage.

---

[44] Rachel Makinson has confirmed that a personal meeting between Clunies Ross and Payne-Scott occurred. Makinson, letter 23 January 2010 to Goss. Clunies Ross had visited RPL for a meeting with professional staff. Payne-Scott told Makinson that she reported to Clunies Ross: "Oh, by the way, did you know that I am married?" Payne-Scott did not provide any other details to Makinson.

Clunies Ross concluded his response to Payne-Scott with a summary of the position of CSIRO:

> You also discuss in your letter the question of the moral obligation people like yourself may feel towards letting us know about such change in circumstances as marriage, and you go on to say that personally you felt no such moral obligation. There can, of course, be two opinions on that point, but I will content myself with pointing out that if everyone thought as you do or acted as you apparently think proper, the administration of C.S.I.R.O. would be greatly complicated, and we would have to introduce a system of rigid scrutiny of the actions of officers instead of relying on their discretion and good sense. You may remember that it was only in December last that we received a notification from you that you had recently opened an account in a Sydney bank in the name of "Ruby Violet Payne-Scott".
>
> The usual procedure in the case of our women officers is, of course, that they are perfectly frank and open about their marriages and in that way help us to administer the law as it affects them and us. I cannot think that your Chief or clerical officers of the Division know of your marriage since they would have felt bound to have acquainted us of this. In this office there was certainly no knowledge of your marriage, nor do we appear to have received the form to which you refer as one which you were asked to complete and return to Head Office, in which your married state is mentioned. [In fact Payne-Scott had submitted this document more than a month earlier, 23 January 1950, to her local CSIRO personnel staff; with the transition from CSIR to CSIRO in mid-1949, the new organisation carried out a survey of all employees as to their marriage status.] I should be grateful if you would let me have details of that form and to whom and when it was submitted.
>
> In conclusion, I think the simplest way of regularizing the whole affair would be for you to tell us the date of your marriage. We will then look into the matter and tell you what should be done in your own and our best interests.

Then on 3 April 1950, Clunies Ross wrote to Payne-Scott again with a complaint about not having heard from her for a month about "the position created by your marriage. I should be glad if you would let me know without delay whether you intend officially to notify us of your marriage, since I do not feel that we can delay much longer in taking action..."

Payne-Scott responded on 12 April 1950, pointing out that she had informed the Head Office earlier and that the forms had been delayed in the RPL office awaiting the collection of all other applications (from all RPL staff). Finally on 24 April 1950 (3 months after Payne-Scott submitted her document to the RPL staff), Bowen did write to Cook (Secretary of CSIRO), with the date of the marriage; apparently the delays were not due to Payne-Scott's reluctance to report her marriage.

At this point, the financial implications of this imbroglio began to be implemented. Cook wrote to both the Superannuation Board and RPL that Payne-Scott was not eligible for the Provident Fund due to her marriage and that a refund of her contributions was required (the CSIR and CSIRO contributions were not to be refunded to her). In the letter to Payne-Scott, the statement was made that she was to be continued as an employee with temporary status, with the requirement for a new appointment each year. The indignity was complete when Cook told the Superannuation Board that the date of the "gazettal" in 1946 was after her marriage in 1944 and thus even the interest on her contributions from 1946 to 1950 was to be disallowed. A dismissive document was prepared by Cook for the CSIRO file that provided a short summary of the whole affair. In an undated handwritten note for the Payne-Scott personnel file he wrote: "This 'marriage' was kept 'secret': for over 6 years. We gazetted RPS after she was married! We registered a protest but apparently glossed over it from then on, for without doubt we wanted employment to continue. There doesn't seem though, as if there was any disciplinary action of any kind."

Perhaps during her period as a CSIR employee (1941–1949), Payne-Scott had become a part of a scientific community that was happy to overlook unjust, bureaucratic fine points in favour of outstanding performance during war time and the immediate post-war period. Apparently only after the administrative overhaul in the post-CSIRO era (mid-1949) did Payne-Scott face renewed scrutiny from CSIRO functionaries, after which her marriage from 6 years earlier became a major issue. It also appears that Payne-Scott, being a stubbornly outspoken person when she considered herself to be in the right, attacked the problem head on by confronting Clunies Ross at the beginning of the transition. Since the laws governing the newly formed CSIRO made no mention of the need to treat women unequally, she tried to clarify that the unfair laws that had been implemented everywhere else need not apply. Clearly, Clunies Ross did not find this to be true, and felt that Ruby's avoidance of injustice not only could no longer be ignored, but must be dealt with, since the legal obligations had been violated.

In the end, the result of this brouhaha was the loss of Payne-Scott's pension contributions that the CSIR and CSIRO had made in the period 1946–1950. She had no superannuation funds at all after her 9 years of service to RPL. Although her continued employment was not in jeopardy, the designation as a "temporary" was quite demoralising.[45] This negative experience may have influenced Payne-Scott's perception of her future at RPL in the course of 1951; she never expressed any interest in returning to CSIRO after the birth of her children.[46] It was not until 1966 that the regulation that married women in the Australian Public Service had to resign their permanent employment on marriage would be rescinded.

Payne-Scott's Resignation in July 1951

Payne-Scott's resignation was quite precipitous as events unfolded in mid-1951. On 18 July 1951, she wrote to Pawsey that she was resigning on the afternoon of 20 July (only 2 days hence) for "private reasons". We know that the reason was her then fifth-month pregnancy, with the birth of Peter expected in November 1951. Undoubtedly, Pawsey knew about the upcoming resignation as F. W. G. White, then the CEO of CSIRO, wrote her a letter on the previous day (17 July 1951):

---

[45] In interviews with Goss in August 2003 and February 2007, Rachel Makinson stated that her similar treatment as a "temporary" at the CSIRO Division of Textile Physics in the years 1953–1982 was quite demoralising.

[46] See footnote 4, chapter 8.

Dear Miss Payne-Scott,
   I have just seen a letter from Dr. Pawsey which tells me that due to the imminent arrival of your baby, you will not be able to carry on with your research work. This event must be giving you a great deal of pleasure but I can well imagine that you regret having to leave off research, at least for the time being. Unfortunately we cannot give a married woman leave without pay [only changed in 1973 when Australian Public Servants were given maternity leave], but I can assure you that I at least would be very pleased to see you return to Radiophysics in due course. I hope the event goes off successfully.

White certainly knew Ruby well as they had both joined RPL at roughly the same time in 1941 and were associated in the research activities of the RPL during the War.

Payne-Scott wrote a poignant letter to White from her temporary residence in the home of Richard and Rachel Makinson [in Nymboida Street Coogee, a Sydney suburb south of the centre of Sydney], while these scientists were on sabbatical in the UK. On 15 August 1951 she wrote to White (personnel file):

I am writing to thank you for the kind letter that you sent me on the eve of my departure from the Radiophysics Laboratory. I am, of course, sorry to give up the research work I have been doing and also to leave the laboratory where I have been so happy and have so many friends. If all goes well I do not expect to be returning to Radiophysics at least for some years, but I hope that I may be able to hear the latest on solar noise at the A.N.Z.A.A. S. [Australia New Zealand Association for the Advancement of Science] and more particularly the U.R.S.I. [International Union for Radio Science] conference in Sydney next year.

An amusing episode, illustrating the ironic nature of the bureaucratic structure of the CSIRO, occurred as Payne-Scott left RPL in 1951. Pawsey wrote to Cook on 24 July 1951, concerning four items that had been checked out to her some years earlier and could not be found after she left on the previous Friday, 20 July 1951. These items were a soldering iron, a small hammer, some pliers and a side cutter (wire cutter). The total replacement cost was less than two Australian pounds. Pawsey wrote Cook: "In view of the length of time [10 years] Miss Payne-Scott was attached to the Division it is recommended that these items now be struck off charge at public expense." Cook agreed to this request on 30 July and Payne-Scott did not suffer the indignity of paying this small sum to CSIRO for the lost items.

The high esteem in which Ruby was held is exemplified by the salary adjustments that were being implemented for the start of the new fiscal year on 1 July 1951, just before her resignation. Already on 27 February 1951, a recommendation had been made that she be promoted to Senior Research Officer, Grade I, at a salary of £A920 per annum, a substantial raise of £A170 per annum.[47] Hers was the second highest salary of any of the scientific staff not in an administrative position; only H. C. Minnett had a higher salary of the scientists who were being reclassified in mid-1951.

---

[47] A memo which accompanied the recommendation for promotion lists eight publications and describes her work and that of A. G. Little on the Potts Hill interferometer in detail: "[It] allows the movement of the sources of radio energy emitted at times of solar flares...to be followed through the solar atmosphere.... Results of great interest and importance are being obtained."

At the farewell party held on 20 July 1951, Dick McGee was present as Pawsey described Payne-Scott as the "best physicist in the lab..." to the attendees. John D. Murray has told us about an amusing event at the farewell. Pawsey made reference to the upcoming birth of her baby (Peter Hall was born exactly 4 months later on 20 November 1951). Pawsey began his summary of Payne-Scott's achievements at RPL: "Miss Ruby Payne-Scott..." and realised that this name was hardly appropriate for someone 5 months pregnant and hastily corrected himself to: "I mean... [long hesitation]...Mrs. Ruby Hall..."

Thus the reason that Ruby resigned in 1951 was not a direct result of the controversy from the previous year concerning her marriage. As a 39 year-old woman, she was much older than average first time mothers in 1951. During or just after the end of the war, Payne-Scott likely had a miscarriage, possibly late 1946. Thus she was quite probably concerned about starting a family. Let us imagine for a moment, what it would have meant to the course of radio astronomy as it developed in its early stages, had Ruby's first pregnancy not ended tragically in a miscarriage. We never would have had her painstaking observations and analysis at the Hornsby site, nor her engineering expertise in the creation of the Potts Hill antenna, nor her mathematical and theoretical contributions to the field over the course of the 5-year period between her first, failed pregnancy, and her second pregnancy, which led to the birth of her son, the mathematician, Peter Hall.

The reason for her resignation was simply that maternity leave was not available in 1951. It is apparent, however, that knowing she would need to leave she did make a valiant attempt to improve the lot of the other female scientists in the new CSIRO by testing the resolve of their employers to punish women for choosing to have professional careers in tandem with marriage, not to mention children.

Certainly the treatment of Payne-Scott by the CSIRO seems draconian in the light of twenty-first century practices. However, her treatment was by no means unusual given the norms of Australian society in 1950. Ruby was the beneficiary of increased opportunities for women scientists at the outbreak of World War II, but it was her skills and experience as a physicist and mathematician that were the true basis of her professional successes in the remarkably productive decade of 1941–1951.

Additional Notes

1. NAA: C3830: B2/2, Part 1.In the letter (NAA: C3830, B2/2, Part 1) written by Payne-Scott to Ryle on 9 December 1948, she described the end of the Hornsby campaign (Chap. 9) and her plans for the new interferometer. "I have now transferred my interests to the type of interferometer that you use, but provided with a phase-changing system so that we can sweep through the complete lobe in 1/25 s instead of relying on the motion of the source, and hence providing a means of measuring the position on the sun of short-period variations of either type [Type I and Type III ?]. I should like to hear your views on the stability of

the calibration of this type of interferometer...." She then described the new Wild and McCready swept-frequency spectrograph (called a "spectrum analyser") to scan the solar spectrum from 50 to 100 MHz. "It should solve many problems that spot-frequency observations leaves open, particularly that of the distribution of intensity with frequency in a burst."

2. Mills told Goss on 1 April 2007 that the exchange between Ruby and Kevin went as follows: Kevin responded, "Well, that's what John Bolton told me." Then Ruby was quoted by Mills as saying, "Well, that doesn't matter if it's wrong." Remarkably, there was extensive correspondence in September 1950 between Bowen and Pawsey about this PC meeting of July 1949 (NAA: C3830, F1/4/PAW/2). Pawsey was in Europe and met Bolton at the URSI Conference in Zurich. There was a conflict brewing between Bolton and Piddington about the interpretation of all sky radio continuum images based on distributions of radio stars pus the contribution of thermal emission throughout the Milky Way. During this discussion, the details of the July 1949 presentations by Bolton and Westfold were reconstructed. Pawsey wrote to Bowen (9 September 1950) about Westfold: "The colloquium took place just before Westfold left [for Oxford] and may be remembered from a criticism by many of us of some other allied data which Westfold put forward and finally withdrew." Bowen replied on 22 September 1950 to Pawsey summarising the colloquium given by Bolton and Westfold: "... Bolton confined himself more or less to a statement of results. Westfold, however, embarked on some ideas about the mechanisms...and ran into some criticisms...These criticisms prevented him finishing what he was going to say, and this appears to have been the crucial point." The Westfold and Bolton papers are summarised in footnote 19 of Piddington's rival paper, *The Origin of Galactic Radio-Frequency Radiation*, 1951.

3. Deery (2000) has reported that Rachel Makinson wrote to him on 8 November 1998 about this meeting. The meeting was probably in the period September to November 1949 in Sydney. Makinson attended a meeting of the Officers Association of the CSIRO in which Tom Kaiser (a CSIR/CSIRO employee on a student assistantship at Oxford) was attacked for engaging in public controversy in London earlier in 1949. Clunies Ross even claimed that the stress of this event had led to his heart attack, an assertion that Makinson thought quite "unfair". Makinson was one of the few attendees who spoke ("cautiously") in support of Kaiser along the grounds of freedom in science and even invoked Sir David Rivett (CSIR CEO from 1927 to 1945 and Chairman from 1946 to the creation of CSIRO in May 1949) in support of this view. Clunies Ross "dismissed us all rather roughly" according to Makinson. At least two other events may have occurred during this meeting: (1) a discussion about equal pay for women at the CSIRO, and (2) a possible private conversation between Payne-Scott and Clunies Ross.

Chapter 11
1952: Ruby Payne-Scott's Last Experience as a Radio Astronomer at the International Union of Radio Science

Payne-Scott experienced a postscript to her career as a radio astronomer in August 1952—13 months after her resignation and less than a year after the birth of her son, Peter Hall. She attended the 10th URSI (International Union of Radio Science) General Assembly at the University of Sydney from 11 to 21 August 1952. URSI was one of the first international scientific congresses to be held outside the US and Europe. The first international congress to be held in Australia was the Second Pan-Pacific Science Congress in Melbourne and Sydney, more than 30 years earlier in August of 1920 (see Additional Notes, No. 1, end of this chapter).

Figure 11.1 shows the arrival in Sydney by ship (the P&O RMS *Strathmore*) of Sir E. V. Appleton, the URSI President and Vice-Chancellor of the University of Edinburgh (third from view's left) being welcomed by the Australian hosts. From the left the others were D. F. Martyn from the Radio Research Board, Canberra (URSI Vice President and Chair of Commission V, Radio Astronomy), Colonel E. Herbays (Brussels, Secretary of URSI), E.G.("Taffy") Bowen (RPL Chief), and J. L. Pawsey (RPL Assistant Chief). Many of the European guests had spent more than a month on board ship across the Indian Ocean (see Additional Notes, No. 2, end of this chapter). Sixty-three overseas delegates came from thirteen countries as well as more than 250 Australians; the conference included tours of Sydney, the local beaches, a harbour cruise and an excursion to Wollongong on the South Coast, which included a tour of the Dapto RPL field station. The two weekends were spent at Jenolan Caves and the Federal Capital, Canberra (with visits to Mt. Stromlo Observatory).

The assembly began on Monday, 11 August 1952 at the University of Sydney. In Fig. 11.2 we see the pin designed for the conference by Ron Bracewell, who was the secretary of the Australian Organising Committee and had only returned from completing his Ph.D. degree at the University of Cambridge in 1949.[1]

---

[1] Interview with Goss, Palo Alto, California, February, 2007. Bracewell reported that his enjoyment of the conference was compromised by a serious illness which forced him to miss many of the scientific sessions. There was a silver lining; he recuperated in the Sydney home of his future wife, Helen.

Fig. 11.1 URSI 1952 began with the arrival of the VIP guests aboard the ship *RMS Strathmore*. We see here the welcome of URSI guests aboard the ship in Sydney. Left to right are D.F. Martyn, Col. E. Herbays (Secretary of URSI, Brussels), Sir Edward Appleton, E. G. "Taffy" Bowen and J. L ("Joe") Pawsey. Based on a letter from Appleton to Bracewell (the Organising Secretary of the Conference) dated 18 June 1952 (Bracewell archive NRAO), the *Strathmore* was predicted to arrive on Friday, 8 August 1952 in the morning. Newspaper accounts from the Sydney newspapers in August 1952 confirm this date. That evening at 8 p.m. Sir Edward gave an address sponsored by the Sydney University Engineering Club (Twenty-Third Annual War Memorial Lecture) in the Great Hall of the University of Sydney, "Science and the Public" (CSIRO Radio Astronomy Image Archive B2842-1)

Fig. 11.2 The URSI pin—a very Australian kangaroo—from 1952, designed by Ron Bracewell. Bracewell gave his pin to Goss in February, 2007. There was also an URSI flag designed by Bracewell and Frank Kerr, which disappeared some years later (Photo by Goss 2007)

At the Administrative Opening Session on 11 August, Appleton gave a lengthy reply to the welcome by Sir John Madsen, the Chairman of the Australian National Committee of URSI. Appleton paid "tribute to the distinction which Australian workers have won for themselves in the field of scientific radio".[2] He detailed the history of the Radio Research Board founded in 1927 under Madsen's leadership and summarised the wartime successes of the Division of Radiophysics Laboratory. He described the recent discovery by Bolton and Stanley of the small angular size of Cygnus A as determined at Dover Heights.[3] He also described David Martyn's work on tides in the earth's ionosphere. The recent decision in the UK to fund the 250 ft. telescope at Jodrell Bank prompted Appleton to say: "[I] would like to see, in due course, a similar instrument at the disposal of your radio astronomers here in the southern hemisphere", a prophetic statement as the Parkes radio telescope was to be opened in less than 10 years. Appleton foresaw the significance of the recently discovered hydrogen 21 cm line, which had been predicted by Henk van de Hulst in 1944, and stressed the importance of frequency protection near 1420.4 MHz, "... so that this new and important discovery may be developed without trouble due to radio traffic interference".[4]

The grand opening of the Assembly was on the following day, with welcoming addresses by the Rt. Hon. R. G. Casey, Minister for External Affairs and the Minister of the CSIRO. Appleton, as President, replied with words of praise for the "young" Australia:

> For in one thing, we can all claim to be Australians ... so we can ourselves believe that our own subject of radio has vast and undreamed-of possibilities of expansion and development in the years which open out before us ... We must be fortified with the pioneering qualities—the Australia qualities—of courage, enthusiasm, and resource.[5]

A highlight of this second day was the Sydney Symphony Orchestra's performance of the very Australian "Corroboree" by John Antill. The bullroarers included in the piece set the stage for Australian cultural experiences to come.[6] The group photograph, shown in Fig. 11.3 was taken on 11 August, 1952.

During the conference there were two excursions: to the RPL field station at Potts Hill and the new solar station at Dapto, where the swept-frequency radio telescope had recently been installed. These stations and other activities of the laboratory were laid out in a delightful sketch map drawn for the booklet, *Research Activities of the Radiophysics Laboratory*, distributed to all USRI participants (Fig. 11.4) (RPL 1952). Seventeen of the 37 pages were devoted to descriptions of the radio astronomy and ionospheric research activities of the laboratory; cloud and rain physics, automatic computation (a computer project) and radio navigation

---

[2] URSI documentation provided by Bracewell in 2006–2007 and by Madam Inge Heleu, Executive Secretary of URSI in February 2007.

[3] In his closing remarks, Appleton confessed that he had only been able to visit the Dapto and Potts Hill field stations. "I am sorry I have not yet been able to visit Dover Heights—which is now an historic site—but I hope to do so before I leave Sydney." It is not known if he visited this scenic sea side site.

[4] URSI, *Proceedings of the General Assembly*, Vol. IX, Fascicule 1, Administrative Proceedings 1952, Brussels, page 9.

[5] Ibid.

[6] Bracewell 1984 (in Sullivan 1984).

Fig. 11.3 Group photograph of URSI participants, on Monday 11 August 1952 outside the Wallace Theatre at the University of Sydney. Brian Robinson (2002) has published a finding chart for 174 of the participants. This is the only photo during the URSI assembly in which Goss have found an image of J. L. Pawsey. Many well known overseas guests attended the Assembly, including Sydney Chapman from Oxford, J. A. Ratcliffe from Cambridge, B. van der Pol from Geneva (CCIR, the International Radio Consultative Committee), A. H. de Voogt from The Hague, S. Silver from the University of California, Berkeley, C. R. Burrows of Cornell, and John Dellinger of RCA in the USA (CSIRO Radio Astronomy Image Archive B2842-15)

were also included. The swept-lobe interferometer was described under the rubric *Locating Sources of Transient Solar Disturbances.* (The instrument was no longer being used in a systematic fashion by August 1952.)

This international conference was a great opportunity for Australian scientists to meet their colleagues from overseas[7]; in addition, the Australians could show the world their successes in radio science, and radio astronomy in particular. Since Australian astronomers outnumbered visitors five to one, the meeting organisers may have limited the slots for Australian presentations, so they did not appear to dominate. Commission V (Radio Astronomy) had six sessions. D. F. Martyn, President of this commission, was responsible for the invitation to URSI to come to Sydney. Nineteen scientific talks were presented (nine of them by Australians) at the four scientific sessions, Of special interest on Monday, 18 August 1952, was the session on "radio emission from interstellar gas", chaired by Professor Schilt from Columbia and Yale (visiting Mt. Stromlo). The newly discovered HI line at 21 cm was the topic of the opening talk by Harold I. ("Doc") Ewen of Harvard. Together with Purcell, Ewen made the initial observation of the line on 25 March 1951 (Ewen and Purcell 1951). The next session, "solar radio noise" was held on 13 August.

---

[7] Due to continued ill will as a result of WWII, the Germans and Japanese were not invited. Martyn did go to Japan after the Assembly, however, to transmit news of the proceedings (Gillmor 1991). Japan submitted a national report in the published URSI proceedings; Germany did not. Also, as the Cold War had begun, there were no participants from the USSR.

Sketch map of the Sydney area, showing the Radiophysics Laboratory and Field Stations at which research is in progress. The two airports are used for experiments in Rain and Cloud Physics and Radio Aids to Navigation. The remaining five field-stations are used for observations in Radio Astronomy.

Fig. 11.4 Sketch map of the Sydney area showing RPL and the field stations (see Fig. 2.4). A booklet was published by RPL for URSI in 1952, *Research Activities of the Radiophysics Laboratory* (RPL Archive, from Christine Vanderleeuw, CSIRO Radiophysics Library, 2007)

Presentations were given by Christiansen (RPL), F. G. Smith (Cavendish Laboratory, Cambridge), Steinberg (CNRS, Paris), Hagen (Naval Research Laboratory, Washington, D.C.), Laffineur (Institut d'Astrophysique, Paris), Piddington (RPL) and Smerd (RPL).

Payne-Scott probably attended this session, but did not give a talk. Here was the opportunity for Payne-Scott to present her new and interesting results. How did it come about that she gave no paper? Pawsey personally asked her to submit a paper. On 11 March 1952, Pawsey sent a memo to all members of the radio astronomy group with a description of the process for making presentations to the URSI Assembly. A multistage selection process required first RPL and then Australian national committee approval. Proposals for papers—a brief half-page outline— were to be given to Bracewell by 18 March. The papers were to represent "an outstanding contribution" and needed to be "reasonably novel". Pawsey sent a copy of the memo and a personal letter to Payne-Scott's home (to Mrs. W. Hall, 120 Woronora Parade, Oatley). He wrote: "In my provisional list, unofficial, I include your work with Alec [Little] as one of our star efforts. Further, if such a paper is to be presented verbally at the Conference, I think you would do it excellently." After a request that she discuss this plan with Alec Little, he asked her to let him know what she would prefer.[8]

---

[8] NAA: C3830, C6/2/4B, correspondence for URSI 1952.

On 20 March 1952 she wrote a letter to Pawsey, including an outline of her planned presentation, "Relation Between Solar Radiation at Metre Wave-Lengths and Other Solar Phenomena", with co-authors Alec Little and Paul Wild. Payne-Scott reported to Pawsey that it would not be possible to have a draft ready by the deadline of 7 April, understandable as she had a 6 month old son and was no longer employed at RPL. She had tried to contact Little, but he was on holiday. She ended the letter by inviting Pawsey and his family to visit her family in Oatley, easier for the Pawseys who had a car while the Halls did not. She signed the letter "Ruby Hall".

The outline written by Payne-Scott discussed "noise storms" (Type I) and "outbursts" (likely Type II and IV), only recogised a few years later. (She did not use the new nomenclature, although Wild had done so in 1952.) The noise storms could be associated with large sunspots and had appreciable amounts of circular polarization, the polarity determined by the associated sunspot. An origin high in the corona was suggested since the radiation was displaced to the limb of the sun when the observations detected a noise storm at times when the sunspot was close to the solar limb. The "outbursts" were usually accompanied by large flares, often associated with radio fadeouts. The events were randomly polarized, with circular polarization occurring towards the end of the event (now known to be the associated Type IVM ("M" for 'moving') events following a Type II event). The movement through the solar atmosphere had been inferred by both the frequency drift (the swept-frequency spectrograph by Wild and McCready) and the direct determination of the motions with the swept-lobe interferometer used by Payne-Scott and Little. An exciting agency moving with the velocity of the "magnetic storm particles" was suggested. There were still questions as to how the two types of evidence (angular motion and frequency drift from high to low frequency) fitted together. The conclusion was that these Type II and IV events were probably caused by particle streams which also caused the magnetic storms on the earth about a day later.

Clearly Ruby Payne-Scott had kept up with developments in the field, even though she was busy as a new mother. Pawsey replied positively in a "Dear Ruby" letter, on 26 March 1952. He had discussed the proposal of a paper with Paul Wild, who "favours the idea". Pawsey wrote:

> This is a proper follow-up and I should like you or Paul [Wild] to attempt it. As I see it, you have the more recent work and hence the right to do it. If you did not feel able to, I should suggest Paul do it. The broad objective is an integrating paper, collecting the known facts and rejecting fallacies, concerning the relations between optical and radio phenomena. It should set this out as an objective.[9]

Pawsey suggested that the three of them meet at Ruby's house to discuss the proposed presentation.

In the end, neither Wild nor Payne-Scott gave a presentation at the URSI Assembly. In the solar session on 13 August 1952, RPL presentations were given only by Christiansen, Piddington and Smerd. The omission of a paper on metre

---

[9] Ibid.

wave solar bursts is surprising. Even if the organisers felt it necessary to limit RPL talks, the leader of the radio astronomy group, Pawsey, viewed the work as one of RPL's "star efforts". Pawsey's objective in pushing for the inclusion of the paper was to elucidate the relationship between optical and radio solar physics. Payne-Scott had always readily faced challenges, but at this time the young mother lacked easy communication with RPL staff and was far from their resources. Also, three authors were involved and the time to decide who would present the paper was running short. In any case, from the modern perspective it is regrettable that Payne Scott did not talk at the URSI conference, thus leaving a gap in the description of recent Australian radio discoveries in the field of solar radio bursts and outbursts.

On Thursday, 14 August 1952, there was an official afternoon tour of the RPL on the University of Sydney grounds, followed by a trip to Potts Hill.[10] Many photographs of the inspection of the various instruments were taken. One example is Fig. 11.5, with Christiansen showing Appleton (dark hat) and van der Pol (lighter coloured hat) the east-west 20 cm solar grating array. The former Chief of RPL, Fred White, who was the CEO of CSIRO by this time, was also at URSI in 1952. He was in the group with Appleton on this tour (Fig. 11.6). A fascinating photo is shown in Fig. 11.7 with Alec Little explaining the workings of the swept-lobe interferometer, the instrument used so effectively by him and Payne-Scott to determine the motions of solar events. Clearly RPL was well prepared for this excursion. It is unlikely that Payne-Scott was present during the tour at Potts Hill field station.

There was also an unscheduled excursion to the Hornsby site, possibly on 14 August. Many of the well known participants were present as judged by existing photographs: Graham-Smith (Cavendish Laboratory), Hanbury Brown (Jodrell Bank of the University of Manchester), Warbarton, Higgins, Hindman, McGee, Mills, Slee and Shain (all RPL).

Several reports about URSI were written in Australian and international publications. Two months before the conference, Bracewell—in his role as Secretary of the URSI Assembly—wrote an article in the *Australian Journal of Science* (21 June 1952). This announcement of the upcoming meeting was full of optimism:

> That URSI has honoured us by its present choice is tangible international recognition of the status of scientific radio in Australia ... it tends to raise Australia's prestige abroad to the level merited ... The Australian National Committee on Radio Science is under an obligation to ensure that foreign delegates to the forthcoming Assembly will find it rewarding, and it is hoped that a successful outcome will lead to fuller Australian participation in these important scientific activities, in radio and in other fields as well.

Bracewell hoped that the Australian prime minister, Sir Roberts Menzies would address the opening session; unfortunately he could not attend.

---

[10] Bolton (1953) described the official field trips to Dapto and Potts Hill. The trip to Hornsby is not mentioned, but Bolton did say that "private visits were made to other field stations engaged on galactic work." Surprisingly, based on the lack of photographic evidence from the CSIRO Radio Astronomy Image Archive, there was likely no tour of the Dover Heights site.

Fig. 11.5 URSI tour of Potts Hill Reservoir Field Station, August 1952. Here Chris Christiansen (*left*) describes the new 32-element east-west grating array, opened November 1951. Sir Edward Appleton, President of URSI, points to one of the 1.7 m antennas. To the extreme right is Prof. Balthasar van der Pol, Director of CCIR in Geneva and prominent Dutch radiophysicst. Note the URSI pin on the jackets of Appleton and Christiansen (Fig. 11.2) (CSIRO Radio Astronomy Image Archive B2842-R6)

Two reviews of radio astronomy at URSI, by Kerr and Bolton, appeared in early 1953. Frank Kerr provided a summary in the popular US astronomy journal, *Sky and Telescope* (January 1953). The Kerr publication contained six figures, including Fig. 11.8. This article described in detail the HI investigations carried out in the Netherlands by J. H. Oort and his colleagues, as well as the new determinations by both Australian and British groups on the sizes of radio sources. Since none of the compact cosmic noise sources had stellar characteristics in terms of size, Kerr proposed using the new terminology: "... the former term *radio stars* is now falling into disuse, in favor of the more general phrase, *discrete sources*". The article gave a short summary of the solar work, concentrating on the new 21 cm continuum imaging of the sun by Christiansen at RPL.[11]

---

[11] Kerr also had doubts about the status of theory in explaining emission processes in radio sources: "The discussion showed that this is a very difficult subject, and a thorough understanding of the nonthermal processes that are important in radio astronomy is still a long way off." Indeed the recognition of the importance of synchrotron emission was not to occur until the mid 1950s. The importance of the publication, *Cosmic Radiation and Radio Stars*, by Alfvén and Herlofson (1950), who proposed synchrotron emission, was not recognised in 1952 (see also Chap. 8). Herlofson (from Stockholm) attended URSI 1952.

Fig. 11.6 Continuation of the tour at Potts Hill during URSI. Again the group is inspecting the 32-element east-west grating array. This 21 cm continuum instrument was used to provide high resolution radio 1-D images of the sun. From right to left: Bowen, Appleton, van der Pol, and Fred White, a former Chief of the Radiophysics Laboratory. At the extreme left is Chris Christiansen (CSIRO Radio Astronomy Image Archive B2842-R65)

Bolton, in the British journal, *Observatory* (February 1953), described the recent Fourier imaging of the quiet sun at Cambridge as well as the results of two eclipse expeditions to the Sudan in 1952 over a wide wavelength range (shortest 8 mm). Bolton gave mixed reviews of the URSI session on "Dynamics of Ionized Media" organised by Professor Harrie Massey: "... [it] gave a very able summary of the present theoretical position ... The writer was, however, left with the impression that a theoretical interpretation of some cosmical phenomena, in particular the radio noise from the disturbed Sun and the galaxy, was far from being achieved, except in the most general terms". As Kerr had done, Bolton criticised the use of the term "radio star": "it seems that the term 'radio star' may be a misnomer". Indeed in a few years the use of this term disappeared.

A group photo of the prominent radio astronomers at URSI was taken, perhaps on one of the days on which Commission V, Radio Astronomy, was scheduled. Since Payne-Scott was present, a possible date was 13 August 1952. In the historic photo shown in Fig. 11.8, Payne-Scott is surrounded by 21 prominent colleagues from Australia, Europe and the US. She is the only woman, and is not wearing the

Fig. 11.7 URSI tour of Potts Hill. Alec Little describes the swept-lobe interferometer that he and Ruby Payne-Scott used from 1949 to 1950. Note the trailer in the background (Fig. 10.11). In the background near the car J. P. Hagen (wearing a hat, from the USA) and H. S. W. Massey (UK) are shown. At the extreme left is C. A. (Lex) Muller from the Netherlands. Payne-Scott was likely not in this tour (CSIRO Radio Astronomy Image Archive B2842-R58)

URSI pin shown in Fig. 11.2.[12] We have been told by Ms. Sally Atkinson that she and a colleague at RPL in the administration, Sylvia Mossom, were the baby sitters this day, looking after the 9 month old Peter Hall while his mother attended the URSI sessions.[13] Sadly, Pawsey was not present for this photograph; Kerr, in the

---

[12] Although Payne-Scott has no pin, she was a registered participant; she appeared in the list of participants with profession "physicist", address 120 Woronora Parade, Oatley. Her name was listed as "Ruby Payne-Scott (Mrs. W. Hall)". Only three women were registered for the URSI Assembly.

[13] Interview with Sally Atkinson, February 1999, Epping, NSW. Sally Atkinson was at the RPL from 1942 to 1971 and secretary to the Chief of RPL, E.G. Bowen from 1946 to 1971. As Honorary Archivist at RPL from 1971 to 1992 she was responsible for transferring over 60 m of files from RPL to the National Archives of Australia. Sylvia Mossom Blackwood (1914–1987) worked in the RPL administration during World War II, starting in 1941; she was also an infrequent babysitter for the Hall family in the 1950s (letter from her husband, Fred Blackwood, July 1999). In the late 1950s, Mossom was a literary assistant to D. P. Mellor as he prepared the influential volume, *The Role of Science and Industry*, in the series *Australia in the War of 1939–1945. Series 4, Civil* (1958), sponsored by the Australian War Memorial.

Fig. 11.8 Most of the radio astronomers at URSI 1952 are shown here. *Front row left to right*: Chris Christiansen, F. Graham-Smith (UK), B. Y. Mills, S.F. Smerd, C. A. Shain, R. Hanbury Brown (UK), Ruby Payne-Scott, A. G. Little, M. Laffineur (France) and J. G. Bolton. *Second row*: J. P. Wild, J. L. Steinberg, J. V. Hindman, F. J. Kerr, C. A. Muller (Netherlands) and O. B. Slee. Third row: C. S. Higgins, J. P. Hagen (USA) and H. I. Ewen (USA). *Back row*: J. H. Piddington, E. R. Hill and L. W. Davies. Individuals with no country designation are Australian. Surprisingly Pawsey, the leader of the radio astronomy group at RPL was not in the photo, likely due to an illness. Likely day for the photo is 13 August 1952 (CSIRO Radio Astronomy Image Archive B2842-43)

Sky and Telescope article, also mentioned Pawsey's absence from the group photo.[14]

An additional photo showing Payne-Scott is shown in Fig. 11.9. This international congress represented the last formal involvement that Payne-Scott had with the Division of Radiophysics and the CSIRO. In later years she did remain in contact with Mills, Christiansen (including participation in anti-Vietnam demonstrations in Sydney in teh late 1960s) and possibly with Pawsey.

Certainly URSI 1952 in Sydney was an encouraging opportunity for the Australian scientific community to show that they were among the world leaders in radio astronomy. A few months later on 27 November 1952, a British reporter for the *Manchester Guardian*, Douglas Wilkie, wrote with sweeping praise:

---

[14] In the closing address to the congress on 21 August 1952, the past President Sir Edward Appleton thanked numerous Australian hosts including Pawsey: "Dr. G. L. (sic) Pawsey, whose absence we deplore". A few months later J.A. Roberts was returned to Australia by ship after finishing a Ph.D. degree at the Cavenish Laboratory in the UK. Although not present for the conference, he had heard that the organisation was outstanding. "I was sorry to hear that you were down with the flu. I gather that the rainmakers [from the RPL Cloud Physics group] turned on a little too much for the conference!" (NAA, C3830, F1/4/ROBE1. Pawsey was present at the opening ceremony. By the last week of August, Pawsey was able to provide tours of the Dapto field station for overseas visitors to the URSI congress (John Murray, interview 2007).

Fig. 11.9 During the URSI Tenth General Assembly a number of photographs were taken of the audience. In a photo of the audience during a lecture by Appleton, Payne-Scott is clearly present. The new mother might have been catching up on sleep during one of the URSI sessions; of course, she could have only been stretching. Her son, Peter, was only about 9 months old. McGee, with a big smile, is seen just to the left of the standing man's nose. The late Brian J. Robinson is just to the left of Ruby's right elbow, in the row behind (CSIRO Archive [Rob Birtles] Canberra 163.108)

> A visiting American scientist remarked that Australia's beer, beaches and radio astronomy were the best in the world ... Why do Australians have a gift of scientific excellence? ... It is certainly at variance with their frequent reputation for cultivating the good-enough as a way of life. The best explanation, it seems, lies in that pioneering heritage ... Now seeking new worlds, they find a challenge in scientific exploration.

Additional Notes

1. A number of authors have asserted that the URSI Assembly of 1952 was the first international meeting outside the US or Europe (e.g., Haynes et al. 1996; Robertson 1992). Already in 1950, Professor F. J. M. Stratton (prominent UK astronomer and the Secretary of the International Council of Scientific Unions) asserted that "no Union so far has ever had a meeting outside Europe and the U.S.A.". (National Archives of Australia-NAA: C3830, C6/2/4A, Part 1). However, there had been earlier meetings of an international nature in Australia. A meeting of the British Association for the Advancement of Science was held in Australia in 1914 at the beginning of World War I (BAAS publication in 1915). This event was a series of lecture tours throughout Australia, with Eddington, Lodge and Rutherford as eminent speakers; the event was essentially British and Australian in character. A more likely candidate for the first international conference in Australia was the 1923 Pan-Pacific Science Congress,

which had more of an international nature with participants from the US, Canada, Britain, Japan, the Netherlands and New Zealand (Hobbs 1923). In addition, the World Radio Conference was held at the University of Sydney in April 1938 with overseas visitors from the UK, the Netherlands (including van derl Pol, Fig. 11.5 and 11.6) and Indonesia (Frater 1982).

2. In fact, Appleton had been persuaded to give a lecture on board, entitled "The Challenge of Scientific Progress". His younger daughter, Rosalind, accompanied Appleton and his wife to Australia; Miss Appleton met the purser of the *Strathmore* on the voyage and they were married later that year in Edinburgh (Clark 1971). During the URSI Assembly, a number of Sydney newspapers carried articles about the Appleton family (including several about the glamorous Rosalind Appleton-Collins [1927–2009], who had taken the stage name Wanda Alpar as a former dancer at the Windmill Theatre in London). On his arrival in Sydney, Appleton was asked by a reporter from *The Daily Telegraph* about the honorary degree that the University of Sydney was to present to him during his visit. When asked about the number of honorary degrees he had received, he responded, 'I don't know—but I've got more degrees than a thermometer' (Bracewell archive, National Radio Astronomy Observatory).

Chapter 12
The Married Life and Motherhood of Ruby Payne-Scott

A Remarkable Family: Bill and Ruby Hall

Ruby Payne-Scott and Bill Hall were happily married for 37 years.[1] Their two children, Peter Gavin Hall and Fiona Margaret Hall, are prominent Australians. It is vital to the understanding of our protagonist that we learn about her family and not just her brief, but vivid career as a physicist and astronomer. The reader must keep in mind that Ruby was forced to make a choice, one that forced her to shear off an entire aspect of her own personality and desire, in favour of another. Ruby made the choice to seek personal happiness in her marriage and children, though doubtless she would have been more than happy to continue her intellectual expansion in the realm of radio astronomy had that been an option. After leaving CSIRO, she put her great breadth and depth of personal knowledge and passion at the disposal of her family; the world profited by gaining two gifted and giving siblings who have fearlessly approached the world on their own terms.

Bill Hall

William "Bill" Holman Hall was born in Inverell, New South Wales, Australia, on 22 August 1911. He died on 21 July 1999, in Wollongong, NSW just before his 88th birthday. As was the case with Ruby, both of Bill's parents came from families that emigrated from the United Kingdom in the late nineteenth century.

---

[1] Based on interviews with Peter Hall in Socorro, New Mexico in May 1998 and with Peter and Fiona Hall in Canberra in February 1999. Additional telephone interviews took place with Peter Hall in Melbourne in February 2007. Extensive interviews in March 2007 in Adelaide were carried out with Fiona, as well as correspondence with both Peter and Fiona in the period 1998–2008. There have also been numerous letters and interviews with Dr. Elizabeth Hall, 1999–2008.

Agnes Paterson, Bill's mother, was born in about 1870 in Glasgow, Scotland and died in Oatley, Sydney, in 1967. She emigrated as a young girl with her family and five siblings. They departed from Glasgow on 15 March 1884 and arrived in Sydney exactly 3 months later.[2] Agnes, then 14, was the eldest child; the others ranged from an infant sister to a brother aged 11.

Bill's father, Sydney Hall, was born in Deal, Kent, close to Dover, in 1872 or 1873. The date of his death is not known but was probably before 1950. Sydney and Agnes were married on 25 August 1902 in Inverell, a town in northern NSW, 230 km west of Grafton, the birthplace of Ruby.

Agnes and Sydney had eight children; Bill was the sixth, born in 1911. Of the six sons, five followed in their father's footsteps to become butchers. There were two sisters. Ivy (1902–1973) never married and lived near Bill and Ruby in Oatley East. She cared for her mother, Agnes, for the last decades of her life. Mary (1908–19??) married name "Pilgrim" was one of the witnesses at Ruby and Bill's "secret" wedding in 1944. Mary remained close to Bill and Ruby's family while the children were young.

Bill Hall's parents moved to Sydney during World War I. Bill remembered his father, Sydney showing him the abattoir (slaughterhouse) at Homebush (later the site of the 2000 Olympic Stadium) where he worked as a butcher. Sydney warned his children to keep away from the abattoir, fearing that they would be distressed by the slaughter of the lambs.

In 1997, 2 years before his death, Bill related a remarkable experience from the early 1930s to his daughter Fiona. Bill was his mother's favourite son, probably because he was one of the younger sons and still at home with her after the older sons had left. In addition, Bill defended his mother against her abusive and probably drunkard husband, Sydney. At the instigation of his mother Agnes, Bill went to Ireland by ship to look for relatives with whom the family had lost contact. This suggests that at least part of the Paterson family originated in Ireland and came to Scotland during the nineteenth century. In any case, Bill had no success and soon returned to Australia. This trip was the first of two overseas trips that Bill made during his lifetime. The second was an extensive, cross-European trip with Ruby in 1976.

The details of Bill's schooling are unknown. Probably he left school in his early teens. He was trained as a French Polisher, a wood finishing technique known to require great patience; this training also gave Bill wide ranging carpentry skills, which he used for the rest of his life. One example is the furniture, including beds that he made for the family home. His children still use some of these pieces. Of

---

[2] 210 passengers were on board the iron sailing ship, *Bann*. Agnes's parents were James and June Paterson from Lanark, Scotland (about 30 km from Glasgow). The family was Presbyterian. The father's occupation was listed as agricultural labourer (later documents listed him as miner and later still as railroad worker). At the time of Agnes's birth, James worked in the Glasgow railway yards. After working in either Inverness or Aberdeen, he decided to emigrate due to the higher wages paid in Australia. Each adult paid £14 for the voyage. Dr. Elizabeth Hall provided many of these details about the family of Bill Hall.

course, there is also the house itself, which Bill and Ruby built in Oatley in 1950–1951. Rachel Makinson recalls letting Bill and Ruby stay in her house while they were still building it, though they also camped on the property while construction progressed and eventually moved in by late August 1951. Their son, Peter, was born a few months later, even though the house had no water or electricity.[3]

Although he was 28 at the start of World War II in 1939, Bill Hall did not go into military service, perhaps because of his colour blindness. In addition, Bill had a "reserved occupation" during the War, working at the Garden Island Naval Base,[4] the principal east coast base of the Royal Australian Navy.

After the war, Bill became a telephone mechanic for the Australian Postmaster General (PMG), the government service that ran the posts, telegraphs and telephones.[5] To qualify as a telephone mechanic, he attended a technical college; Ruby helped him with mathematics.

Friends and neighbours remembered Bill as a "good solid man",[6] who spent lots of time with his children while they were growing up. With no car and no television, the parents created many opportunities for free and creative learning for their son and daughter as they grew up in the 1950s and 60s.

Fiona remembers her father with admiration. Bill was always proud of his wife's illustrious career as a radio physicist, from 1941 to 1951 at the RPL, CSIR and later CSIRO. Bill told Fiona that her mother's troubles with CSIR and CSIRO were due to the fact that she had the "wrong gender". In 1987 he told Fiona, "Doing well is tough, even more so for women. I saw what your mother went through as she battled discrimination."[7] Bill was a strong supporter of sexual equality. Fiona has stated that her father did not have "a sexist nerve in his body".

A key aspect of Bill's life that led to his meeting Ruby Payne-Scott, was his avid participation in bush walking (rambling in the UK or hiking in the US). On 6 November 1936, at age 25, he joined the Sydney Bush Walkers (SBW), an organisation founded in 1927 by Jack Debert.[8] Bill was clearly an accomplished

---

[3] National Archives of Australia-NAA: A1/1/1, Part 6.

[4] It is not known what work Bill did at Garden Island Naval Base. We assume he worked on maintenance of ships, perhaps as a carpenter. A request to the National Archives of Australia in September 2012 yielded no information.

[5] Merle Watman (letter to E. Hall 1999) remembered that Bill worked at the Sutherland PMG exchange, while Arthur Gilroy (letter to E. Hall 1999) suggested that Bill was in charge of the Blakehurst exchange. Peter Hall remembered that in the late 1950s his father occasionally worked on a night shift, leaving home at 6pm. The increased wages were a welcome supplement to the family income.

[6] Betty Hall interview, February 2007.

[7] This conversation occurred when Fiona's career was blossoming. At a time when she had been invited to visit Japan to exhibit her work in an art gallery, her brother had just received a prestigious prize from a French mathematical society

[8] Butler has described this group in some detail in the delightful reminiscence, *The Barefoot Walker: A Remarkable Story of Adventure, Courage and Romance*, written by Dorothy Butler 1991. Butler (1911–2008) joined the SBW in 1931 and made a number of major ascents in the Blue

Fig. 12.1 Dot English Butler and Bill Hall; a large brown snake had been killed by Bill on a Sydney Bush Walkers (SBW) event—late 1930s or early 1940s (Bill Hall family collection, used by permission of Peter Hall)

bush walker. In this period just prior to WWII, he was a member of the "Tigers", a group of 10–12 high-powered walkers which included Dot (*née* English) Butler, Alex Colley, Hilma Galliott Colley, Max Gentle and the founder, Debert. Figure 12.1 is an adventuresome photograph of Dot (Dorothy) and Bill holding a large brown snake killed by Bill; the brown snake is one of the more deadly Australian snakes with a potent venom. To qualify for this group, the applicant had to prove herself or himself over a 70 mile, 3-day hike in the Blue Mountains with an over 9,000 ft. (2,700 m) change in elevation. Butler has written about herself, "There were those of mighty statute and physique among the Tigers, but also among them was this small, neat girl who, once the going became really tough, could out walk and out climb all of them." The group broke up in WWII, since many members were in the armed forces.

Ruby Payne-Scott joined the SBW on 10 January 1941,[9] probably meeting Bill soon afterwards. It was customary to be subjected to a 3-month probationary period; the initiate was required to satisfactorily take part in 31 day test walks and a full weekend test walk. She would have been judged on walking ability and on a general capacity to "fit-in". By Easter, 13 April 1941, Payne-Scott was on a bush walk with Bill in a large group on the south coast of NSW near Pigeon House Mountain. Alex Colley, a well known environmentalist who joined the SBW in 1936 and was also a "Tiger", has written about how these dissimilar individuals could only have met on a bush walk:

Mountains, west of Sydney. The term 'bush walking' had only been invented in the late 1920s. Later in her life she went to New Zealand and made a number of ascents, including Mount Cook.

[9] Dates of Bill's and Ruby's joining the SBW provided by Bill Holland (SBW archivist) to E. Hall, 19 March, 1999.

> I knew Bill quite well... His background was very different to that of Ruby... later [he] became a telephone mechanic. He stuttered and looked a bit of a rough diamond—we called him Ben Hall—the bushranger [the infamous Australian outlaw of the 19th century]. He was a very hardy bush walker and a very popular member of the club. In normal circumstances Bill and Ruby would never have met, or contemplated marriage. But on a bush walk you share a range of experiences, some very challenging...and form close, often life long friendships, irrespective of educational, occupational or social background. Bill and Ruby were very devoted to each other.[10]

Members of the SBW remember that, despite the differences in their circumstances, Ruby and Bill were fortunate to find one another.[11] Betty Hall, who became close with Ruby in her later years, and raised children in the same neighbourhood of Oatley, in fact met Ruby and Bill in 1946 through the SBW. Betty offers a sweet impression of Ruby and Bill's relationship:

> Like the rest of us, Ruby may have had her faults, but I have the warmest recollections of her. Although she was not in general demonstrative it still moves me to tears to remember how she referred to Bill as "my Billy" and how her face softened when she spoke to him.

Ruby apparently made some walks with the "Tigers", but was never an official member. Figure 12.2 is a photograph of Bill and Ruby in 1947 on a Christmas trip to Pretty Plains near Mt. Kosciusko; Ruby is washing Bill's hair. In Fig. 12.3, Ruby, Bill and a colleague are shown on the same Christmas excursion. A marvellous photo of Ruby and her close friend Betty Hall at the summit of Mt. Twynam in the Mt. Kosciusko National Park is shown in Fig. 12.4. In Fig. 12.5, Bill, Ruby and Val Gilroy are studying a map in Tasmania on the Overland Track in Cradle Mountain National Park (Tasmania) between Windermere and Pelion West in 1949; the weather was clearly wet and cold.

R. W. Younger, a teenage member of the SBW during WWII, remembers that Ruby and Bill went out of their way to help new members make friends and adjust to the "onerous activities associated with bush walking. Ruby became a member of the [SBW] committee and took part in discussions at general meeting".[12]

During the 1946 Christmas period, Betty Hall went on one of the group's regular Mt. Kosciusko trips before she was a member.[13] She wrote:

> On this first trip I took the wrong kind of billycan—a tin one which makes the tea taste terrible—I was teased by Ruby for taking "Clive of India" brand curry powder, thus

---

[10] Letter to Goss, 18 March 1999, from Alex Colley. Colley also wrote: "Bush walkers come from all walks of life, but we are all equal on walks".

[11] Elizabeth ("Betty") Hall has written (19 March 1999) that the SBW "was also a very successful marriage bureau... There is nothing like a hard walk to bring out the best and the worst in people; after you have walked with them a few times you have no illusions about your chosen partner!"

[12] Letter to Goss, 15 March 1999.

[13] Betty began bush walking with another young woman. On one of their first trips in the Burragorang Valley they returned quite late, missing the last bus from the top of the track. They were rescued by a man in a truck, who told them about the Sydney Bush Walkers; since she knew she needed to develop a network of friends, she became a prospective member in December 1946 and a full member on 30 April 1947.

Fig. 12.2 Bill Hall and Ruby Payne-Scott. Pretty Plains, Mt. Kosciusko National Park, Christmas 1947. Bill has his hair washed by Ruby (Bill Hall family collection, used by permission of Peter Hall)

Fig. 12.3 Bush walking in Mount Kosciusko National Park. A trip organised by Ruby Payne-Scott and Bill Hall at Christmas 1947 to Mount Tate: Ruby, Bill and possibly Bill Caw (Photo by Betty Hall. Bill Hall family collection, used by permission of Betty Hall)

betraying my Imperialist tendencies, and my sleeping-bag was too thin for Kosciusko weather. Bill and Ruby forgave me the billycan, used the curry-powder, insisted that I sleep between them to keep warm, and during the trip taught me many of the things a bushwalker in Australia needed to know. Listening to their talk around the campfire gave me my first introduction to Australian left-wing politics. Ruby was a forceful arguer and antagonised some people, but she was not an intellectual snob. Although I was no match for her, she never made me feel like a fool.

Fig. 12.4 A marvellous photo of Betty Hall (*sitting*) and Ruby Payne-Scott (*standing and holding on to the marker for the "peak" of the mountain*) on Mount Twynam in Mount Kosciusko National Park (Photo with Betty Hall's camera by an unknown companion. Bill Hall family collection, used by permission of Betty Hall)

Fig. 12.5 Overland Track in Tasmania 1949. Bill Hall, Ruby Hall and Val Gilroy. Between Windermere and Pelion West, Cradle Mountain National Park (From Bill Hall family collection. Photo by Arthur Gilroy, used by permission of Lindsay Baudinet, his daughter)

After the probationary hikes, Betty came before the full SBW committee to be considered for membership. There was some doubt about her because she had been hitchhiking, an activity frowned upon; however, Ruby spoke up for her saying that Betty "was good at washing up". Betty has stated that "the friends that I made in the SBW, including Bill and Ruby, influenced my decision to stay in Australia" (see Additional Note, end of this chapter).

Betty also reports that, prior to their marriage in September 1944, Ruby and Bill were living "in sin" in a house in Ashfield, a less common occurrence in 1944 than

Fig. 12.6 (**a** and **b**) The new house of Ruby and Bill Hall, likely in mid-1951. 120 Woronora Parade, Oatley, NSW, Sydney. Bill and Ruby camped out during the construction of the new house. In Fig. 12.6b Bill can be seen (Photo by Ruby Hall. Bill Hall family collection, used by permission of Peter Hall)

60 years later. The marriage certificate listed her address as 5 Fairleigh Street Ashfield, Sydney, while his address was listed as 6 Stanley Street, Arncliffe, which was the existing home of his family. In late August or early September 1951, Bill and Ruby moved to Oatley,[14] where they designed and built their house, camping out while the construction was going on. They lived elsewhere during the week and camped on the building site during the weekend while working on the

---

[14] NAA; C3830, A1/1/1 Part 6. Payne-Scott wrote to Pawsey on 18 October 1951 (a month before Peter was born): "We moved in here about 6 week ago without either light or water, but things are gradually straightening out, although I still seem to entertain sundry tradesmen most days of the week."

Fig. 12.7 A young Peter Hall in Oatley (Photo by Ruby Hall. Bill Hall family collection, used by permission of Peter Hall)

house (Fig. 12.6a, b show the house at 120 Woronora Parade, Oatley during this period; Bill can be seen faintly in the background as Ruby took the picture). Peter and Fiona grew up in this house, from the time they were born, in late 1951 and late 1953, respectively.

Peter Hall: Mathematician

Peter was born on 20 November 1951, 4 months after his mother left CSIRO. An early photograph, probably taken by his mother, shows Peter playing in the garden of their home in Oatley (Fig. 12.7).

His mother encouraged Peter to seek an advanced education by suggesting a variety of educational and career choices. However, it was probably her passion for the sciences that influenced him most. Peter remembers his mother telling him about her great excitement when she observed the radio emission of the sun at sun rise, as shown by interference fringes from the sea-cliff interferometer in 1946. He recounts:

[She was excited] by the realisation that the radio emission was associated with sunspots; quite late in her life, the excitement was still with her.[15]

As a child, Peter found in the Oatley home tables of solar azimuth and elevation for different times of the year as a function of local time. These were probably left

---

[15] From the interview on 12 February 2007. It is possible that Payne-Scott was recalling her memories of the sun rise on 26 January 1946 (Australia Day) when interference fringes from the radio sun were observed for the first time.

over from the period when his mother was observing the sun at Dover Heights in 1946–1947. Peter was a keen photographer and used these tables to determine the earliest times in the afternoon when the sun would be best for taking photos near their house, thus utilizing a complex set of scientific data to guide his hobby.

Payne-Scott's experience as a mathematics teacher made a lasting impression on Peter. In the 1960s, while she was the Science Mistress at Danebank School, she was also an adult education instructor at St. George's Technical College in Sydney. She told Peter that success in mathematics was as much a matter of intuition as it was knowledge of the rules. As an example of this she told the story of a butcher who was a student in her adult education class. This man could neither add nor subtract large numbers, nor even work out simple algebra. He was, however, adept in the accurate estimation of fractions. Payne-Scott asked him the source of his proficiency. His answer was simple:

> Well, we have a carcass of meat. One third is gristle and bone and will be thrown away. Two-thirds is the saleable portion, on which I make a profit. I've done this so often that I always work out my profit margin on two-thirds of what I buy from the wholesaler.

Peter also remembered how absent-minded his mother was when she was concentrating on mathematics problems or lesson planning for the next day. This inattention could lead to over-cooked, if not burned food in the kitchen. Because lamb was cheap in the 1960s, a frequent evening meal at the Hall house consisted of lamb cutlets or chops. When her mind was too preoccupied with other considerations, Ruby would forget to check the oven, with disastrous results. As smoke appeared, she would utter her self-invented swear word, "gordy buggers". Though Ruby's cuisine was often not of top quality, Bill was a better chef. Once when Peter was a child, and his mother was in hospital for several days, Peter remembers the improved meals at home as Bill looked after the cooking. While Ruby was in hospital, Bill would even bring home a store-bought pie or cake on the way from work, something Payne-Scott would never do. When Peter's mother returned from hospital, he noticed the quality of the cuisine deteriorated again.

Poor cooking skills aside, Ruby Payne-Scott was clearly an inspiring force in her son's life. Peter had initially planned to study physics at university:

> ... when I finished [Sydney Technical High School], I was determined to become a physicist and I couldn't have given any better reason than that my mother ... showed me that physics was a wonderful and honourable profession.

When he began at Sydney University in 1970, the first physics lecture was given by Professor Robert May.[16]

> ... but he only gave the first lecture to the advanced class in physics and then he disappeared to do something more uplifting. And from that point on, things went downhill. If the other physics lecturers had been as good as [May], I probably would have been a physicist.

---

[16] Robert May (now Lord May) was Professor of Theoretical Physics at Sydney University and later President of the Royal Society from 2000 to 2005.

Instead, Peter Hall became a mathematician with an Honours Degree in Mathematical Statistics from the University of Sydney in 1974. Peter then spent 2 years, 1974–1976, in the UK at Oxford. In 1976 he received an M.Sc. from the Australian National University, Canberra and a DPhil from Oxford.

Peter Hall has had a remarkable career with numerous honours, establishing himself as a leading mathematical statistician in Australia. After a short period as a Lecturer in Statistics at the University of Melbourne (1976–1978), he went to the Australian National University from 1978 to 2006, becoming a Professor in 1988. In 2006, he became a Federation Fellow of the Australian Research Council at the University of Melbourne. He was elected a Fellow of the Australian Academy of Science in 1987 and a Fellow of the Royal Society of London in 2000. On Australia Day 2013, he was honoured as an Officer in the Order of Australia. His publication record is prolific, including four monographs, and a book on a single highly specialised subject—at the time of this book's publishing he was a listed contributor to 179 publications. He has received numerous honours and awards and was President of the Australian Mathematical Society (2006–2008). Peter Hall married Jeannie Jean Chien on 15 April 1977. In Fig. 12.8, Professor Peter G. Hall is shown on a visit to the NRAO, P. V. Dominici Science Center, Socorro, New Mexico, on 12 May 1998.

Fiona Hall: Artist

Fiona Hall, who became a well-known and much acclaimed Australian artist (Fig. 12.9), was born on 16 November 1953, almost exactly 2 years after her brother Peter, who we can see giving his little sister a ride on a tricycle in Fig. 12.10. The two are shown in a formal portrait in Fig. 12.11. A more playful image (Fig. 12.12), shows Fiona and Peter with their father on holiday at Jamberoo, NSW, about 100 km south of Oatley.

From an early age of two or three, Fiona was given materials for simple art work by her mother. Fiona described the educational philosophy of her parents[17]:

> I think [their] philosophy was you educate your children by example, not telling them what's what. I think I knew as a child subconsciously that I could do whatever I wanted to. There were no expectations about what a girl's later role in life should be.

Fiona Hall has described semi-austere conditions while she was growing up. The family ate dried fruit with a few store-bought biscuits as a snack. Payne-Scott sewed for her daughter, making frilly party dresses for her birthdays. The Hall household was full of books and discussions about theatre, the visual arts, literature and especially public affairs and politics. There was little music, however, in their home.

---

[17] Karen Pakula in *The Sydney Morning Herald*, 1 March 2008.

Fig. 12.8 Professor Peter Hall on a visit to the National Radio Astronomy Observatory Domenici Science Center Operations Center, New Mexico, USA, on 12 May 1998 (Photo by Goss)

Fig. 12.9 Fiona Hall. A formal portrait taken by Greg Weight in 1995. The fabric, woven by Fiona from coke cans is the same as that shown in Fig. 12.16. Used on the dust jacket of Ewington, 2005 (Image courtesy the artist and Roslyn Oxley9 Gallery, Sydney)

Fig. 12.10 The Hall children in the 1950s. Peter is seen giving his little sister a ride on a tricycle (Photo by Ruby Hall. Bill Hall family collection, used by permission of Peter Hall)

Fig. 12.11 Formal portrait of the Hall children, 1950s (Bill Hall family collection, used by permission of Peter Hall)

An early, but pivotal event in Fiona's life occurred in 1967 when she was 14; her mother took her to the Art Gallery of New South Wales to see the landmark exhibition "Two Decades of American Painting". Their "outing [was] typical of the expeditions organised by Ruby to develop the children's interests that leaves a lasting impression".[18]

There were sometimes conflicts with her mother while Fiona was growing up; both were remarkably single minded. Fiona was very close to her father; on weekends the two would walk through the bush to visit her paternal grandmother and her Aunt Ivy who lived with Agnes. Her father called Fiona "my little shadow".

---

[18] Julie Ewington, *Fiona Hall*, 2005, page 180. In this book, Ewington has given a thorough description of Fiona Hall's artistic career.

Fig. 12.12 Bill, Fiona and Peter Hall on a family holiday at Jamberoo, New South Wales (Photo by Ruby Hall. Bill Hall family collection, used by permission of Peter Hall)

A running joke, originating with Bill, was: "Who shall go first through the jungle so the tigers will get you [Fiona] or me [Bill] first?" Fiona had little affinity with her aged grandmother, who was 83 years older than Fiona and showed more affection for her grandsons. Agnes Hall was, however, the only grandparent the Hall children knew; she died in 1967 when Fiona was 10. In contrast, Fiona was close to her Aunt Ivy, who died in 1973 at age 71. The grandmother and aunt were suspicious of Bill and Ruby Hall; in their house in the late 1950s percolated coffee instead of tea was prevalent—"perhaps the drink of the radical left!"[19]

Fiona attended Oatley West Primary School and then Penshurst Girls' High School, where she had a reputation for being quite self-assured.[20] While she was in high school, Fiona considered architecture as a career. This is a fascinating connection, as her mother had long been interested in architecture as well.[21] But in the last year of high school, Fiona decided to attend art school. In the early 1970s, she studied at East Sydney Technical College, where she focused initially on painting

---

[19] Quote from Fiona Hall to Libby and W.M. Goss at her home, Adelaide, 9 March 2007.

[20] Telephone call from the late Marie Alewood, a neighbour of Payne-Scott's, 12 February 2007.

[21] When Peter Hall went through the house in 1999 (after the death of his father), he found numerous architectural books. Though Ruby's interests ranged from mathematics and physics through particularly biology, a special interest of hers was architecture. She told Betty Hall that if she could have a second life she would return as an architect.

and then photography.[22] No stranger to photography, she had started experimenting with a simple Kodak camera when she was 11. She and her brother would compete for the use of the family laundry "darkroom" for the development of film. In her late 20s, Fiona Hall studied at the Visual Studies Workshop in Rochester, New York, between 1978 and 1982, with visits back to Australia before the death of her mother in 1981 (Ewington 2005).

Since the early 1980s, Fiona Hall has branched into many new art forms, often using everyday objects such as soap, Tupperware, video tapes and legal tender from numerous countries. Some of her most famous objects have consisted of finely cast metal sculptures set inside sardine tins. The book by Ewington and the accompanying volume from the 2008 exhibition by MacGregor, Savage, Webb, O' Brien and Hall, provide masterful overviews of the oeuvre of Fiona Hall (see Hall 2008).

Figures 12.13, 12.14, 12.15, and 12.16 show some the striking examples of Fiona Hall's early work which illustrate the eclectic nature of her art. In these pieces, Fiona took famous paintings from the past and "reconstructed" them out of odd bits of everyday flotsam. In Fig. 12.13, created in 1981, we see images that suggest a perhaps subconscious connection with her mother's astronomical research work from 1945 to 1951. These reconstructed paintings are "notable for the obvious pleasure Hall took in multiple slippages of meaning between historical templates and contemporary life. She substitutes contemporary artefacts for depicted objects..."[23] In Hall's *Starry Night*,[24] electrical cords swirl in the night sky as in van Gogh's famous *Starry Night*. In Fig. 12.13, a related study shows the *Aurora Borealis/Aurora Australis*, a study remarkably related to solar bursts and the ionosphere through magnetism (note the magnet at the bottom). With some imagination the Van Allen belts of the earth can be discerned in this inventive creation. This work is from the Antipodean Suite show of 1981, the year of Ruby Payne-Scott's death. Whether intentional or through the odd workings of inspiration upon which an artist touches, these pieces Fiona created near the time of her mother's death have clear echoes of Ruby's astronomical work. The connection between Type II bursts (related to aurora), first detected at Dover Heights in March 1947, and the *Aurora Australis* is striking.

Fern Garden, located in the inhospitable[25] courtyard of the National Gallery of Australia in Canberra was a commissioned work completed in 1998. The details of the garden are visible in Fig. 12.14.

---

[22] In *Fiona Hall* (2005, page 27), Ewington has written, "In 1974 it seemed that Fiona Hall emerged fully formed as an artist." At this time she began exhibiting photographs, which were collected and published.

[23] Ewington (2005), page 42.

[24] An image of this work was used as the cover for Goss and McGee's *Under the Radar—the First Woman in Radio Astronomy: Ruby Payne-Scott*.

[25] Ewington (2005), page 118: "In a seemingly inhospitable courtyard, formed by the junction of the original National Gallery building of 1982 and the extensions opened in 1998, Hall sited 58 mature ... giant tree ferns that are amongst the most ancient plants in Australia."

Fig. 12.13 Reconstructed painting, *Aurora Borealis/Aura Australis* in the Antipodean Suite series from 1981, by Fiona Hall. The viewer can see a relationship to solar radio emission and even the ionosphere (Image courtesy the artist and Roslyn Oxley9 Gallery, Sydney)

Fig. 12.14 Details of the "Fern Garden", the National Gallery of Australia in Canberra. Fifty-eight *Dicksonia antarctica* large tree ferns are located in this garden which can be viewed from windows inside the museum (Image courtesy the artist and Roslyn Oxley9 Gallery, Sydney)

In 2000, Fiona Hall designed *A Folly for Mrs. Macquarie* (Fig. 12.15), located in the Royal Botanic Gardens in Sydney, a place she often walked through with her father Bill in the last decade of his life. The death of her father in 1999 is commemorated in the iron work; her father's name is inscribed on the scythe.[26] The site provides a remarkable view of Sydney Harbour.

---

[26] Fiona has told Goss (18 April 2009) that she often walked through the Royal Botanic Garden (the location of the sculpture) with her father in the early 1990s; she would meet him near Sydney Harbour and they would walk through the Domain to Kings Cross railway station. Bill Hall would then take the train back to his home in Oatley. Fiona thinks that her father was attracted to this location as it was close to Garden Island where he worked at the naval dockyard during World War II.

Fig. 12.15 *A Folly for Mrs. Macquarie*, completed in 2000. A public art work for the Sydney Sculpture Walk, commissioned by the City of Sydney. The *Folley* is located in the Royal Botanic Gardens, in front of Government House with a sensational view of Sydney Harbour (Image courtesy the artist and Roslyn Oxley9 Gallery, Sydney)

The challenging installation, *Give a Dog a Bone*, was shown in three versions during 1996 and 1997. The installation had a larger than life sized photograph of her father (Fig. 12.6a), centrally placed and surrounded by a shrine of boxes, each piled with items Fiona had carved out of soap (Fig. 12.16b, c). The central portrait was the last photograph of her father taken by Fiona Hall for public display; Bill Hall died about 3 years later. The ironic nature of the installation is obvious:

> Hall required a king for this castle of worthless wealth [supermarket objects carved from soap]. The children's song which gave the works its title mentions 'this old man comes rolling home' ... and Hall commissioned a larger than life-size photograph of her elderly father, William Holman Hall, naked except for a fabulous full-length cloak knitted from aluminium Coca-Cola tins.[27]

Fiona Hall has clearly been influenced by her mother[28]:

> If I've inherited anything from my mother, it's applying a logic [as she describes her eclectic combinations of concept and medium]... I will come up with something that's

---

[27] Ewington, page 138. The knitted cloak photograph has the title "The Social Fabric" (1996); Greg Weight was the photographer. Note the same fabric in Fig. 12.10. Fiona Hall has told Goss that she was apprehensive about asking her father to pose in this semi-nude fashion. She was pleased that he was totally at ease during the photographic session.

[28] Brook Turner *The Australian Financial Review Magazine* 27 April 2012.

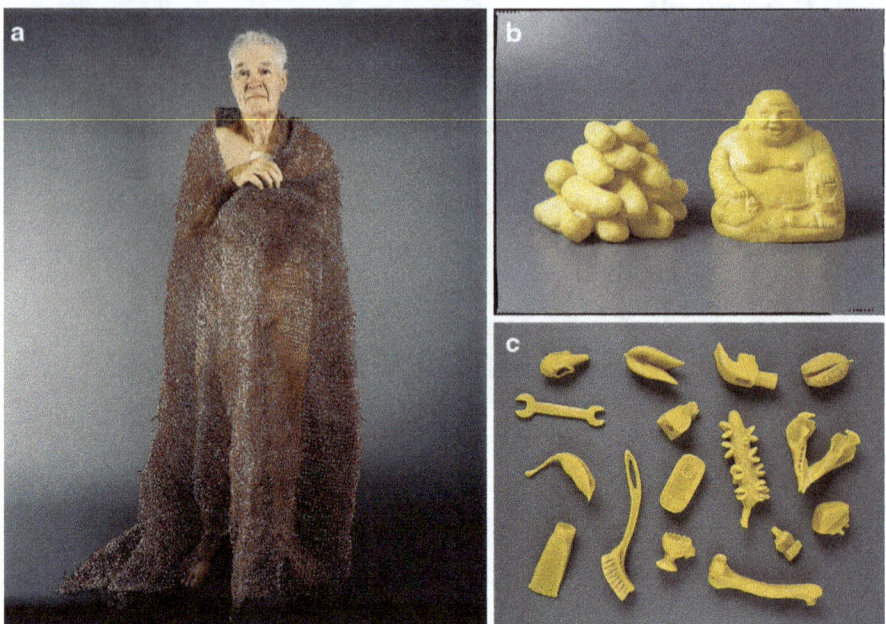

Fig. 12.16 (a) From *Give a Dog a Bone*. This portrait was the last photograph of her father taken by Fiona Hall for public display. The fabric of his cape, woven by Fiona from coke can strips, is also seen in Fig. 12.11 (Image courtesy the artist and Roslyn Oxley9 Gallery, Sydney). (b) From *Give a Dog a Bone*. Buddha and pellets carved from soap by Fiona Hall (Image courtesy the artist and Roslyn Oxley9 Gallery, Sydney). (c) From *Give a Dog a Bone*. Sixteen items carved from soap by Fiona Hall (Image courtesy the artist and Roslyn Oxley9 Gallery, Sydney)

> quite illogical and pushes things that are...very disparate, and then try and mesh them materially and conceptually..... [I doubt myself] constantly. I think for a lot of people who work across various creative disciplines, including science, it's not that you start out so much with confidence as faith that the hunch you've got is worth embarking upon.

In 2008, more than 100,000 people visited the exhibit "Fiona Hall: Force Field" at the Museum of Contemporary Art in Sydney. To date this has been the most popular Australian exhibition, based on the number of visitors.[29] The exhibition was a retrospective on the work of Fiona Hall since the 1970s including photography, sculpture and installation art. In mid-2008 the exhibition moved to New Zealand, first to the Wellington City Gallery and then to Christchurch in late 2008 (Hall 2008).

Major exhibits have continued in 2011 and 2012. In November 2011, the exhibit "Shot Through" opened on 3 November 2011[30] at the Rosyln Oxley9 Gallery in Paddington (Sydney) Australia. This exhibition "features creatures from the

---

[29] *Wentworth Courier*, Sydney, 18 June 2008. Drew Sheldrick.

[30] Goss met Fiona Hall the following day at the exhibit which ran to 26 November 2011.

International Union for Conservation's Red List of endangered species, painstakingly rendered in camouflage fabric of the countries from which they come".[31] An example is a 2.3 by 1.4 by 0.75 m California condor, wearing a flattened cola can around its neck. This piece was part of the dOCUMENTA(13) exhibition in Kessel, Germany, in mid-2012, an international show every 5 years dedicated to "fundamental research on multi-species cohabitation in the world".[32]

Tim Lloyd wrote in 2009[33]:

> She's considered by some to be Australia's finest living artist.... She's best described as a sculptor, although her mediums are incredibly varied. Hall makes art from the everyday: weaving beads into sea creatures, painting leaves on banknotes, making bird's nests from shredded banknotes, erotic images from sardine cans, woven heads from videotapes. Political themes suffuse her work.

Fiona Hall has been recognised with two major awards. In September 2009, she received South Australia's Ruby Award, the Premier's Award for Lifetime Achievement.[34] In September 2011, she received the Artist Award in the Melbourne Art Foundation Awards for the visual Arts to recognise her 35-year contribution to visual arts.

The children of Ruby Payne-Scott and William H. Hall continue to have remarkable careers, clearly influenced by two gifted and loving parents who made their children's development a lifetime priority.

Additional Note

1. Elizabeth Kate Hall, née Hurley (no relation to Bill Hall), was born in London on 5 June 1919 and grew up over a florist shop run by her parents, who had inherited it from Betty's grandparents. As an only child, she became an orphan after her father's death when she was seven and her mother's when she was 17; she ran the florist shop for several years with the help of the manager. She was conscripted into the Royal Navy in early 1944 when she was 25 and the shop was then closed. In the Women's Royal Naval Service (WRNS), she became a transport driver. She volunteered for overseas service and was sent to Australia just as the War was ending in 1945. In Sydney she was attached to the civil engineers as a driver and saw a lot of the city. She was discharged on 5 July 1946, and having decided to stay in Australia, found a job while she lived in a single room in Bondi, an eastern beach suburb in Sydney. With no ties back in London, Betty remained in Australia.

---

[31] Brook Turner *The Australian Financial Review Magazine* 27 April 2012.
[32] Ibid.
[33] Tim Lloyd *The Advertiser* (Adelaide Australia) 12 September 2009.
[34] The Ruby award is named after Dame Ruby Beatrice Litchfield (1912–2001), a community and arts leader in South Australia.

Chapter 13
1963–1974: Employment at Danebank School

In 1963, after her children had reached the ages of 12 (Peter) and 10 (Fiona), Ruby Payne-Scott took a position as a part-time science and senior mathematics teacher at the Danebank Church of England School for Girls in Hurstsville,[1] only 5 km distant from her home in Oatley. Hurstville was easily accessible by train from Oatley and Ruby's children were, of course, at local schools. At Danebank School, she was known as Mrs. Ruby Hall. Two photographs of the school in early 2007 are shown in Fig. 13.1a, b.

Ruby had prior teaching experience, having taught at Woodlands School for a year and a half in 1938–1939, and though this new experience started off in a promising fashion, it became quite disappointing after a few years. A major reason was the debilitating effect of the possible onset of Alzheimer's disease.

In 1963, Danebank was a small school and had no specialist scientist teacher. The Wyndham Scheme had been introduced into NSW schools in 1962, and science in Years 7–10 was compulsory for all students, leading to optional multidisciplinary sciences in Years 11 and 12. Biology was the only science taught at Danebank. Mrs Joyce Cowell, Principal of Danebank School, was impressed by Ruby's academic qualifications, but there was little or no discussion of Ruby's earlier achievements. At this time, teaching was viewed by women with young families as an ideal profession, as it was possible to combine school hours and holidays with the duties of raising a family. Teachers at private schools did not have to be academically qualified, and their salaries were lower than those of teachers in state schools. Private schools, including Danebank, therefore often had a reputation of offering an inferior standard of education. Some parents would send difficult or poorly achieving children to a private school in the hope that smaller classes and a different philosophy might bring about improvements in attitude and academic progress. Facilities at Danebank were basic, with only one small room for science. Payne-Scott was responsible for organising the new laboratories and equipment essential

---

[1] I am indebted to Carolyn Little, retired Science Coordinator, Danebank School, for details of Payne-Scott's career at Danebank; most of the text in this chapter has been contributed by Little.

Fig. 13.1 (**a** and **b**) Two views of Danebank School in February 2007 (Images obtained by Goss with the assistance of Jan Christensen and Sue Brian, Betty Hall's daughters)

for the new State and Commonwealth requirements for Science education. A striking positive comment about Payne-Scott appears in a school history from 1983: "When she commenced at Danebank there were no real laboratory facilities...she planned the new laboratories and ordered and installed the new equipment needed for the new science syllabus. She left behind...a very fine Science Department."[2] She remained effectively in charge of science and senior mathematics at the school until her retirement in 1974.

Payne-Scott never achieved any real intimacy with other members of the school staff. Relationships at the school were formal—first names were never used. Ruby never mentioned her previous career, other than to mention that she had spent time at the CSIRO. She spoke, however, with great love and affection of her family, and came across as a typical mother with domestic concerns which were shared with other staff members. She made generous offers of clothing to other teachers' children. Despite her frustration with the progress of many students, she expressed a deep concern for each of them individually, displaying a surprising knowledge of their backgrounds as a source of their problems. However, as she aged and was probably affected by the onset of Alzheimer's disease, her stories deteriorated into

---

[2] Joyce C. Cowell, *Danebank Church of England School for Girls, The First Fifty Years*, 1983. The school was founded in 1933.

longwindedness and her ability to cope also declined. At her retirement in 1974, her farewell speech occupied more time than that of Cowell, who was retiring at the same time as Principal.

The history of Payne-Scott's relations with students is troubling. Past students have commented that she did not engage well with many of them, even though the classes were small. She appeared remote and was unduly stern in her treatment of the pupils, and could be dismissive of their academic standards. From the mid-1960s students found her behaviour eccentric. Her class management skills were poor, with girls walking in and out of the class without being noticed; perhaps her poor eyesight contributed to this inattention. On occasion she called girls by the wrong names. She would ring a handbell to attract attention, leading to the nickname "Tinker Bell". This was in contrast to her teaching at Woodlands School in 1938–1939; Kate Foy, at this time a mathematics teacher at Danebank, had been her student at Woodlands in Adelaide in the late 1930s and had memories of an effective and tolerant teacher.[3]

The brighter students, however, fared well under her instruction. She clearly "knew her stuff" according to one student whom Ruby taught in the early 1970s; she brought books from home for this girl and favoured her with a gift at Christmas. Some students understood that she was clever and had had a prior career in physics, but the extent of that career was never discussed.

These contradictory impressions were probably caused by her illness. Because of her deteriorating condition in the early 1970s, the School Council introduced a policy of compulsory retirement in order to force her to leave. Payne-Scott did not maintain contact with the School after 1974, and the lingering memory of her there is that of an eccentric individual.

When Goss began gathering material on Payne-Scott in 1997, he contacted the Principal of Danebank School who was surprised to hear about her past achievements. The Science Coordinator, Carolyn Little, was herself a physics teacher and had an interest in radio astronomy and the history of science. Little decided that the life of such a significant former member of the Danebank staff should be celebrated by the School and even by the wider community.[4] The School initiated an annual lecture, the Ruby Payne-Scott Lecture, to be presented at the School by successful women scientists. The first lecture in 1999 was given, appropriately, by Dr Anne Green,[5] a radio astronomer at the University of Sydney, on the topic, "Microwaves in Space". Peter and Fiona Hall have both supported the

---

[3] Communication from Foy to Carolyn Little in 1999; communication to Dick McGee in June 2007.

[4] Joyce C. Cowell wrote Goss in 1997, "She was a dedicated and caring teacher for 12 years of her service."

[5] In 2008, Professor Anne Green (née Barwick) was the Head of the School of Physics at the University of Sydney. She is probably the third woman radio astronomer in Australia following Payne-Scott and Professor Beverly Wills (née Harris) of the University of Texas, Austin. McGee, co-author of Under the Radar, was in the audience for this inaugural lecture. The lectures have continued through 2012. Four of the subsequent lecturers have been radio astronomers from the

lecture series; they continue to do so, attending a number of these events. In 2012, Dr. Lisa Harvey-Smith of the Australia Square Kilometre Project (CSIRO) was the lecturer.

During the years of the Ruby Payne-Scott Lecture series, they have been presented by outstanding women scientists in a variety of fields. The guest speaker has generally given an account of her area of research and achievement, together with some insight into difficulties she might have encountered along the way. Each lecture is preceded by biographical information by students and visitors who had some connection with Ruby. On occasions, this has included a short dramatisation of events in Ruby's life, interviews with former students, and excerpts from radio and TV programs. In 1999, for example, a theatrical presentation, entitled "Who Was Ruby Payne-Scott? A Reflection on the Life and Work of Ruby Payne-Scott" was presented by the Year 11 Drama and Physics classes with the role of Ruby played by Catherine Bond. The Danebank audience has been joined by students from local schools such as Penshurst Girls' High School (Fiona Hall's old school), St George Girls' High School, Georges River College, Oatley and St George Christian School, among others.

Sadly, Payne-Scott's experience at Danebank School was not a happy one. But as her daughter, Fiona, remarked: "Danebank School is a minor part of her life. Radar, radio astronomy and our family were the major events. Mum's constant battling against adversity and her lucky breaks, such as coming to Sydney from the country as a young [child], characterise her life."[6]

CSIRO: Jessica Chapman, 2000; Kate Brooks, 2005; Ilana Feain, 2009; and Lisa Harvey-Smith, 2012. Both Brooks and Feain are also recipients of the Payne-Scott Award, given by the CSIRO.

[6] Interview March 2007, Adelaide.

Chapter 14
The Last Years and Legacy of Ruby Payne-Scott

Last Years

When Payne-Scott left her position at Danebank School in late 1974, her mental condition was deteriorating. Her conflicts with colleagues at Danebank were possibly due to the first signs of Alzheimer's disease, which accelerated until her death only 6 years later in 1981. She aged rapidly as indicated in the passport photograph taken of her in 1976 before the trip to Europe (see Fig. 14.1). This photo shows an aged individual, yet her daughter, Fiona, has little memory of her appearing as an old person—the transition occurred rapidly.

The accompanying mental deterioration can be judged by a letter she wrote to Joan Freeman Jelley on 16 February 1976. The letter was to congratulate Freeman on the receipt of the Rutherford Prize. She explained to Freeman that her name was now Ruby Hall, because she had adopted Bill's surname when the children were born in the 1950's. She wrote about her teaching and was effusive about her children's successes. At this time Peter was at Oxford working on a DPhil in probability theory. Ruby wrote about Fiona:

> Fiona is our surprise. She is a very good artist and her favourite medium is photography... [she] is now using her savings to explore Europe.

The striking aspect of this letter is the poor quality of the handwriting when compared to her handwritten letters and documents from earlier periods. Probably the poor quality of her penmanship can be associated with the mental degradation that was beginning at the age of only 64.

An ambitious overseas trip was undertaken by Bill and Ruby Hall in 1976. Peter was finishing his degree at Oxford, which he had begun in 1974. Peter remembers that she seemed in a reasonable mental state in 1974 although he later remembered that she had occasionally been confused before he left Australia. Bill and Ruby Hall flew to Japan and then took a boat to Vladivostok for the start of the Trans-Siberian

Fig. 14.1 Ruby Hall's passport photo before the trip she and Bill took to Europe via Japan and then Vladivostok (Trans Siberian Express). Her aging was apparent in this photo from 1976, 5 years before her death (Bill Hall family collection, used by permission of Peter Hall)

Fig. 14.2 The "Trans Siberian Group" at a hotel in Japan; Bill and Ruby Hall are in the *front row*, extreme right (Bill Hall family collection, used by permission of Peter Hall)

Express to Europe via Moscow.[1] Figure 14.2 shows the Australian group that undertook this trip. After some sightseeing in Moscow and a visit to Paris, they met Peter in London in the summer of 1976. He noticed immediately that his mother was confused and showed signs of physical deterioration; his father had written to him earlier that his mother was quite unwell and that Peter should not be surprised by her worsening state. In fact, Bill Hall took his wife to a medical specialist in Harley Street (the area with many medical clinics), but this consultation was not at all helpful. Peter returned to Australia to begin his University of Melbourne appointment and met his parents when they returned to Australia by air at the end of 1976; he spent Christmas with his parents in Oatley. Clearly the cruel nature of Alzheimer's disease was beginning to take a toll.

---

[1] Peter and his mother had discussed the possibility of taking this train trip across Russia together; however, he did the trip by himself on his way to Oxford in 1974.

Woody Sullivan talked to Payne-Scott on the telephone on 3 March 1978; she was rather confused with failing memory. On 30 March 1981—about 2 months before her death—Sullivan phoned her house and likely spoke with Bill Hall; he told Sullivan that Ruby was in a nursing home with no memory of the past.

As Ruby's condition worsened, her husband, Bill looked after her with admirable dedication. They had been members of the Hughes League of Health for many years; even when her illness was well advanced, Bill took her to classes where she was able to join in the discussions in a limited fashion. He also took her on short bush walks near their home. As she became more handicapped, for the last year or so of her life she was moved to a nursing home in Mortdale (a suburb immediately to the north of Oatley). A few years earlier she had taught her husband to cook some of her favourite foods, which he would take to the nursing home. Ruby earlier had told her daughter, Fiona, that as she aged she would be able to tolerate the loss of physical mobility, but could not accept any loss of her mental faculties.

After her death on 25 May 1981—3 days before her 69th birthday—Bill Hall distributed her ashes on the grounds of their house in Oatley which they had designed and built in 1949–1950 at 120 Woronora Parade. He told Peter that his mother was now "at peace". In a discussion with a representative of the funeral company, Bill and Peter were faced with a dilemma as to what religion to put on a form. Bill said that Ruby Hall would have preferred "atheist"; in the end this section was left blank.

Ruby Payne-Scott's Legacy; Why Did She Not Return to RPL in the 1960s?

Ruby Payne-Scott played a key role in the development of Australian radio astronomy in the decisive period March 1944–1951. Pawsey relied on her for many aspects of the new research programme that was to set the course of Australian astronomy until the end of the twentieth century. She had the requisite engineering experience based on the radar work of 1941–1945 and especially the mathematical skills to formulate the concept of radio interferometry and Fourier synthesis in radio astronomy. Her contributions to solar physics were also important. She was without doubt the discoverer of Type III bursts[2] at Dover Heights and Hornsby, participated in the discovery of the rare Type II bursts in 1947 (and ironically ceased to believe in the reality of this important classification in late 1947), participated in the discovery of Type I bursts at Dover Heights in 1945 and 1946, and even observed Type IV bursts in 1946 and 1949 without realising the uniqueness of these events. These bursts were only recognised to be a new type of solar radio emission event in 1957 by Boischot and Denisse. They suggested that the Type IV emission was caused by synchrotron emission from relativistic electrons. Payne-Scott and Little observed the motions of these Type IVM

---

[2] Her preferred terminology was "unpolarized" bursts while Pawsey preferred "isolated" bursts. Paul Wild invented the terminology used today, Type III.

Fig. 14.3 Photograph of a group visiting Mt. Field National Park near Hobart during the January 1949 ANZAS conference. From left: Frank Kerr (RPL), Kathleen Kerr, H. L. Humphries (then in the Cloud Physics group at RPL), Don Yabsley (RPL, wearing a fedora), John Jaeger (University of Tasmania, wearing a cap), Pawsey (RPL, wearing a scarf) and Ruby Payne-Scott (RPL, in a rain poncho). She is eating, as is often the case in group photos. (Identifications provided by the late Don Yabsley.) In 2002, Jim Lovell of the University of Tasmania found the cabin near Wombat Moor near Lake Dobson. The cabin is named, "Telopea", the botanical name for the well-known Australian flowering plant, the Waratah (Bill Hall family collection, used by permission of Peter Hall)

('moving') events directly for the first time using the swept lobe interferometer at Potts Hill in 1949–1951; Payne-Scott played a major role in the design of this innovative instrument. She presented these groundbreaking results at a conference of the Australian and New Zealand Association for the Advancement of Science in Hobart in 1949. Typical of Ruby, she was making good use of her time by eating during the photo session shown in Fig. 14.3. Payne-Scott and Pawsey set the stage for the monumental solar research carried out by Wild and colleagues using the Dapto swept frequency spectrograph in the 1950s, leading to the construction and opening of the Culgoora Radioheliograph in 1967, an instrument that dominated solar metre wave radio astronomy until the instrument was closed in 1984 to make way for the Compact Array of the Australia Telescope opened during the Australian Bicentenary in 1988.

It is tempting to ask the question: What would Payne-Scott's role have been had she returned to CSIRO in the early 1960s, at the time she began her teaching career again? Probably the transition would have been difficult for her; the solar group had made major advances and the large scale Radioheliograph was under construction. Perhaps she would have joined the 21 cm hydrogen group of Frank Kerr and colleagues Jim Hindman, John Murray, Brian Robinson and Dick McGee. In the late 1940s until her resignation in July 1951, she had been quite interested in the HI line, urging a dedicated search for the line.[3] The excitement of the confirmation of the HI line in Australia by Christiansen and Hindman on 6 July (Pawsey 1951) preceded her resignation by exactly 2 weeks, following the initial detection of the line by Ewen and Purcell at Harvard on 25 March 1951 (Ewen and Purcell 1951; Muller and Oort 1951).[4]

How did she feel about the end of her professional career in 1951, when she was only 39? We will never know. We can only hazard a guess, based on two newspaper interviews she gave in 1952, close to the time of the URSI conference of 11–21 August 1952. The first interview occurred before the conference. The date from the archival newspaper article is missing as well as the name of the newspaper; a likely date was the week of 4 August. The news article was titled "Scientist Gives Up Her Career". Ruby was described as a shy, blonde from Oatley, formerly Ruby Payne-Scott, MSc. She told the reporter that she might take part in the URSI conference but:

> Obviously, I can't do two things at once.... I can't look after a young child and do my job properly. Peter is just eight months old and I would like to have another child.... I was in the physics department as a demonstrator for two years.... Then I decided there was not much opening for women in that field.... I worked on high frequency radar during the war when I joined [RPL]. At our laboratory, after the war, we started research—concerning radio waves emitted from the sun and stars. Then I became interested in disturbances in the sun that seemed to be connected with flares—eruptions of bright light. These flares were associated with radio fadeouts and magnetic storms on the earth.

This summary of her career as both a physicist and a mother is short but comprehensive. We can detect a sense of frustration, or perhaps regret that she had no choice but to leave her physics profession behind in order to have children.[5]

However, it is clear that Payne-Scott had no real desire to return to RPL.[6] In her earlier career she had been motivated mainly by the engineering challenges of the early solar research. Astronomy was not the driving force, although she quickly learned many aspects of astronomy in the years following the end of WWII. In the 1960s, she realised that she would be returning to a new RPL, a research establishment made up of teams working in the sphere of "big science". By then, RPL had become an *astronomy* institute, consisting of the Parkes 64 m telescope opened in 1961 as well as the Culgoora Radioheliograph completed in 1967. Her role, as a

---

[3] Reported by John Murray in a letter from 24 January 2004.

[4] Wendt et al. (2008b) have described the early Australian confirmation of the HI line by Christiansen and Hindman.

[5] The other interview was after the conference in the Sydney Morning Herald of 24 August 1952. Much of this interview was a confused discussion with a young woman who was an observing assistant (Pamela White of Glebe) under the direction of Alec Little at Potts Hill; the text about Payne-Scott (also called Mrs. W.H. Hall in the interview) points out that she had been in charge of the Potts Hill station from 1948 to 1951. She now had a 9 month old baby. She and her husband had just built their own house in Oatley, having moved in about August 1951. She and Bill had ceased being active in the Sydney Bush Walkers. She and her husband did discuss the Post Master General's "new methods. But I know absolutely nothing about telephones."

[6] Discussion with Peter Hall, 12 February 2007.

member of a small team working on isolated experiments with a handful of individuals designing, building and then using the apparatus, as had occurred in 1950, was no longer feasible. Also her main collaborators, Mills, Christiansen and Little, had left to start new departments at the University of Sydney. Payne-Scott decided that she would be happier spending a few years as a school teacher. As Claire Hooker has so aptly said (Hooker 2004):

> By her own desire, in the second half of her life [Payne-Scott] stuck to the women's roles of mother, wife and teacher. Yet if she was a brief feminine flare in radio astronomy, she was a bright one, both as a physicist and as a woman.

Perhaps the forgotten contributions of remarkable women such as Ruby Payne-Scott will become a phenomenon of the past. Payne-Scott had the scientific background, the drive and insight that enabled her to become a leader in the new field of radio astronomy that developed in rapid fashion as World War II ended. She, along with a handful of colleagues, lead the way to Australian prominence in radio astronomy that continues into the 21th century.

Bibliography for Making Waves

S 1982- see Sullivan's collection from 1982: *Classics in Radio Astronomy*

Adam-Smith, P.: Australian Women at War. Thomas Nelson Australia, Melbourne (1984)
Alexander, E.: The sun's radio energy. Radio Electr. (NZ) **1**, 16 (1946)
Alexander, E.: Report on the Investigation of the Norfolk Island Effect. Radio Development Laboratory, Department of Scientific and Industrial Research, Wellington (1945). RD1/518
Alfvén, H., Herlofson, N.: Cosmic radiation and radio stars. Phys. Rev. **78**, 616 (1950). S 1982
Allen, C.W.: Solar radio noise of 200 Mc./s. and its relation to solar observations. Mon. Not. R. Astron. Soc. **107**, 386 (1947)
Allen, C.W.: The variation of decimetre-wave radiation with solar activity. Mon. Not. R. Astron. Soc. **117**, 174 (1957)
Allen, N.: Australian women in science—a comparative study of two physicists. Metascience **8**, 75 (1990)
Allen, N.: Textile physics and the wool industry: an Australian woman scientist's contribution. Agric. Hist. **67**, 67 (1993)
Allen, C.W., Gum, C.S.: Survey of galactic radio-noise at 200 Mc/s. Aust. J. Sci. Res. A **3**, 224 (1950)
Appleton, E.V.: Departure of long-wave solar radiation from black-body intensity. Nature **156**, 534 (1945)
Appleton, E.V., Hey, J.S.: Solar radio noise-I. Philos. Mag. (Ser. 7) **37**, 73 (1946a)
Appleton, E.V., Hey, J.S.: Circular polarisation of solar radio noise. Nature **158**, 339 (1946b)
Baade, W., Minkowski, R.: Identification of the radio sources in Cassiopeia, Cygnus a, and Puppis a. Astrophys. J. **119**, 206 (1954). S 1982
Barnard, M.: One Single Weapon. Unpublished manuscript, CSIR, Division of Radiophysics. (1946)
Bayne, M. (ed.): Australian Women at War. Research Group of the Left Book Club of Victoria, Melbourne (1943)
Bird, T.S.: Going strong after 50 years—antenna R&D at the CSIRO division of radiophysics. IEEE Antenn. Propag. Mag. **35**, 39 (1993)
Boischot, A., Denisse, J.F.: Les émssions de type IV et l'origine des rayons cosmiques associés aux éruptions chromosphériques. C.R. Acad. **245**, 2194 (1957)
Bolton, J.G.: Radio astronomy at U.R.S.I. Observatory **73**, 23 (1953)
Bolton, J.G.: Sky and telescope, Vol. 12. January 1953. Cover image showing J.G. Bolton and the 4.9 metre parabolic radio telescope at Dover Heights Sydney. (1953b)
Bolton, J.G.: Radio astronomy at Dover Heights. Proc. Astron. Soc. Aust. **4**, 349 (1982)
Bolton, J.G., Slee, O.B.: Galactic radiation at radio frequencies. V. The sea interferometer. Aust. J. Phys. **6**, 420 (1953)
Bolton, J.G., Westfold, K.C.: Galactic radiation at radio frequencies. I. 100 Mc/s. Survey. Aust. J. Sci. Res. A **3**, 19 (1950a)

Bolton, J.G., Westfold, K.C.: Galactic radiation at radio frequencies. III. Galactic structure. Aust. J. Sci. Res. A **3**, 251 (1950b)
Bolton, J.G., Westfold, K.C.: Galactic radiation at radio frequencies. IV. The distribution of radio stars in the galaxy. Aust. J. Sci. Res. A **4**, 476 (1951)
Bolton, J.G., Stanley, G.J., Slee, O.B.: Galactic radiation at radio frequencies. VIII. Discrete sources at 100 Mc/s between declinations +50° and −50°. Aust. J. Phys. **7**, 110 (1954)
Bowen, E.G.: Radar in war. Aust. J. Sci. **VIII**, 33 (1945)
Bowen, E.G.: The origins of radio astronomy in Australia. In: Sullivan III, W.T. (ed.) The Early Years of Radio Astronomy: Reflections Fifty Years after Jansky's Discovery, p. 85. Cambridge University Press, Cambridge (1984)
Bowen, E.G.: From wartime radar to postwar radio astronomy in Australia. J. Electr. Electron. Eng., Aust.—IE Aust. IREE Aust. **8**, 1 (1988)
Bowen, E.G. (ed.): A Textbook of Radar: A Collective Work by the Staff of the Radiophysics Laboratory C.S.I.R.O., Australia, 1st edn. Angus and Robertson, Sydney (1947), 2nd edn. Cambridge University Press, Cambridge (1954)
Bracewell, R.N.: An instrumental development in radio astronomy. Observatory **70**, 185 (1950)
Bracewell, R.N.: An honour to Australian science: The U.R.S.I. Assembly. Aust. J. Sci. **14**, 173 (1952)
Bracewell, R.N.: Early work on imaging theory in radio astronomy. In: Sullivan III, W.T. (ed.) The Early Years of Radio Astronomy: Reflections Fifty Years after Jansky's Discovery, p. 167. Cambridge University Press, Cambridge (1984)
Briton, J.N.: Lightweight air warning and G.C. I. radar in Australia. J. Inst. Eng. Aust. **18**, 121 (1947)
Brown, R.H.: Boffin: A Personal Story of the Early Days of Radar, Radio Astronomy and Quantum Optics. Adam Hilger, Bristol (1991)
Brown, L.: Technical and Military Imperatives: A Radar History of World War II. Taylor and Francis, New York (1999)
Brown, R.H., Minnett, H.C., White, F.W.G.: Edward George Bowen 1911–1991. Hist. Records Aust. Sci. **9**, 151 (1992)
Buderi, R.: The Invention that Changed the World. How a Small Group of Radar Pioneers Won the Second World War and Launched a Technological Revolution. Simon and Shuster, New York (1996)
Butler, D.: The Barefoot Bush Walker: The Remarkable Story of Adventure, Courage and Romance. Australian Broadcasting Corporation, Sydney (1991)
Carey, J.: Departing from their sphere? Australian women in science, 1880–1960. In: Pons, X. (ed.) Departures: How Australia Reinvents Itself. Melbourne University Press, Melbourne (2002)
Carlslaw, H.S., Jaeger, J.C.: Conduction of Heat in Solids, 1st edn. Clarendon, Oxford (1947)
Christiansen, W.N., Hindman, J.V.: Report of a cable from J.P. Pawsey. Nature **168**, 358 (1951). S 1982
Clark, R.W.: Sir Edward Appleton. G.B.E., K.C.B., F.R.S. Pergamon Press, Oxford (1971)
Conklin, N.D.: Two Paths to Heaven's Gate. National Radio Astronomy Observatory, Charlottesville (2006)
Cowell, J.C.: The Danebank Church of England School for Girls, The First Fifty Years. Danebank School, Sydney (1983)
Deery, P.: Scientific freedom and post-war politics: Australia, 1945–1955. Hist. Records Aust. Sci. **13**, 1 (2000)
Eddington, A.S.: The Internal Constitution of Stars. University of Cambridge Press, Cambridge (1926)
Eldershaw, M.B. (actually Marjorie Barnard and Flora Eldershaw): Tomorrow and Tomorrow and Tomorrow. Virago Press (1947, 1983). A House is Built (1929)
Evans, W.F.: History of the Radiophysics Advisory Board 1939–1945. Commonwealth Scientific and Industrial Research Organization, Australia, Melbourne (1970)
Ewen, H.I., Purcell, E.M.: Observation of a line in the galactic radio spectrum: radiation from galactic hydrogen at 1,420 Mc./sec. Nature **168**, 356 (1951). S 1982
Ewington, J.: Fiona Hall. Piper Press, Sydney (2005)

Fielder-Gill, W., Bennett, J., Davidson, J.A., Pollard, A., Porter, F.H., Slayter, R.T.: The "Bailey Boys": the University of Sydney and the training of radar officers. In: MacLeod, R. (ed.) The "Boffins" of Botany Bay: Radar at the University of Sydney, 1939–1945. Hist. Records Aust. Sci. **12**, 469 (1999)
Frame, T., Faulkner, D.: Stromlo: An Australian Observatory. Allen and Unwin, Sydney (2003)
Fränz, K.: Messung der emfängerempfindlichkeit bei kurzen elektrischen wellen. Hochfrequenztech. Electroakustik **59**, 143 (1942)
Frater, R.H.: Radio science in Australia 1932–1982. Golden Jubilee Publication of the IREE, p. 15. Sydney (1982)
Frater, R.H., Ekers, R.D.: John Paul Wild 1923–2008. Hist. Records Aust. Sci. **23**, (2012)
Freeman, J.: A Passion for Physics: The Story of a Woman Physicist. The Institute of Physics Publishing, Bristol (1991)
Friis, H.T.: Noise figures of radio receivers. Proc. Inst. Radio Eng. **32**, 419 (1944)
Gillmor, C.S.: Ionospheric and radio physics in Australian science since the early days. In: Home, Kohlstedt (eds.) International Science and National Scientific Identity. Kluwer, Dordrecht (1991)
Ginzburg, V.L.: Cosmic rays as the source of galactic radio emission. Dokl. Akad. Nauk SSSR **76**, 377 (1951). S 1982
Golub, L., Pasachoff, J.M.: Nearest Star: The Surprising Science of Our Sun. Harvard University Press, Cambridge (2001)
Goss, W.M., McGee, R.X.: Under the Radar, the First Woman in Radio Astronomy: Ruby Payne-Scott. Springer, Heidelberg (2009)
Green, A.L.: Superheterodyne tracking charts-I. AWA Tech. J. **5**, 77 (1941)
Greenstein, J.L., Henyey, L.G., Keenan, P.C.: Interstellar origin of cosmic radiation at radio-frequencies. Nature **157**, 805 (1946)
Guerlac, H.E.: Radar in World War II. American Institute of Physics/Tomash Publishers, Los Angeles (1987)
Hall, F.: Force Field. (exhibition guide with contributions from Fiona Hall, Elizabeth Ann Macgregor, Vivienne Webb, Paula Savage and Gregory O'Obrien). Museum of Contemporary Art (Sydney) and City Gallery (Wellington), Sydney and Wellington (2008)
Hamersley, H.: Cancer, physics and society: interactions between the wars. In: Home, R.W. (ed.) Australian Science in the Making. Cambridge University Press, Cambridge (1988)
Haynes, R., Haynes, R., Malin, D., McGee, R.X.: Explorers of the Southern Sky: A History of Australian Astronomy. University of Cambridge Press, Cambridge (1996)
Hey, J.S.: Solar radiation in the 4 to 6 metre wavelength band on 27th and 18th February 1942. Army Operations Research Group Report No 275. (1942)
Hey, J.S.: Solar radiation in the 4–6 metre radio wave-length band. Nature **157**, 47 (1946)
Hey, J.S.: Solar radio eclipse observations. Vistas Astron. **1**, 521 (1955)
Hey, J.S.: The Evolution of Radio Astronomy. Science History Publications, New York (1973)
Hey, J.S., Parsons, S.J., Phillips, J.W.: Fluctuations in cosmic radiation at radio frequencies. Nature **158**, 234 (1946). S 1982
Hey, J.S., Parsons, S.J., Phillips, J.W.: Some characteristics of solar radio emissions. Mon. Not. R. Astron. Soc. **108**, 354 (1948)
Higgins, C.S., Shain, C.A.: Observations of cosmic noise at 9.15 Mc/s. Aust. J. Phys. **7**, 460 (1954)
Hindman, J.V., 39 other names—including Miss R. Payne-Scott: Letter to the editor about defence research in the CSIR. Sydney Morning Herald and Daily Telegraph (Sydney) (1948)
Holman, G.D.: Solar eruptive events. Physics Today **65**(4), 56 (2012)
Hobbs, W.H.: The second pan-pacific science congress. Science **58**, 342 (1923)
Hooker, C.: Irresistible Forces: Australian Women in Science. University of Melbourne Press, Melbourne (2004)
Jaeger, J.C., Westfold, K.C.: Transients in an ionized medium with applications to bursts of solar noise. Aust. J. Sci. Res. A **2**, 322 (1949)
Jansky, K.G.: Directional studies of atmospherics at high frequencies. Proc. Inst. Radio Eng. **20**, 1920 (1932)

Jansky, K.G.: Electrical disturbances apparently of extraterrestrial origin. Proc. Inst. Radio Eng. **21**, 1387 (1933). S 1982

Jones, R.V.: Most Secret War: British Scientific Intelligence 1939–1945. Hamish Hamilton Limited, London (1978)

Kai, K., Melrose, D.B., Suzuki, S.: Storms. In: McLean, D.J., Labrum, N.R. (eds.) Solar Radiophysics. Cambridge University Press, Cambridge (1985)

Kellermann, K.I., Moran, J.M.: The development of high-resolution imaging in radio astronomy. Annu. Rev. Astron. Astrophys. **39**, 457 (2001)

Kellermann, K.I., Orchiston, W., Slee, O.B.: Gordon James Stanley and early development of radio astronomy in Australia and the United States. Publ. Astron. Soc. Aust. **22**, 13 (2005)

Kerr, F.J.: Radio astronomy at the URSI assembly. Sky Telescope **12**, 59 (1953a)

Kerr, F.J.: Radio astronomy at Potts Hill. Sydney Water Board J. **2**(4), 123 (1953b)

Kerr, F.J., Shain, C.A., Higgins, C.S.: Moon echoes and penetration of the ionosphere. Nature **163**, 310 (1949)

Kerr, F.J., Hindman, J.F., Robinson, B.J.: Observations of the 21 cm line from the magellanic clouds. Aust. J. Phys. **7**, 297 (1954)

Kipenheurer, K.O.: Cosmic rays as the source of general galactic radio emission. Phys. Rev. **79**, 738 (1950). S 1982

Lehany, F.J., Yabsley, D.E.: A solar noise outburst at 600 Mc/s. and 1,200 Mc/s. Nature **161**, 645 (1948)

Lehany, F.J., Yabsley, D.E.: Solar radiation at 1200 Mc/s., 600 Mc/s., and 200 Mc/s. Aust. J. Sci. Res. **A2**, 48 (1949)

Little, A.G., Payne-Scott, R.: The position and movement on the solar disk of sources of radiation at a frequency of 97 Mc/s. I. Equipment. Aust. J. Sci. Res. A **4**, 489 (1951)

Lloyd, T.: Hall of fame. SA Weekend, The Advertiser, Adelaide, p. 6. (12 September). (2009)

Lovell, A.C.B.: Joseph Lade Pawsey 1908–1962. Biograph. Mem. F. R. Soc. **10**, 229 (1964)

Lovell, A.C.B.: Impact of world war II on radio astronomy. In: Kellermann, K.I., Sheets, B. (eds.) Serendipitous Discoveries in Radio Astronomy. Green Bank West Virginia (National Radio Astronomy Observatory), West Virginia (1983)

Lovell, A.C.B., Banwell, C.J.: Abnormal solar radiation on 72 megacycles. Nature **158**, 517 (1946)

MacLeod, R.: Introduction: Revisiting Australia's Wartime Radar Programme. In The "Boffins" of Botany Bay: Radar at the University of Sydney, 1939–1945. Hist. Records Aust. Sci. **12**, 411 (1999)

Makinson, R.E.B., Somerville, J.M., Makinson, K.R.: Magnetically controlled gas discharges. CSIR Radiophysics Laboratory Report RP222/1. (1944)

Martyn, D.F.: Polarization of solar radio-frequency emissions. Nature **158**, 308 (1946a)

Martyn, D.F.: Temperature radiation from the quiet sun in the radio spectrum. Nature **158**, 632 (1946b). S 1982

McCready, L.L., Pawsey, J.L., Payne-Scott, R.: Solar radiation at radio frequencies and its relation to sunspots. Proc. R. Soc. A, Math. Phys. Sci. **190**, 357 (1947). S 1982

McGee, R.X., Bolton, J.G.: Probable observation of the galactic nucleus at 400 Mc./s. Nature **173**, 985 (1954)

Mellor, D.P.: The Role of Science and Industry. Australia in the war of 1939–1945. Series 4- Civil. Australian War Memorial, Canberra (1958)

Melrose, D.B., Minnett, H.C.: Jack Hobart Piddington 1910–1997. Hist. Records Aust. Sci. **12**, 229 (1998)

Mills, B.Y.: Scanning considerations in LW/AWH Mk. II. CSIR Radiophysics Laboratory Report RP TI 137/4. (1945)

Mills, B.Y.: The positions of six discrete sources of cosmic radiation. Aust. J. Sc. Res. A **5**, 456 (1952)

Mills, B.Y., Little, A.G.: A high-resolution aerial system of a new type. Aust. J. Phys. **6**, 272 (1953)

Mills, B.Y., Thomas, A.B..: Observations of the sources of radiation-frequency radiation in the constellation of Cygnus. Aust. J. Sci. Res. A **4**, 158 (1951)

Minnett, H.C., Robertson, R.: Frederick William George White 1905–1994. Hist. Rec. Aust. Sci. **11**, 239 (1996)
Minnett, H.C.: The radiophysics laboratory at the University of Sydney. In: MacLeod, R. (ed.) The "Boffins" of Botany Bay: Radar at the University of Sydney, 1939–1945. Hist. Records Aust. Sci. **12**, 419 (1999)
Minnett, H.C., Alexander, T.B., Bullock, E., Day, G., Fielder-Gill, W., Mills, B.Y., Richardson, R.C.: Light-weight air warning radar. In: MacLeod, R. (ed.) The "Boffins" of Botany Bay: Radar at the University of Sydney, 1939–1945. Hist. Records Aust. Sci. **12**, 457 (1999a)
Minnett, H.C., Alexander, T.B., Cooper, B.F.C., Porter, F.H.: Radar and the bombing of Darwin. In: MacLeod, R. (ed.) The "Boffins" of Botany Bay: Radar at the University of Sydney, 1939–1945. Hist. Records Aust. Sci. **12**, 429 (1999b)
Morrell, P.: A history of homeopathy in Britain. http://www.homeoint.org/morrell/articles/pm_brita.htm (1998)
Muller, C.A., Oort, J.H.: The interstellar hydrogen line at 1420 Mc./sec. and an estimate of galactic rotation. Nature **168**, 357 (1951). S 1982
Newton, H.W.: The lineage of the great sunspots. Vistas Astron. **1**, 666 (1955)
Norman, L.: The Brown and the Yellow: Sydney Girls' High School 1883–1983. Oxford University Press, Melbourne (1983)
O'Dea, M.C.: Ian Clunies Ross: A Biography. Hyland House, Melbourne (1997)
Orchiston, W.: Dr. Elizabeth Alexander: first female radio astronomer. In: Orchiston, W. (ed.) The New Astronomy: Opening the Electromagnetic Window and Expanding Our View of Planet Earth. Springer, Dordrecht (2005)
Orchiston, W., Slee, O.B.: The radiophysics field stations and the early development of radio astronomy. In: Orchiston, W. (ed.) The New Astronomy: Opening the Electromagnetic Window and Expanding Our View of Planet Earth. Springer, Dordrecht (2005)
Orchiston, W., Slee, O.B., Burman, R.: The genesis of solar radio astronomy in Australia. J. Astron. Hist. Her. **9**, 35 (2006)
Pasachoff, J.M.: A new understanding of our sun. In: 1989 Britannica Yearbook of Science and the Future. Encyclopedia Britannica, Chicago (1989)
Paterson, M.S.: John Conrad Jaeger 1907–1979. Hist. Records Aust. Sci. **5**, 64 (1982)
Pawsey, J.P.: Directional errors of an array employing beam swinging due to a scattered signal from an object to one side. CSIR Radiophysics. Laboratory Report RP158. (1942)
Pawsey, J.L.: Solar radio-frequency radiation. Proc. Inst. Electr. Eng. Part III **97**, 290 (1950)
Pawsey, J.L.: Possible use of solar radiation to check sensitivity of microwave radio receiving equipment. CSIRO Radiophysics Laboratory Report RPR 111. (1950b)
Pawsey, J.L.: Cable from Pawsey to nature confirming the 21 cm line 12 July 1951. Nature **168**, 358 (1951). see Chritriansen and Hindman, 1951. (S 1982)
Pawsey, J.L., Smerd, S.F.: Solar radio emission. In: Kuiper, G.P., Middlehurst, B.M. (eds.) The Sun. University of Chicago Press, Chicago (1953)
Pawsey, J.L., Yabsley, D.E.: Solar radio-frequency radiation of thermal origin. Aust. J. Sci. Res. A **2**, 198 (1949)
Pawsey, J.L., Payne-Scott, R.: Measurements of the noise level picked up by an s-band aerial. CSIR Radiophysics Laboratory Report RP 209. (1944)
Pawsey, J.L., Payne-Scott, R., McCready, L.L.: Radio-frequency energy from the sun. Nature **157**, 158 (1946)
Payne-Scott, R.: The Chronicle (Sydney Girls High School). June, p. 51, November, p. 28. (1927)
Payne-Scott, R.: Relative intensity of spectral lines in indium and gallium. Nature **131**, 365 (1933)
Payne-Scott, R.: The wavelength distribution of the scattered radiation in a medium traversed by a beam of x or gamma rays. Br. J. Radiol. **10**, 850 (1937)
Payne-Scott, R.: A note on the design of iron-cored coils at audio frequencies. AWA Tech. J. **6**, 91 (1943)
Payne-Scott, R.: S band signal generator. CSIR Radiophysics Laboratory Report PD (preliminary draft) 30. (1943b)

Payne-Scott, R.: A thermal noise generator for absolute measurement of receiver noise factor at 10 cms. CSIR Radiophysics Laboratory Report RP 211. (1944a)

Payne-Scott, R.: The present position of low-power s-band measurements in the radiophysics laboratory. CSIR Radiophysics Laboratory Report TI 121/1. (1944b)

Payne-Scott, R.: Effect of summer temperatures on s band crystals. CSIR Radiophysics Laboratory Report TI 80/2. (1944c)

Payne-Scott, R.: Present position of fundamental R.F. [Radio Frequency] measurements in the radiophysics laboratory. CSIR Radiophysics Laboratory TI 191/1. (1945a)

Payne-Scott, R.: Ultimate visibility of signals on a PPI display and the effect of electrical parameters on visibility. CSIR Radiophysics Laboratory Report RP 252/1. (1945b)

Payne-Scott, R.: Solar and Cosmic radio frequency radiation; survey of knowledge available and measurements taken at radiophysics laboratory to Dec. 1, 1945. CSIR Radiophysics Laboratory Report SRP 501/27. (1945c)

Payne-Scott, R.: A study of solar radio frequency radiation on several frequencies during the sunspot of July–August, 1946. CSIR Radiophyscis Laboratory Report, RPL 9. (1947)

Payne-Scott, R.: The visibility of small echoes on radar RRP displays. Proc. Inst. Radio Eng. **36**, 180 (1948)

Payne-Scott, R.: Solar noise records taken during 1947and 1948. CSIR Radiophysics Laboratory Report RPL 30. (1948b)

Payne-Scott, R.: Bursts of solar radiation at metre wavelengths. Aust. J. Sci. Res. A **2**, 214 (1949)

Payne-Scott, R.: Women's rates of pay. October, p. 3. Women's Pay. December, p.5. CSIRO Officers Association Bulletin. (1949b)

Payne-Scott, R.: Some characteristics of non-thermal solar radiation at metre wave-lengths. J. Geophys. Res. **55**, 203 (1950). In collection of papers Summary of Proceedings of Australian National Committee of Radio Science, URSI, Sydney, 16–20 January, 1950

Payne-Scott, R., Green, A.L.: Superheterodyne tracking charts-II. AWA Tech. Rev. **5**, 251 (1941). reprinted in 'Wireless Engineer', July, 1942

Payne-Scott, R., Little, A.G.: The positions\ and movement on the solar disk of sources of radiation at a frequency of 97 Mc/s II. Noise storms. Aust. J. Sci. Res. A **4**, 508 (1951)

Payne-Scott, R., Little, A.G.: The positions and movement on the solar disk of sources of radiation at a frequency of 97 Mc/s. III. Outbursts. Aust. J. Sci. Res. A **5**, 32 (1952)

Payne-Scott, R., Love, W.H.: Tissue cultures exposed to the influence of a magnetic field. Nature **137**, 277 (1936)

Payne-Scott, R., Mills, B.Y.: Notes on interferometer errors. CSIRO Radiophysics Laboratory unpublished report. (1949)

Payne-Scott, R., Yabsley, D.E., Bolton, J.G.: Relative times of arrival of bursts of solar noise on different radio frequencies. Nature **160**, 256 (1947)

Penzias, A.A., Wilson, R.W.: A measurement of excess antenna temperature at 4080 Mc/s. Astrophys. J. **142**, 419 (1965)

Pfeiffer, J.: The Changing Universe. Random House, New York (1956)

Pick, M., Vilmer, N.: Sixty-five years of solar radio astronomy: flares, coronal mass ejection, sun-earth connection. Astron. Astrophys. Rev. **16**, 1 (2008)

Piddington, J.H.: The origin of galactic radio-frequency radiation. Mon. Not. R. Astron. Soc. **111**, 45 (1951)

Piddington, J.H., Minnett, H.C.: Solar radiation of wavelength 1.25 centimetres. Aust. J. Sci. Res. A **2**, 539 (1949)

Radiophysics Laboratory: Research Activities of the Radiophysics Laboratory. URSI, Sydney (1952)

Reber, G.: Cosmic static. Proc. Inst. Radio Eng. **28**, 68 (1940a). S 1982

Reber, G.: Notes: cosmic static. Astrophys. J. **91**, 621 (1940b). S 1982

Reber, G.: Cosmic static. Astrophys. J. **100**, 279 (1944). S 1982

Reber, G., Greenstein, J.L.: Radio-frequency investigations of astronomical interest. Observatory **67**, 15 (1947)

Roberts, J.A.: Solar radio bursts of spectral type II. Aust. J. Phys. **12**, 327 (1959)

Robertson, P.: Beyond Southern Skies: Radio Astronomy and the Parkes Telescope. Cambridge University Press, Cambridge (1992)

Robinson, R.D.: Flare continuum. In: McLean, Labrum (eds.) Solar Radiophysics. Cambridge University Press, Cambridge (1985)

Robinson, B.J.: Recollections of the URSI Tenth General Assembly, Sydney Australia, 1952. Radio Science Bulletin, No. 300, March 2002, p. 22. (2002)

Rowe, A.P.: One Story of Radar. Cambridge University Press, Cambridge (1948)

Ryle, M.: The generation of radio-frequency radiation in the sun. Proc. R. Soc. A **195**, 82 (1948)

Ryle, M.: A new radio interferometer and its application to the observation of weak radio stars. Proc. R. Soc. **A211**, 351 (1952). S 1982

Ryle, M., Vonberg, D.D.: Solar radiation on 175 Mc./s. Nature **158**, 339 (1946). S 1982

Schedvin, C.B.: Shaping Science and Industry: A History of Australia's Council for Scientific and Industrial Research, 1926–1949. Allen and Unwin, Sydney (1987)

Shain, C.A., Higgins, C.S.: Observations of the general background and discrete sources of 18.3 Mc/s cosmic noise. Aust. J. Phys. **7**, 130 (1954)

Sheldrick, D.: Record Numbers Visit MCA (Museum of Contemporary Art) for Hall Retrospective. Wentworth Courier, Sydney (2008). 18 June

Simmonds, E.: More Radar Yarns. Radar Returns, Hampton, Victoria (1992)

Simmonds, E., Smith, N.: Radar Yarns. Radar Returns, Hampton, Victoria (1991)

Simmonds, E., Smith, N.: Echoes Over the Pacific. Radar Returns, Hampton, Victoria (1995)

Slee, O.B.: Some memories of the Dover Heights field station, 1946–1954. Aust. J. Phys. **47**, 517 (1994)

Smart, W.M.: Text-Book on Spherical Astronomy, 1st edn. Cambridge University Press, Cambridge (1931)

Smerd, S.F., Dulk, G.A.: The solar type II burst of October 13, 1969. Aust. J. Phys. **24**, 185 (1971)

Smith, F.G.: An accurate determination of the positions of four radio stars. Nature **168**, 555 (1951). S 1982

Smith, F.G. (ed.): The determination of the position of a radio star. Mon. Not. R. Astron. Soc. **112**, 497 (1952)

Smith, F.G. (ed.): The Determination of Positions of Discrete Radio Sources. Unpublished report, Cavendish Laboratory, Cambridge (1948)

Southworth, G.C.: Microwave radiation from the sun. J. Frankl. Inst. **239**, 285 (1945). S 1982

Stanley, G.J., Slee, O.B.: Galactic radiation at radio frequencies. II. The discrete sources. Aust. J. Sci. Res. A **3**, 234 (1950)

Stewart, R.T.: Moving type IV bursts. In: McLean, Labrum (eds.) Solar Radiophysics. Cambridge University Press, Cambridge (1985)

Sullivan III, W.T. (ed.): Classics in Radio Astronomy. Reidel, Dordrecht (1982). the S 1982 volume

Sullivan III, W.T. (ed.): The Early Years of Radio Astronomy: Reflections Fifty Years after Jansky's Discovery. Cambridge University Press, Cambridge (1984)

Sullivan III, W.T.: Early years of Australian radio astronomy. In: Home (ed.) Australian Science in the Making. Cambridge University Press, Cambridge (1988)

Sullivan III, W.T.: Some highlights of interferometry in early radio astronomy. In: Cornwell, Perley (eds.) Radio Interferometry: Theory, Techniques, and Applications, IAU Colloquium No 131. Astronomical Society of the Pacific, San Francisco (1991)

Sullivan III, W.T.: Cosmic Noise: A History of Early Radio Astronomy. Cambridge University Press, Cambridge (2009)

Suzuki, S., Dulk, G.A.: Bursts of type III and type V. In: McLean, Labrum (eds.) Solar Radiophysics. Cambridge University Press, Cambridge (1985)

Thomas, B.M., Robinson, B.J.: Harry Clive Minnett 1917–2003. Hist. Records Aust. Sci. **16**, 199 (2005)

Thomson, J.: The WAAAF [Women's Auxiliary Australian Air Force] in Wartime Australia. Melbourne University Press, Melbourne (1991)

Torokfalvy, P., Armstrong, B.: Homoeopathy in Australia- a brief history. http://www.homeopathyoz.org/downloads/HomHistoryAUSTRALIA_full_.pdf (2010)

Townes, C.H.: Interpretation of radio radiation from the Milky Way. Astrophys. J. **105**, 235 (1947)

Turner, B.: The alchemist. Financial Review Magazine, Sydney. 27 April. (2012)

URSI: Proceedings of the General Assembly, (Sydney) vol. IX. URSI, Brussels (1952)

Vonwiller, O.U.: Intensity measurements in the arc spectrum of thallium. Phys. Rev. **35**, 802 (1930)
Vonwiller, O.U.: Cancer research in the University of Sydney: a review. Journal of Cancer Research Committee of the University of Sydney, October, 1938, **80**, 69 (1938)
Wendt, H.: The contribution of the division of radiophysics Potts Hill and Murraybank field stations to international radio astronomy. PhD thesis, James Cook University (2008)
Wendt, H., Orchiston, W., Slee, O.B.: W.N. Christiansen and the development of the solar grating array. J. Astron. Hist. Her. **11**, 173 (2008a)
Wendt, H., Orchiston, W., Slee, O.B.: W.N. Christiansen and the initial Australian investigation of the 21 cm hydrogen line. J. Astron. Hist. Her. **11**, 185 (2008b)
Westfold, K.C.: The wave equations for electromagnetic radiation in an ionized medium in a magnetic field. Aust. J. Sci. Res. A **2**, 169 (1949)
Wild, J.P.: Observations of the spectrum of high-intensity solar radiation at metre wavelengths. II. Outbursts. Aust. J. Sci. Res. A **3**, 399 (1950a)
Wild, J.P.: Observations of the spectrum of high-intensity solar radiation at metre wavelengths. III. Isolated bursts. Aust. J. Sci. Res. A **3**, 541 (1950b)
Wild, J.P.: The spectrum analysis of solar bursts at metre wavelengths. J. Geophys. Res. **55**, 205 (1950c). In collection of papers Summary of Proceedings of Australian National Committee of Radio Science, URSI, Sydney, 16–20 January, 1950
Wild, J.P.: Observations of the spectrum of high-intensity solar radiation at metre wavelengths. IV. Enhanced radiation. Aust. J. Sci. Res. A **4**, 36 (1951)
Wild, J.P.: Observational radio astronomy. Adv. Electron. Electr. Phys. **7**, 299 (1955)
Wild, J.P.: Origin of radio-astronomy in CSIRO. National Archives of Australia, D12/1/5 (1965)
Wild, J.P.: The exploration of the sun by radio. Aust. Phys. **5**, 117 (1968) (Fourth Pawsey Memorial Lecture, 30 April, 1968. Brisbane)
Wild, J.P.: A new look at the sun. Highlights Astron. (International Astronomical Union) **3**, 3 (1974)
Wild, J.P.: The beginnings. In: McLean, Labrum (eds.) Solar Radiophysics. Cambridge University Press, Cambridge (1985)
Wild, J.P.: The beginnings of radio astronomy in Australia. Proc. Astron. Soc. Aust. **7**, 95 (1987)
Wild, J.P., McCready, L.L.: Observations of the spectrum of high-intensity solar radiation at metre wavelengths. I. The apparatus and spectral types of solar burst observed. Aust. J. Sci. Res. A **3**, 387 (1950)
Wild, J.P., Murray, J.D., Rowe, W.C.: Evidence of harmonics in the spectrum of a solar radio outburst. Nature **172**, 533 (1953)
Wild, J.P., Murray, J.D., Rowe, W.C.: Harmonics in the spectra of solar radio disturbances. Aust. J. Phys. **7**, 439 (1954)
Wild, J.P., Sheridan, K.V., Trent, G.H.: The transverse motions of the sources of solar radio bursts. In: Bracewell, R., IAU/URSI Symposium (eds.) Paris Symposium on Radio Astronomy. Stanford University Press, Stanford (1959a)
Wild, J.P., Sheridan, K.V., Neylan, A.A.: An investigation of the speed of the solar disturbances responsible for type III radio bursts. Aust. J. Phys. **12**, 369 (1959b)
Wild, J.P.: The exploration of the sun by radio. Australian Physicist, August 1968, p. 117. (Fourth Pawsey Memorial Lecture from 30 April 1968, Brisbane) (1968)
Wilde, S.: Unions in CSIRO: Part of the Equation. Hyland House, Melbourne (1998)
Wilkie, D.: The scientist's place in Australia. Manchester Guardian (27 November). (1952)
Woodlands School: Woodlands 1923–1973. The First Half Century of Woodlands Glenelg Church of England Girls' Grammar School, Inc., South Australia. Rigby Limited, Adelaide (1973)
Woodlands School: Woodlands Reflections. Memories of Woodlands School over 75 Years as Recalled by Some of the Old Scholars. Alpha Visuals, Adelaide (1999)
Zhao, J.H., Morris, M.R., Goss, W.M., An, T.: Dynamics of ionized gas at the galactic center: very large array observations of the three-dimensional velocity field and location of the ionized streams in Sagittarius A west. Astrophys. J. **699**, 186 (2009)

Biographical Sketch of the Author

W. Miller Goss received his undergraduate degree in astronomy from Harvard in 1963 and a Ph.D. from the University of California (Berkeley) in 1967, while working on the newly discovered OH radio frequency line at the Hat Creek Observatory. He then moved to Australia to the CSIRO Division of Radiophysics, first as a postdoctoral fellow and later as a staff member. In 1976, he was the recipient of the Pawsey Medal of the Australian Academy of Science. From 1977 to 1986, he was on the staff of the Kapteyn Astronomical Institute of the University of Groningen, the Netherlands, and then professor from 1980 to 1986. In 1986, he moved to the National Radio Astronomy Observatory (NRAO) in Socorro New Mexico (USA). He was director of the Very Large Array and the Very Long Baseline Array from 1988 to 2002. He and Dick McGee began working on *Under the Radar, the First Woman in Radio Astronomy: Ruby Payne-Scott*, in 1997. It was published in 2009 with book launches in Sydney and Canberra. At present, W.M. Goss and Claire Hooker are working on a joint biography of J.L. Pawsey and J.G. Bolton. W.M. Goss' scientific interests are radio astronomical studies of the interstellar medium in the Milky Way and nearby galaxies. W.M. Goss is currently on the scientific staff of NRAO and is the author of over 500 astronomical publications.

Biographical Sketch of the Author

W.M. Goss at the National Radio Astronomy Observatory, Socorro, New Mexico, USA

Index

Note: "n" following page numbers indicate endnotes and "f" following page numbers indicate figure or figure caption. *passim* indicates numerous, scattered mentions within page range.

A
Adam-Smith, Patsy, 68
Alexander, Elizabeth, 91, 94, 95
Allen, C.W. ("Cla"), 15, 98, 106, 107, 115, 124n, 129, 132, 135n, 142
Allen, Nessy
   Freeman-Makinson comparison, 73n
   Makinson study, 75n
Alzheimer's disease, 6, 235, 236, 239, 240
Amalgamated Wireless Australasia (AWA), 64–65, 71, 72, 76, 82f, 95n, 148n
Andrews, M., 191
Apollo 11 lunar mission, Parkes radio telescope, 38
Appleton, Sir Edward, 94–146 *passim*, 202–213 *passim*
   Rosalind (daughter), 213n
Army Operational Research Group, UK (AORG), 97, 111, 129, 146n
Aspray, W., 86n
Atkinson, Sally, 210
Aurora, 14, 19f, 20, 25f, 138f, 139, 186, 229, 230f
Australia Day, 26 January 1946 observations Dover Heights, 4, 26, 29, 223n
Australian Women's Army Services (AWAS), 69
Australian Broadcasting Corporation (ABC), 8
Australian Council of Trade Unions (ACTU), 81n
Australian New Zealand Association for the Advancement of Science (ANZAAS), 126
Australian Security Intelligence Organisation (ASIO, earlier CIS, Commonwealth Investigation Service), viii, 3, 6, 8, 48, 191
Award, Payne-Scott career, viii, 9, 238n

B
Bailey boys, after Victor A. Bailey, 70n, 86n
Baldwin, John E., 127n, 162
Barnard, Marjorie, 74, 86n
Bastian, T.S. ("Tim"), 42f
Bayne, M., 67, 68, 81n
Bell Laboratories, 77n, 80, 89n, 93
Bethe, H., 13
Big Bear Observatory New Jersey Institute of Technology, 20f
Boischot, A., 4, 186, 241
Bolton, John G., ix, x, 6, 8, 29–31, 31f, 38, 98, 113n, 124n, 126, 135–142 *passim*, 142n, 144–145, 148n, 150–152, 157n, 159n, 164, 165n, 178, 199n, 203, 207n, 208–209, 211f
   "bust up" with Payne-Scott, 150
   Controversy *Nature* paper 1947, 141–142
   refraction, 126
Bolton-Stanley New Zealand Expedition (Cosmic Noise Expedition to NZ 1948), 144n, 165n
Bowen, E.G. (Taffy), 3, 10n, 69–85 *passim*, 86n, 91–99 *passim*, 100f, 101n, 104–112 *passim*, 126n, 140–161 *passim*, 171, 195, 199n, 201, 202f, 209f, 210n
Bracewell, R.N. ("Ron")
   Observatory publicity reports, 187
   solar bursts from aircraft, Ryle, 163
Brian, S. ("Sue"), 9n, 54n, 236n
Briggs, Edna, 57
Briton, J.N., 82, 94, 99, 100f, 112
Brookman, Elizabeth, 62, 63f
Brooks, Kate, 238n
Brown, H.J., 73, 76
Brown, Lyn, 149

Brown, R.Hanbury, 69, 112n, 207, 211f
Buchwalter, Louise, 86n
Burgmann, Victor, 93
Bursts. *See also* Type III; Type I, storm bursts
   isolated bursts, 133f, 155, 175, 179, 182, 241n
   unpolarized bursts, 140, 151, 154–156, 159–165, 178, 179n, 182, 241n
Bush walking, 1, 2, 6, 7f, 217–220, 241
Butler, D. ("Dot") (née English), 217, 218, 218f

C

Cambridge University UK
   collaboration and conflict with RPL 1948–1951, 161
   RPL and Cambridge reports on position determinations, 169
Cancer Research Committee, University of Sydney (CRC), 58–61
Carrington, R., 1859 solar flare discovery, 20
Carslaw, H.S., 57
Carter, A.W.L. ("Alan"), 189f
Casey, R.G., Minister in Charge of CSIRO 1950–60, 203
Cathode ray tube (CRT), 85, 85n, 144, 155, 168
Chain Home Overseas Low (COL, radar), Australia and New Zealand, 93–97 *passim*
Chain Home, (CH, radar), UK, 69
Chapman, Jessica, 238n
Christiansen, W.N. ("Chris"), 35, 36, 164, 178–211 *passim*
   Payne-Scott interaction, 4, 64, 88n, 170, 172–173, 211, 244
Chromosphere, 20–23, 21f, 42f, 123, 164
Clark, Marie, 64, 65, 148, 152, 155, 190
Cleveland Street School, Sydney, 54
Clunies Ross, Sir W. Ian, 5, 81n, 192–196, 199n
   letter to Payne-Scott 3 March 1950 about her marriage, 192, 194–196
   meeting regarding marriage of Payne-Scott, 5, 194
Collaroy, NSW, RAAF radar station, 96, 98
Colley, Alex, 218, 219n
Commonwealth Investigation Service (CIS, later ASIO), 191
Commonwealth Scientific and Industrial Research Organisation (CSIRO) Formation 1949, 192

Commonwealth Solar Observatory (CSO), at Mt Stromlo, Australian Capital Territory, 15, 104n, 106n, 108f, 129, 146n
Communist Party of Australia (CPA), 6, 190, 191
Cook, G.A., 112, 195, 197
Coonabarabran, NSW, 50
Coronal holes, 23, 25f
Coronal loops, 23, 24f, 25f
Coronal mass ejection (CME), 20, 23, 24, 26f
Coulson, R.B., 10, 149, 190
Cowell, Joyce, 235–237
Crater, solar, 107, 109
Critchley, Laurie, 8
CRT. *See* Cathode ray tube (CRT)
CSIR. *See* Council for Scientific and Industrial Research (CSIR), 1926–1949
CSIR(O)A (Officer's Association)
   contributions Payne-Scott to Bulletin, 81, 191, 194
   women's pay equality, 81–85
Culgoora Radioheliograph. *See* Radioheliograph (Culgoora)
Curran, Joan, (née Strothers), 85n
Cygnus, A, optical identification, 124n, 126, 144n, 150n, 169, 182, 203
   Potts Hill observations, 35, 170, 177n

D

Davies, Pauline N., 8
Davis, Wendy, 66n
Deas-Thomson Scholarship, University of Sydney, 57
Debert, J., 217, 218
Decibel (dB), 77–80, 82, 85n
Dedman, J.J. Minister for CSIR, 112
Deery, P., 199n
Denisse, J.F., 4, 186, 241
Department of Scientific and Industrial Research (DSIR) New Zealand, 72, 94, 101n
Dish, The, Australian movie by Rob Sitch, 38
Drakeford, A.S., Minister for Air, 69, 70
DSIR. *See* Department of Scientific and Industrial Research (DSIR) New Zealand

E

Eastman, E., 74
Eclipses. *See* Solar eclipses
Eddington, Sir Arthur, 90, 212n

Index 257

Edwards, Glenys, 66n
Eldershaw, Flora, 86n
Enhanced radiation. *See* Type I
Ewen, H.I., 204, 211f, 243
Ewington, Julie, 226, 227, 229, 231

F

Fadeouts, in the ionosphere (SID-sudden ionospheric disturbance), 131, 139, 145n
Fairbank, Jane, 86
Fairweather, G., 189
Feain, Ilana, 238n
Flares, solar, 15, 18f, 19–21, 41, 132, 154, 158, 186, 187, 198
Fleurs field station RPL, 34f, 36
Fourier synthesis, 104, 122, 123, 241
Foy, Kate, 64, 237
Fraunhofer, 13
Freeman, Joan, (Jelley), 2, 6, 11n, 57n, 67, 73–75, 86n, 99n, 134n, 136, 149, 196n, 239
Frequency Agile Solar Radiotelescope (FASR), 41–42
Frequency response, 27
Friis, H., 77n

G

Gary, Dale, 138n
Georges Heights Field Station of RPL, 82f, 83n, 109, 141, 170, 172, 173
Geyer, Dawn, 63f, 66n
Gilroy, A., 217n, 219, 221f
Google, Doodle, 9, 9f, 38, 106
Grafton, NSW, 47f, 48–50, 216
Graham-Smith, Sir Francis, 161, 169, 205, 207, 211f
Grating array, solar, 33f, 35, 173f, 207, 208f, 209f
Green, Anne, 237
Green, A.L., 64, 65
Greenstein, Jesse L., 110n, 140
G2V-stellar type of the sun, 16
Guerlac, H.E., 85n

H

Hall, Agnes, (née Paterson), Bill's mother, 228
Hall, Elizabeth ("Betty" née Hurley), 43n, 216n, 219–221, 233n
  CPA and Payne-Scott, 191
  Payne-Scott, first meeting, 219

Hall, Fiona-daughter of Ruby Payne-Scott, 6, 215, 217, 225–233
  interactions with father, 227, 228, 230
  mother's influence on, 227, 231, 232
Hall, Ivy, 216, 227
Hall, Peter-son of Ruby Payne-Scott, 6, 215, 217, 223–225, 226f, 227f, 228f, 239–241, 243
  memories of his mother, 224
Hall, Sydney (Bill's father), 216
Hall, William ("Bill"-husband of Ruby Payne-Scott), 2, 6, 96, 134, 172, 192, 215–222, 218f, 220f, 221f, 222f, 228f, 230n, 231, 240–241
Hamersley, H., 58–60
Harvey-Smith, Lisa, 238
Hayward, R.H. ("Bob"), 77n
Heliosphere, 23
Herbays, C.E., 201, 202f
Hey, J.S., 90, 93, 94, 97, 108, 111, 114, 115, 124, 124n, 129, 145n, 146n, 165
Higgins, C., 152, 153, 207, 211f
Higgs, A., 192, 193f
Hinode ("Sunrise") Japanese space satellite, 18f
History Detectives, 8
Hodgson, R., 1859 discovery of solar flare, 20
Hole in the Ground antenna, Dover Heights, 31f, 33
Holt, H. Minister for CSIR, 72
Hooker, Claire, 1, 8, 244, 253
Houses in Australia, Scott and Payne-Scott families, 49, 50f, 56f
Hughes League of Health, 241
Hush-Hush (musical revue RPL), 10n

I

Interferometer, first in radio astronomy, 168
International Union of Radio Science (URSI)
  arrival of guests in Sydney, 201, 202f
  congress 1952, 201
  excursions, 203, 207
  one of the first international conferences in Australia, 201
  opening ceremony, 203
  post conference publicity, 211–212
Inverell, NSW, 215, 216
Ionosphere, research, 93, 95

J

Jaeger, J.C., 57, 158, 160, 161, 242f
Jamberoo, NSW, 225, 228f

Jansky, Karl J., 27, 77n, 89, 90, 93, 106, 109–111, 125
Jansky Very Large Array (NRAO, Socorro, New Mexico USA). *See* Very Large Array (VLA)
Jodrell Bank, University of Manchester, 33, 129, 161n

K

Kaiser, T.R.("Tom"), 100f, 199n, 200n
Kerr, Frank J., 31f, 73n, 94, 95, 98, 150, 152, 153, 165n, 171n, 202f, 208–211, 242
Kosciusko, Mt., 219–221

L

Lightfoot, G., 72
Light Weight Air Warning, Height radar (LW/AWH), 82–85
   Mills' contributions, 83
Little, Alec, 3–4, 4f, 5, 35, 87, 168–189 *passim*, 189f, 190n, 198n, 205–206, 210f, 211f, 241, 243n, 244
Little, Carolyn
   Payne-Scott lecture series, 237
Lodge, Sir Oliver, 212n
Loran navigation, 95
Lovell, J.E.J. ("Jim"), 242f
Lovell, Sir Bernard (A.C.B.), 88n, 129

M

Mack, F., 49
Madsen Building University of Sydney (RPL from WWII), 90n
Madsen, Sir John, 202–203
Makinson, Rachel, 67n, 68, 74–75, 86n, 100f, 101, 134, 190–200 *passim*, 217
Makinson, R.E.B. ("Dick"), 86n
Manchester, R.N. ("Dick"), 38
Manchester, University, 15
Marsden, Sir Ernest, 72n
Martyn, D.F. 70–72, 101n, 110, 129, 139, 140, 146n, 158, 159n, 201–204
   hires Payne-Scott 1941, 71, 72
Massey, H., 209, 210f
May, R., 224
McCready, Lindsay L., 64, 71, 76, 85n, 94–128 *passim*, 135–173 *passim*, 190n, 199n, 206
   letters to Pawsey 1947–1948, 157
McGee, R.X. ("Dick"), 8, 31f, 126, 141, 191, 198, 207, 212f, 237n, 242

McKenzie, Florence V., 68
Mellor, D.P., 210n
Melrose, D.B. ("Don"), 6n
Michelson interferometer, 27, 32f, 34–35, 120, 126, 167–170 *passim*
   Potts Hill interferometer, 169
Milky Way, 13, 15, 16, 32, 33, 36, 38, 40f, 80, 95, 97, 98, 109–111, 113, 125, 178, 199n
   structure, 41
Mills, B.Y. ("Bernie"), 3, 36, 76–85 *passim*, 127n, 134, 149, 157n, 169–192 *passim*, 199n, 211f, 244
Mills Cross, 34f, 36, 173
Minkowski, Rudolph, 142n, 177n
Minnett, H.C. ("Harry"), 6n, 68–101 *passim*, 113, 114f, 135, 136, 149, 190, 198
Miscarriage, Ruby Payne-Scott, 99, 103, 133–137, 198
Montgomery, Dorothy, 86
Moppett, W., 59
Mossom, Sylvia, 210
Mount Stromlo Observatory (MSO), Canberra, 104
Muller, C.A., 210f, 211f, 243
Murphy, Peter, 191
Murray, Joan, 134
Murray, John D., 88n, 148, 165n, 198, 211n, 242, 243
Museum of Contemporary Art (MCA), Sydney, 232

N

National Gallery of Australia, 229, 230f
National Standards Laboratory (CSIR and CSIRO), 100f
Neale, Amy-(mother of Ruby Payne-Scott), 48, 49, 53, 56
New Zealand, 30, 93–95, 97, 101n, 111, 144n, 165n, 213n, 218
Nicol, Phyllis, 57
Noble, Grace, 65
Noise Storms. *See* Type I
Norfolk Island, 93, 94, 97, 105

O

Oatley, NSW, 134n, 172, 216–225 *passim*
Officer's Association of CSIR and CSIRO (OA), 81
Oort, Jan H., 208
Orchiston, Wayne, 87–98 *passim*

Index   259

Outbursts. *See also* Type IV, Type II
  giant Type II burst of March 8, 1947, 137
Oxley, R. (Roslyn Oxley9 Gallery,
    Sydney), 232

P

Page, Earle, Prime Minister of Australia, Ruby
  Payne-Scott at birth, physician in
  Grafton, NSW, Australia, 49
Parkes radio telescope,-The Dish, 38, 39f,
  101n, 172, 203, 243
Pawsey, Joseph L., vii, viii, 3, 7, 16, 29, 31f,
  35, 73–77 *passim*, 80, 83, 87–92 *passim*,
  94–96 *passim*, 99, 100f, 103–105
  passim, 111, 113, 114, 118, 122–128
  passim, 129, 135, 136, 139, 141–144
  passim, 146–148 *passim*, 151–152,
  155–159 *passim*, 161–166 *passim*,
  168–174 *passim*, 184, 185, 187, 189,
  189f, 197, 198, 199n, 202f, 204f,
  205–207 *passim*, 210, 211, 222, 241,
  242f, 243
  assessment solar group late 1947, 143
  review paper of 1950, 150n, 158
  trip to USA, Canada, UK and Europe
    1947–1948, 142–144 *passim*, 162n,
    167, 170, 190
Payne-Scott, Amy (née Neale, mother of Ruby
  Payne-Scott). *See* Neale, Amy
Payne-Scott, Cyril (father of Ruby), 45–50
  passim, 56
Payne-Scott, Henry (brother of Ruby), 49,
  53, 56
Payne-Scott, Marguerita (aunt of Ruby), 44–48
  passim
Payne-Scott, Ruby
  absence from RPL late 1946, 99, 103,
    133–135 *passim*
  Adelaide 1938–1939, 56, 61–64 *passim*, 66
  ancestors, 43–48
  aperture synthesis (Fourier synthesis),
    Pawsey, Payne-Scott and McCready, 5,
    41, 87, 104, 122–124
  ASIO, 6, 191
  AWA, 64–65, 71n, 72, 76, 82f, 95n, 149n
  award, "Ruby Payne-Scott" CSIRO, viii,
    237n
  Barnard, Marjorie, 6, 150–151, 199n
  Bolton, conflict, 6, 150–151, 199n
  Bolton, praise of Payne-Scott, ix
  childhood, 49–51, 53–56
  confusion (radio source confusion), ix, 5,
    85, 119

CSIR/CSIRO Officers Association, 81, 134,
  191, 194, 199n
death, 229, 239, 241
Europe trip, 239–240
farewell celebration from RPL
  (20 July 1951), 197–198
first interferometry in radio astronomy,
  117, 120
first radio astronomy observation in
  Australia (1944), 88–91
first woman radio astronomer, 91
Google Doodle, 9f, 38, 106
HI line, Payne-Scott interest in, 242–243
loyalty oath, 192, 193f
LW/AWH, 82f, 82–83
marriage, vii, 2, 5, 9, 56, 81n, 96, 127n, 134,
  192–196
miscarriage, 99, 103, 133–135, 198
PC, 87, 88n, 95n, 106n, 115n, 117n, 124n,
  129, 134–135, 144, 145n, 165n, 167n,
  172–175 *passim*, 199n
personnel files CSIR/CSIRO, 72n, 112,
  134–136 *passim*, 192, 193f, 196–197
PPI, 3, 9f, 83–85
provident fund (pension), 112, 127n, 195
refraction, 119f, 119, 125–126, 170, 182n
resignation from CSIRO, 5, 179, 197–198
RPL employment in 1941, 71–72
Sullivan interview with Payne-Scott, 241
summary paper by Payne-Scott in
  December 1945, 3, 96–99, 105–110
  passim, 125
superannuation (pension), viii, 2, 5, 72, 112,
  127n, 194–196
Sydney Girls High School, 51n, 54, 65, 68n
Sydney Teachers College, 47, 60–61
secret research in post war CSIRO, 190
Payne-Scott, Valerie (aunt of Ruby), 43–49
  passim
Pearcey, T., 145n
Pedler, Jocelyn ("Jock", née Britten-Jones),
  61, 63f, 66n
Pfeiffer, J., 8
Piddington, Jack, 6n, 76, 80, 95, 127, 135, 157,
  164, 199n, 205, 207, 211f
Pilgrim, Mary, 192, 216
Plan Position Indicator (PPI), 3, 9, 83–85, 168
Potts Hill field station of RPL
  97 MHz swept-lobe interferometer, design,
    3, 4f, 5, 35, 87, 151, 156, 167–176, 174f,
    176f, 178–189 *passim*, 181f, 204, 206,
    207, 210f
  Paper II, 182–185

Potts Hill field station of RPL (*cont.*)
  Paper III, 185–189
  Paper I of Payne-Scott and Little, 180–182
  testing of swept-lobe interferometer, 174–179
  URSI tour 1952, 208f, 209f, 210f
Potts Hill Reservoir, 4f, 28f, 32, 35, 167n, 171, 173f, 208f
PPI. *See* Plan Position Indicator (PPI)
Propagation Committee at RPL (PC) to 1949, 129, 144, 145, 165n, 173–175, 199
  meeting nature, 88n, 112, 116n, 134, 158, 178
  minutes, 87, 95, 124, 135, 167, 172
  name change to Radio Astronomy Committee, 87
  Payne-Scott colloquium on velocities Type III bursts, 132, 140
Pulsars, 38
Purcell, E.M., 86, 204, 243

R
Radar School, Richmond NSW Aerodrome, 70
Radar, WWII, 15, 69, 85
Radiation Laboratory of the Massachusetts Institute of Technology WWII, 76, 85n, 86n
Radio Astronomy Committee, from 1949, 87, 164, 177, 179
Radio Astronomy, origin of name, 164
Radio Australia, 139, 152, 153f
Radio Detection and Ranging (RADAR), origin of acronym, 68
Radio Direction Finding (RDF), early British term for radar, 68
Radioheliograph (Culgoora), 36–37, 37f, 92n, 185n, 242, 243
Radiophysics Laboratory (RPL) of CSIR and CSIRO
  competition with UK colleagues, 161, 162
  formation in 1939, 68
  post war activities, 91–94
  postwar programme, 112n
  staff early in WWII, 73n
Radio reflector (paraboloid), 32–34
Radio Research Board (RRB) of CSIR and CSIRO, 64n, 68, 157n, 203
Radio stars, 15, 31, 118f, 137f, 150, 151n, 162, 169, 178, 208, 209

Ratcliffe, J.A., 72n, 142n, 143, 161, 163, 164, 169, 170, 187, 204f
Reber, Grote, 89, 90, 93, 97, 98, 109–111, 125, 140, 142n
Refraction, 125
Rivett, Sir A.C. David, 81, 92, 125, 190, 200n
Roberts, J.A. ("Jim"), 138, 211n
Robinson, Brain J., 31f, 204f, 212f, 242
Ross. *See* Clunies Ross
Rowe, A.P., 69n
Royal Australian Air Force (RAAF), 69–71, 84, 86n, 105. *See also* Women's Auxiliary Australian Air Force (WAAAF)
Royal Botanic Gardens, Sydney, 230, 231f
Royal Society, Proceedings, 127n
Rutherford, Sir Ernest, 212n
Ryle, Sir Martin, 16, 35, 122n, 123, 143, 146n, 147, 157–177 *passim*, 184, 199n
  solar bursts from aircraft, 163

S
Sarkissian, John, 38
Schmidt, Brain, 9n
Schwabe, Samuel Heinrich, discovery of solar cycle, 19
Science Show, ABC radio, 8
Scott, Agnes (grandmother of Ruby), 44–48 *passim*
Scott, Hubert (grandfather of Ruby), 44–47 *passim*
Scott, Henry Thomas (great uncle of Ruby), 44, 46
Scott, John (great grandfather of Ruby), 44
Sea-cliff interferometer (Lloyd's mirror), 165n
Secret research, CSIRO, viii, 190
Shain, C. Alexander ("Alex"), 152, 153, 207, 211f
Shore defence radar (ShD), 29, 95, 113, 114f, 130, 131f
Slee, O. Bruce, 8, 30f, 31, 91, 113n, 126n, 135, 144n, 145, 148n, 150, 207, 211f
Smart, W.M., 91
Social Education and Research Concerning Humanity Foundation (SEARCH), Sydney, 191n
Solar and Heliosphere Observatory (SOHO), 26f
Solar eclipse, 15, 24f
Solar grating array (Potts Hill), 33f, 35, 173f, 207

Index 261

Solar images, movies, 5, 27, 35, 168, 182
Solar outburst, March 1947, 137
Solar wind, 23, 24, 25f
Southwest Pacific Area (SWPA), WWII, 69, 71
Southworth, G.C., 80, 89–94 *passim*, 111, 116, 127n
Stanley, Gordon J., 29, 30, 31, 32f, 100f, 101, 113n, 124n, 126n, 141, 144, 145, 203, 222
Steel, W.H. ("Beattie"), 100f, 101
Stratton, F.J., 212
Strong, Virginia P., 86n
Struve, Otto, 142n
Sullivan, W.T. III ("Woody"), 8, 80–94 *passim*, 97, 110n, 117, 122n–123n, 161, 175n, 184n, 241
Sun
  chromosphere, 20–23, 21f
  convection cells, granulation, 17
  corona, 13, 20, 23–26, 97, 123
  3-D model, 13, 14f
  filaments, 21f, 22f, 23
  flares, 19, 20
  G2V- stellar type of sun, 16
  heliosphere, 23
  interior and photosphere, 16–17
  observations by X-ray, 23
  plages, 21
  prominences, 21
  solar cycle, 19, 23
Sun, bursts. *See* Bursts; Outbursts
Sun, quiet, 27, 35, 41, 80, 81, 120, 124, 130, 131n, 144n, 145, 169, 180, 209
Sunspots, 17–19
Sutherland, Dame Joan A., 10n, 11n
Swept lobe interferometer, 4f, 5, 35, 87, 151, 156, 167–176, 178, 180, 181f, 187–189, 204, 206, 207, 210f, 242
Sydney Bush Walkers (SBW), 6, 65, 149, 153, 191n, 217, 218f, 219n, 221, 243n
  The Tigers, 219
Sydney Girls High School, 48, 51n, 54, 65, 68n, 86n
Sydney, map of radio astronomy sites, 10f, 205f
Sydney Morning Herald, 46n, 48n, 54, 70, 190, 243n
Sydney teachers college, 47n, 60, 61
Sydney University. *See* University of Sydney

Sydney Water Board, 170–172
Synchrotron emission, 110, 125, 178, 186n, 208n, 241

T
Tasmania, 143n, 219, 221f, 242f
Tiverton, UK, 43, 44
Townes, C.H., 97n
Type I. *See* Bursts Type I
Type II. *See* Outbursts Type II
Type III. *See* Bursts Type III

U
Unicorn, HMS, 6
University of Sydney
  appointments board (WWII), 67
  Moppett effect-CRC, 59
  Payne-Scott, 57–61

V
Vandals at Dover Heights, 130
van de Hulst, H., 203
Very High Frequency (VHF), 27
Very Large Area (VLA), National Radio Astronomy Observatory, 5, 38–41
Vonwiller, O.U., 57–60

W
Watheroo Observatory, 187
Watman, Merle, 153, 158n, 217n
Watson-Watt, R.A., 69, 112
Wendt, Harry, 172, 173f, 173n, 243n
Westfold, Kevin C., 123, 158, 160, 161, 163n, 178, 199n
White, Sir Frederick W.G. ("Fred"), 72, 76, 93, 101n, 104, 105, 111n, 114f, 197, 207, 209f
White, Stephen, 41
Wilde, Sally, 81n
Wild, J. Paul, 8, 37f, 92, 104, 119, 122, 130, 136, 138, 140, 145n, 147, 148, 155, 156, 159, 161, 164, 168, 169, 177, 179, 180, 182, 185, 186, 188–190, 199n, 206, 211f, 241n, 242
Williams, Robyn, 8
Wills, Beverley, (née Harris), 237n
Window, Project (UK) (chaff in the US), 85n

Women, Australian armed forces WWII, 67
Women, Australia work force WWII, 67–68
Women's Auxiliary Australian Air Force (WAAAF), 69–71
Women's Royal Australian Naval Service (WRANS), 67n, 68
Women's Royal Naval Service (WRNS, UK), 233n
Woodlands Church of England Girls' Grammar School, Glenelg, 61–63, 66n, 235, 237
Woolley, Sir Richard, 142

Y
Yabsley, D.E. ("Don"), 8, 88n, 95, 113, 136–142 *passim*, 170–171, 173, 178, 190n, 242f
Yagi antenna, (invented by S. Uda and H. Yagi), 28
Y factor of receivers, 78, 79
Younger, Robert and Christa, 191n, 219

GPSR Compliance

The European Union's (EU) General Product Safety Regulation (GPSR) is a set of rules that requires consumer products to be safe and our obligations to ensure this.

If you have any concerns about our products, you can contact us on

ProductSafety@springernature.com

In case Publisher is established outside the EU, the EU authorized representative is:

Springer Nature Customer Service Center GmbH
Europaplatz 3
69115 Heidelberg, Germany

www.ingramcontent.com/pod-product-compliance
Lightning Source LLC
LaVergne TN
LVHW010338260326
834688LV00036B/766

*9 7 8 3 6 4 2 3 5 7 5 1 0 *